Springer Series in Statistics

Springer Series in Statistics

Alho/Spencer: Statistical Demography and Forecasting.
Andersen/Borgan/Gill/Keiding: Statistical Models Based on Counting Processes.
Atkinson/Riani: Robust Diagnostic Regression Analysis.
Atkinson/Riani/Ceriloi: Exploring Multivariate Data with the Forward Search.
Berger: Statistical Decision Theory and Bayesian Analysis, 2nd edition.
Borg/Groenen: Modern Multidimensional Scaling: Theory and Applications, 2nd edition.
Brockwell/Davis: Time Series: Theory and Methods, 2nd edition.
Bucklew: Introduction to Rare Event Simulation.
Cappé/Moulines/Rydén: Inference in Hidden Markov Models.
Chan/Tong: Chaos: A Statistical Perspective.
Chen/Shao/Ibrahim: Monte Carlo Methods in Bayesian Computation.
Coles: An Introduction to Statistical Modeling of Extreme Values.
Devroye/Lugosi: Combinatorial Methods in Density Estimation.
Diggle/Ribeiro: Model-based Geostatistics.
Efromovich: Nonparametric Curve Estimation: Methods, Theory, and Applications.
Eggermont/LaRiccia: Maximum Penalized Likelihood Estimation, Volume I: Density Estimation.
Fahrmeir/Tutz: Multivariate Statistical Modeling Based on Generalized Linear Models, 2nd edition.
Fan/Yao: Nonlinear Time Series: Nonparametric and Parametric Methods.
Ferraty/Vieu: Nonparametric Functional Data Analysis: Theory and Practice.
Fienberg/Hoaglin: Selected Papers of Frederick Mosteller.
Frühwirth-Schnatter: Finite Mixture and Markov Switching Models.
Ghosh/Ramamoorthi: Bayesian Nonparametrics.
Glaz/Naus/Wallenstein: Scan Statistics.
Good: Permutation Tests: Parametric and Bootstrap Tests of Hypotheses, 3rd edition.
Gouriéroux: ARCH Models and Financial Applications.
Gu: Smoothing Spline ANOVA Models.
Györfi/Kohler/Krzyżak/Walk: A Distribution-Free Theory of Nonparametric Regression.
Haberman: Advanced Statistics, Volume I: Description of Populations.
Hall: The Bootstrap and Edgeworth Expansion.
Härdle: Smoothing Techniques: With Implementation in S.
Harrell: Regression Modeling Strategies: With Applications to Linear Models, Logistic Regression, and Survival Analysis.
Hart: Nonparametric Smoothing and Lack-of-Fit Tests.
Hastie/Tibshirani/Friedman: The Elements of Statistical Learning: Data Mining, Inference, and Prediction.
Hedayat/Sloane/Stufken: Orthogonal Arrays: Theory and Applications.
Heyde: Quasi-Likelihood and its Application: A General Approach to Optimal Parameter Estimation.
Huet/Bouvier/Poursat/Jolivet: Statistical Tools for Nonlinear Regression: A Practical Guide with S-PLUS and R Examples, 2nd edition.
Ibrahim/Chen/Sinha: Bayesian Survival Analysis.
Jiang: Linear and Generalized Linear Mixed Models and Their Applications.
Jolliffe: Principal Component Analysis, 2nd edition.
Knottnerus: Sample Survey Theory: Some Pythagorean Perspectives.
Küchler/Sørensen: Exponential Families of Stochastic Processes.
Kutoyants: Statistical Inference for Ergodic Diffusion Processes.

(continued after index)

Jiming Jiang

Linear and Generalized Linear Mixed Models and Their Applications

Jiming Jiang
Department of Statistics
University of California, Davis
Davis, CA 95616
jiang@wald.ucdavis.edu

Library of Congress Control Number: 2006935876

ISBN-10: 0-387-47941-4 e-ISBN-10: 0-387-47946-5
ISBN-13: 978-0-387-47941-5 e-ISBN-13: 978-0-387-47946-0

Printed on acid-free paper.

9 8 7 6 5 4 3 2 1

springer.com

To my children

Preface

Over the past decade there has been an explosion of developments in mixed effects models and their applications. This book concentrates on two major classes of mixed effects models, linear mixed models and generalized linear mixed models, with the intention of offering an up-to-date account of theory and methods in the analysis of these models as well as their applications in various fields.

The first two chapters are devoted to linear mixed models. We classify linear mixed models as Gaussian (linear) mixed models and non-Gaussian linear mixed models. There have been extensive studies in estimation in Gaussian mixed models as well as tests and confidence intervals. On the other hand, the literature on non-Gaussian linear mixed models is much less extensive, partially because of the difficulties in inference about these models. However, non-Gaussian linear mixed models are important because, in practice, one is never certain that normality holds. This book offers a systematic approach to inference about non-Gaussian linear mixed models. In particular, it has included recently developed methods, such as partially observed information, iterative weighted least squares, and jackknife in the context of mixed models. Other new methods introduced in this book include goodness-of-fit tests, prediction intervals, and mixed model selection. These are, of course, in addition to traditional topics such as maximum likelihood and restricted maximum likelihood in Gaussian mixed models.

The next two chapters deal with generalized linear mixed models. These models may be regarded as extensions of the Gaussian mixed models. They are useful in situations where responses are correlated as well as discrete or categorical. McCullagh and Nelder (1989) introduced one of the earlier examples of such models, a mixed logistic model, for the infamous salamander mating problem. Since then these models have received considerable attention, and various methods of inference have been developed. A major issue regarding generalized linear mixed models is the computation of the maximum likelihood estimator. It is known that the likelihood function under these models may involve high-dimensional integrals that cannot be evaluated analytically.

Therefore, the maximum likelihood estimator is difficult to compute. We classify the methods of inference as likelihood-based, which include the Bayesian methods, and nonlikelihood-based. The likelihood-based approaches focus on developing computational methods. Markov chain Monte Carlo methods have been used in both the likelihood and the Bayesian approaches to handle the computations. The nonlikelihood-based approaches try to avoid the computational difficulty by considering alternative methods. These include approximate inference and estimating equations. Another challenging problem in this area is generalized linear mixed model selection. A recently developed method, called *fence*, is shown to be applicable to selecting an optimal model from a set of candidate models.

There have been various books on mixed effects models and related topics. The following are the major ones published in the last ten years.

1. Demidenko (2004), *Mixed Models—Theory and Applications*
2. McCulloch and Searle (2001), *Generalized, Linear and Mixed Models*
3. Sahai and Ageel (2000), *Analysis of Variance: Fixed, Random and Mixed Models*
4. Verbeke et al. (2000), *Linear Mixed Models for Longitudinal Data*
5. Pinheiro and Bates (2000), *Mixed-Effects Models in S and S-Plus*
6. Brown and Prescott (1999), *Applied Mixed Models in Medicine*
7. Rao (1997), *Variance Components Estimation: Mixed Models, Methodologies and Applications*
8. Littell et al. (1996), *SAS System for Mixed Models*

The latest publication, book 1, provides a quite comprehensive introduction on Gaussian mixed models and applications in problems such as tumor regrowth, shape, and image. Book 2 emphasizes application of the maximum likelihood and restricted maximum likelihood methods for estimation. Book 3 intensively studies the ANOVA type models with examples and computer programs using statistical software packages. Books 4, 5, 6 and 8 are mainly application-oriented. Many examples and case studies are available, and computer programs in SAS and S-Plus as well. Book 7 presents a comprehensive account of the major procedures of estimating the variance components, and fixed and random effects. These books, however, do not discuss non-Gaussian linear mixed models, neither do they present model diagnostics and model selection methods for mixed models. As mentioned, non-Gaussian linear mixed models, mixed model diagnostics, and mixed model selection are among the main topics of this book.

The application of mixed models is vast and expanding so fast as to preclude any attempt at exhaustive coverage. Here we use a number of selected real-data examples to illustrate the applications of linear and generalized linear mixed models. The areas of application include biological and medical research, animal and human genetics, and small area estimation. The latter has evolved quite rapidly in surveys. These examples of applications are considered near the end of each chapter.

The text is supplemented by numerous exercises. The majority of the problems are related to the subjects discussed in each chapter. In addition, some further results and technical notes are given in the last section of each chapter. The bibliography includes the relevant references. Three appendices are attached. Appendix A lists the notation used in this book. Appendix B provides the necessary background in matrix algebra. Appendix C gives a brief review of some of the relevant results in probability and statistics.

The book is aimed at students, researchers, and other practitioners who are interested in using mixed models for statistical data analysis or doing research in this area. The book is suitable for a course in a MS program in statistics, provided that the sections of further results and technical notes are skipped. If the latter sections are included, the book may be used for two courses in a PhD program in statistics, perhaps one on linear models and the other on generalized linear models, both with applications. A first course in mathematical statistics, the ability to use a computer for data analysis and familiarity with calculus and linear algebra are prerequisites. Additional statistical courses, such as regression analysis, and a good knowledge of matrices would be helpful.

A large portion of the material in this book is based on lecture notes from two graduate courses, "Linear Models" and "Generalized Linear Models", that the author taught at Case Western Reserve University between 1996 and 2001. The author would like to thank Prof. Joe Sedransk for initiating these courses. Grateful thanks are also due, in particular, to Prof. Partha Lahiri for his role in drawing the author's attention to application of mixed models to small area estimation, and to Prof. Rudy Beran for his advice and encouragement. My special thanks also go to Prof. J. N. K. Rao, who reviewed an earlier version of this book and made valuable suggestions. A number of anonymous reviewers have reviewed the earlier drafts of this book, or chapters of this book. Their comments led to major changes, especially in the organization of the contents. The author wishes to thank them for their time and effort that have made the book suitable for a broader audience. Two graduate students have helped with the data analyses that are involved in this book. In this regard, my sincere thanks go to Ms. Zhonghua Gu and Ms. Thuan Nguyen. In addition, the author thanks Ms. Qiuyan Xu and Ms. Zhonghua Gu for proofreading the book and providing solutions to some of the exercises. Finally, the author is grateful for Dr. Sean M. D. Allan for his comments that helped improve the presentation of this preface.

Jiming Jiang
Davis, California
July, 2006

Contents

1

Linear Mixed Models: Part I

1.1 Introduction

The best way to understand a linear mixed model, or mixed linear model in some earlier literature, is to first recall a linear regression model. The latter can be expressed as $y = X\beta + \epsilon$, where y is a vector of observations, X is a matrix of known covariates, β is a vector of unknown regression coefficients, and ϵ is a vector of (unobservable random) errors. In this model, the regression coefficients are considered fixed. However, there are cases in which it makes sense to assume that some of these coefficients are random. These cases typically occur when the observations are correlated. For example, in medical studies observations are often collected from the same individuals over time. It may be reasonable to assume that correlations exist among the observations from the same individual, especially if the times at which the observations are collected are relatively close. In animal breeding, lactation yields of dairy cows associated with the same sire may be correlated. In educational research, test scores of the same student may be related.

Now, let us see how a linear mixed model may be useful for modeling the correlations among the observations. Consider, for example, the example above regarding medical studies. Assume that each individual is associated with a random effect whose value is unobservable. Let y_{ij} denote the observation from the ith individual collected at time t_j, and α_i the random effect associated with the ith individual. Assume that there are m individuals. For simplicity, let us assume that the observations from all individuals are collected at a common set of times, say, $t_1, \dots, t_k$. Then, a linear mixed model may be expressed as $y_{ij} = x_{ij}'\beta + \alpha_i + \epsilon_{ij}$, $i = 1, \dots, m$, $j = 1, \dots, k$, where x_{ij} is a vector of known covariates; β is a vector of unknown regression coefficients; the random effects $\alpha_1, \dots, \alpha_m$ are assumed to be i.i.d. with mean zero and variance σ^2; the ϵ_{ij}s are errors that are i.i.d. with mean zero and variance τ^2; and the random effects and errors are independent. It follows that the correlation between any two observations from the same individual is $\sigma^2/(\sigma^2 + \tau^2)$, whereas observations from different individuals are uncorrelated. This model is a special case of the so-called longitudinal linear mixed model, which is discussed in much detail in the sequel. Of course, this is only a simple model and it may not capture all the correlations among the observations. Therefore, we would like to have a richer class of models that allows further complications.

A general linear mixed model may be expressed as

$$y = X\beta + Z\alpha + \epsilon, \tag{1.1}$$

where y is a vector of observations, X is a matrix of known covariates, β is a vector of unknown regression coefficients, which are often called the fixed effects, Z is a known matrix, α is a vector of random effects, and ϵ is a vector of errors. Both α and ϵ are unobservable. Compared with the linear regression model, it is clear that the difference is $Z\alpha$, which may take many different forms, thus creating a rich class of models, as we shall see. A statistical model must come with assumptions. The basic assumptions for (1.1) are that the random effects and errors have mean zero and finite variances. Typically, the covariance matrices $G = \mathrm{Var}(\alpha)$ and $R = \mathrm{Var}(\epsilon)$ involve some unknown dispersion parameters, or variance components. It is also assumed that α and ϵ are uncorrelated.

We conclude this section with three examples illustrating the applications of linear mixed models, with more examples to come later.

1.1.1 Effect of Air Pollution Episodes on Children

Laird and Ware (1982) discussed an example of analysis of the effect of air pollution episodes on pulmonary function. About 200 school children were examined under normal conditions, then during an air pollution alert, and on three successive weeks following the alert. One of the objectives was to determine whether the volume of air exhaled in the first second of a forced exhalation, denoted by FEV_1, was depressed during the alert.

Note that in this case the data were longitudinally collected with five observational times common for all the children. Laird and Ware proposed the following simple linear mixed model for analysis of the longitudinal data: $y_{ij} = \beta_j + \alpha_i + \epsilon_{ij}$, $i = 1, \dots, m$, $j = 1, \dots, 5$, where β_j is the mean FEV_1 for the jth observational time, α_i is a random effect associated with the ith child, ϵ_{ij} is an error term, and m is the total number of children. It is assumed that the random effects are independent and distributed as $N(0, \tau^2)$, the errors are independent and distributed as $N(0, \sigma^2)$, and the random effects and errors are independent. It should be mentioned that some measurements were missing in this study. However, the above model can be modified to take this into account. In particular, the number of observations for the ith individual may be denoted by n_i, where $1 \leq n_i \leq 5$.

Based on the model, Laird and Ware were able to analyze the data using methods described in the sequel, with the following findings: (i) a decline in mean FEV_1 was observed on and after the alert day; and (ii) the variances and covariances for the last four measurements were larger than those involving the baseline day. The two authors further studied the problem of identification of sensitive subgroups or individuals most severely affected by the pollution episode using a more complicated linear mixed model.

1.1.2 Prediction of Maize Single-Cross Performance

Bernardo (1996) reported results of best linear unbiased prediction, or BLUP, of single-cross performance using a linear mixed model and genetic relationship among parental inbreds. Grain yield, moisture, stalk lodging and root lodging data were obtained for 2043 maize single crosses evaluated in the multilocation testing program of Limagrain Genetics, from 1990 to 1994. The objective of the study was to investigate the robustness of BLUP for identifying superior single crosses when estimates of genetic relationship among inbreds are erroneous.

In analysis of the single-cross data, Bernardo proposed the following linear mixed model: $y = X\beta + Z_0c + Z_1g_1 + Z_2g_2 + Zd + e$, where y is a vector of observed performance for a given trait (i.e., hybrid by multilocation trial means); β is a vector of fixed effects of multilocation yield trials; c is a vector of check effects; g_j is a vector of general combining ability effects of group j, $j = 1, 2$; d is a vector of specific combining ability effects; e is a vector of residual effects; and X, Z_j, $j = 0, 1, 2$ and Z are (known) incidence matrices of 0s and 1s. Here c, g_1, g_2, and d are treated as random effects. It is assumed that c, g_1, g_2, d, and e are uncorrelated with their covariance matrices modeled according to the genetic relationship.

Based on the linear mixed model, Bernardo was able to predict the performance of untested single crosses given the average performance of the tested ones, using the BLUP method (see Section 2.3.1 for details). The results indicated that BLUP is robust when inbred relationships are erroneously specified.

1.1.3 Small Area Estimation of Income

Large-scale sample surveys are usually designed to produce reliable estimates of various characteristics of interest for large geographic areas. However, for effective planning of health, social, and other services, and for apportioning government funds, there is a growing demand to produce similar estimates for smaller geographic areas and subpopulations for which adequate samples are not available. The usual design-based small-area estimators (e.g., Cochran 1977) are unreliable because they are based on a very few observations that are available from the area. This makes it necessary to "borrow strength" from related small areas to find indirect estimators that increase the effective sample size and thus precision. Such indirect estimators are typically based on linear mixed models or generalized linear mixed models that provide a link to a related small area through the use of supplementary data such as recent census data and current administrative records. See Ghosh and Rao (1994) for a review.

Among many examples of applications, Fay and Herriot (1979) used a linear mixed model for estimating per-capita income (PCI) for small places from the 1970 Census of Population and Housing. In the 1970 Census, income was collected on the basis of a 20% sample. However, of the estimates required, more than one-third, or approximately 15,000, were for places with populations of fewer than 500 persons. With such small populations, the sampling error of the direct estimates is quite significant. For example, Fay and Herriot estimated that for a place of 500 persons, the coefficient of variation of the direct estimate of PCI was about 13%; for a place of 100 persons, that coefficient increased to 30%. In order to "borrow strength" from related places and other sources, Fay and Herriot proposed the following linear mixed model, $y_i = x_i'\beta + v_i + e_i$, where y_i is the natural logarithm of the sample estimate of PCI for the ith place (the logarithm transformation stablized the variance); x_i is a vector of known covariates related to the place; β is a vector of unknown regression coefficients; v_i is a random effect associated with the place; and e_i represents the sampling error. It is assumed that v_i and e_i are distributed independently such that $v_i \sim N(0, A)$, $e_i \sim N(0, D_i)$, where A is unknown but D_is are known.

1.2 Types of Linear Mixed Models

There are different types of linear mixed models, and different ways of classifying them. One way of classification is according to whether or not a normality assumption is made. As will be seen, normality provides more flexibility in modelling, while models without normality are more robust to violation of distributional assumptions.

1.2.1 Gaussian Mixed Models

Under Gaussian linear mixed models, or simply Gaussian mixed models, both the random effects and errors in (1.1) are assumed to be normally distributed. The following are some specific types.

1. ANOVA model. As usual, ANOVA refers to analysis of variance. Some of the earliest (Gaussian) mixed models in the literature are of ANOVA type. Here we consider some simple cases.

Example 1.1 (One-way random effects model). A model is called a random effects model if the only fixed effect is an unknown mean. Suppose that the observations y_{ij}, $i = 1, \dots, m$, $j = 1, \dots, k_i$ satisfy $y_{ij} = \mu + \alpha_i + \epsilon_{ij}$ for all i and j, where μ is an unknown mean; α_i, $i = 1, \dots, m$ are random effects that are distributed independently as $N(0, \sigma^2)$; ϵ_{ij}s are errors that are distributed independently as $N(0, \tau^2)$; and the random effects are independent of the errors. Typically, the variances σ^2 and τ^2 are unknown. To express the model in terms of (1.1), let $y_i = (y_{ij})_{1 \le j \le k_i}$ be the (column) vector of observations from the ith group or cluster, and similarly $\epsilon_i = (\epsilon_{ij})_{1 \le j \le k_i}$. Then, let $y = (y_1', \dots, y_m')'$, $\alpha = (\alpha_i)_{1 \le i \le m}$, and $\epsilon = (\epsilon_1', \dots, \epsilon_m')'$. It is easy to show that the model can be expressed as (1.1) with $\beta = \mu$ and suitable X and Z (see Exercise 1.1), in which $\alpha \sim N(0, \sigma^2 I_m)$ and $\epsilon \sim N(0, \tau^2 I_n)$ with $n = \sum_{i=1}^m k_i$. One special case is when $k_i = k$ for all i. This is called the balanced case. It can be shown that, in the balanced case, the model can be expressed as (1.1) with $X = 1_m \otimes 1_k = 1_{mk}$, $Z = I_m \otimes 1_k$, and everything else as before (with $k_i = k$), where $\otimes$ denotes the Kronecker product, defined in Appendix B. Note that in this case $n = mk$.

Example 1.2 (Two-way random effects model). For simplicity, let us consider the case of one observation per cell. In this case, the observations y_{ij}, $i = 1, \dots, m$, $j = 1, \dots, k$ satisfy $y_{ij} = \mu + \xi_i + \eta_j + \epsilon_{ij}$ for all i, j, where μ is as in Example 1.1; ξ_i, $i = 1, \dots, m$, η_j, $j = 1, \dots, k$ are independent random effects such that $\xi_i \sim N(0, \sigma_1^2)$, $\eta_j \sim N(0, \sigma_2^2)$; and ϵ_{ij}s are independent errors distributed as $N(0, \tau^2)$. Again, assume that the random effects and errors are independent. This model can also be expressed as (1.1) (see Exercise 1.2). Note that this model is different from that of Example 1.1 in that, in the one-way case the observations can be divided into independent blocks, whereas no such division exists in the two-way case.

A general ANOVA model is defined by (1.1) such that

$$Z\alpha = Z_1\alpha_1 + \cdots + Z_s\alpha_s, \tag{1.2}$$

where $Z_1, \dots, Z_s$ are known matrices; $\alpha_1, \dots, \alpha_s$ are vectors of random effects such that for each $1 \le i \le s$, the components of α_i are independent and distributed as $N(0, \sigma_i^2)$. It is also assumed that the components of ϵ are independent and distributed as $N(0, \tau^2)$, and $\alpha_1, \dots, \alpha_s, \epsilon$ are independent. For ANOVA models, a natural set of variance components is $\tau^2, \sigma_1^2, \dots, \sigma_s^2$. We call this form of variance components the original form. Alternatively, the Hartley–Rao form of variance components (Hartley and Rao 1967) is: $\lambda = \tau^2, \gamma_1 = \sigma_1^2/\tau^2, \dots, \gamma_s = \sigma_s^2/\tau^2$.

A special case is the so-called balanced mixed ANOVA models. An ANOVA model is balanced if X and Z_i, $1 \le i \le s$ can be expressed as $X = \bigotimes_{l=1}^{w+1} 1_{n_l}^{a_l}$, $Z_i = \bigotimes_{l=1}^{w+1} 1_{n_l}^{b_{i,l}}$, where $(a_1, \dots, a_{w+1}) \in S_{w+1} = \{0, 1\}^{w+1}$, $(b_{i,1}, \dots, b_{i,w+1}) \in S \subset S_{w+1}$. In other words, there are w factors in the model; n_l represents the number of levels for factor l $(1 \le l \le w)$; and the $(w+1)$st factor corresponds to "repetition within cells." Thus, we have $a_{s+1} = 1$ and $b_{i,s+1} = 1$ for all i. For example, in Example 1.1, the model is balanced if $k_i = k$ for all i. In this case, we have $w = 1$, $n_1 = m$, $n_2 = k$, and $S = \{(0, 1)\}$. In Example 1.2 the model is balanced with $w = 2$, $n_1 = m$, $n_2 = k$, $n_3 = 1$, and $S = \{(0, 1, 1), (1, 0, 1)\}$ (Exercise 1.2).

2. Longitudinal model. These models gain their name because they are often used in the analysis of longitudinal data. See, for example, Diggle, Liang, and Zeger (1996). One feature of these models is that the observations may be divided into independent groups with one random effect (or vector of random effects) corresponding to each group. In practice, these groups may correspond to different individuals involved in the longitudinal study. Furthermore, there may be serial correlations within each group which are in addition to the random effect. Another feature of the longitudinal models is that there are often time-dependent covariates, which may appear either in X or in Z (or in both) of (1.1). Example 1.1 may be considered a simple case of the longitudinal model, in which there is no serial correlation within the groups. The following is a more complex model.

Example 1.3 (Growth curve model). For simplicity, suppose that for each of the m individuals, the observations are collected over a common set of times $t_1, \dots, t_k$. Suppose that y_{ij}, the observation collected at time t_j from the ith individual, satisfies $y_{ij} = \xi_i + \eta_i x_{ij} + \zeta_{ij} + \epsilon_{ij}$, where ξ_i and η_i represent, respectively, a random intercept and a random slope; x_{ij} is a known covariate; ζ_{ij} corresponds to a serial correlation; and ϵ_{ij} is an error. For each i, it is assumed that ξ_i and η_i are jointly normally distributed with means μ_1, μ_2, variances σ_1^2, σ_2^2, respectively, and correlation coefficient ρ; and ϵ_{ij}s are independent and distributed as $N(0, \tau^2)$. As for the ζ_{ij}s, it is assumed that they satisfy the following relation of the first order autoregressive process, or AR(1): $\zeta_{ij} = \phi\zeta_{ij-1} + \omega_{ij}$, where ϕ is a constant such that $0 < \phi < 1$, and ω_{ij}s are independent and distributed as $N\{0, \sigma_3^2(1 - \phi^2)\}$. Furthermore, the three random components (ξ, η), ζ, and ϵ are independent, and observations from different individuals are independent. There is a slight departure of this model from the standard linear mixed model in that the random intercept and slope may have nonzero means. However, by subtracting the means and thus defining new random effects, this model can be expressed in the standard form (Exercise 1.3). In particular, the fixed effects are μ_1 and μ_2, and the (unknown) variance components are σ_j^2, $j = 1, 2, 3$, τ^2, ρ, and ϕ. It should be pointed out that an error term, ϵ_{ij}, is included in this model. Standard growth curve models do not include such a term.

Following Datta and Lahiri (2000), a general longitudinal model may be expressed as

$$y_i = X_i\beta + Z_i\alpha_i + \epsilon_i, \qquad i = 1, \dots, m, \tag{1.3}$$

where y_i represents the vector of observations from the ith individual; X_i and Z_i are known matrices; β is an unknown vector of regression coefficients; α_i is a vector of random effects; and ϵ_i is a vector of errors. It is assumed that α_i, ϵ_i, $i = 1, \dots, m$ are independent with $\alpha_i \sim N(0, G_i)$, $\epsilon_i \sim N(0, R_i)$, where the covariance matrices G_i and R_i are known up to a vector θ of variance components. It can be shown that Example 1.3 is a special case of the general longitudinal model (Exercise 1.3). Also note that (1.3) is a special case of (1.1) with $y = (y_i)_{1\le i\le m}$, $X = (X_i)_{1\le i\le m}$, $Z = \text{diag}(Z_1, \ldots, Z_m)$, $\alpha = (\alpha_i)_{1\le i\le m}$, and $\epsilon = (\epsilon_i)_{1\le i\le m}$.

3. Marginal model. Alternatively, a Gaussian mixed model may be expressed by its marginal distribution. To see this, note that under (1.1) and normality, we have

$$y \sim N(X\beta, V) \quad \text{with} \quad V = R + ZGZ'. \tag{1.4}$$

Hence, without using (1.1), one may simply define a linear mixed model by (1.4). In particular, for the ANOVA model, one has (1.4) with $V = \tau^2 I_n + \sum_{i=1}^{s} \sigma_i^2 Z_i Z_i'$, where n is the sample size (i.e., the dimension of y). As for the longitudinal model (1.3), one may assume that $y_1, \ldots, y_m$ are independent with $y_i \sim N(X_i\beta, V_i)$, where $V_i = R_i + Z_i G_i Z_i'$. It is clear that the model can also be expressed as (1.4) with $R = \text{diag}(R_1, \ldots, R_m)$, $G = \text{diag}(G_1, \ldots, G_m)$, and X and Z defined below (1.3).

A disadvantage of the marginal model is that the random effects are not explicitly defined. In many cases, these random effects have practical meanings and the inference about them may be of interest. For example, in small area estimation (e.g., Ghosh and Rao 1994), the random effects are associated with the small area means, which are often of main interest.

4. Hierarchical models. From a Bayesian point of view, a linear mixed model is a three-stage hierarchy. In the first stage, the distribution of the observations given the random effects is defined. In the second stage, the distribution of the random effects given the model parameters is defined. In the last stage, a prior distribution is assumed for the parameters. Under normality, these stages may be specified as follows. Let θ represent the vector of variance components involved in the model. Then, we have

$$\begin{aligned} y|\beta, \alpha, \theta &\sim N(X\beta + Z\alpha, R) \quad \text{with} \quad R = R(\theta), \\ \alpha|\theta &\sim N(0, G) \quad \text{with} \quad G = G(\theta), \\ (\beta, \theta) &\sim \pi(\beta, \theta), \end{aligned} \tag{1.5}$$

where π is a known distribution. In many cases, it is assumed that $\pi(\beta, \theta) = \pi_1(\beta)\pi_2(\theta)$, where $\pi_1 = N(\beta_0, D)$ with both β_0 and D known, and π_2 is a known distribution. The following is an example.

Example 1.4. Suppose that, (i) conditional on the means μ_{ij}, $i = 1, \ldots, m$, $j = 1, \ldots, k_i$, and variance τ^2, y_{ij}, $1 \le i \le m$, $1 \le j \le n_i$ are independent and distributed as $N(\mu_{ij}, \tau^2)$; (ii) conditional on β and σ^2, $\mu_{ij} = x_{ij}'\beta + \alpha_i$, where x_{ij} is a known vector of covariates, and $\alpha_1, \ldots, \alpha_m$ are independent and distributed as $N(0, \sigma^2)$; (iii) the prior distributions are such that $\beta \sim$

$N(\beta_0, D)$, where β_0 and D are known, σ^2 and τ^2 are both distributed as inverted gamma with pdfs $f_1(\sigma^2) \propto (\sigma^2)^{-3} e^{-1/\sigma^2}$, $f_0(\tau^2) \propto (\tau^2)^{-2} e^{-1/\tau^2}$, and β, σ^2, and τ^2 are independent. It is easy to show that (i) and (ii) are equivalent to the model in Example 1.1 with μ replaced by $x'_{ij}\beta$ (Exercise 1.4). Thus, the difference between a classical linear mixed model and a Bayesian hierarchical model is the prior.

1.2.2 Non-Gaussian Linear Mixed Models

Under non-Gaussian linear mixed models, the random effects and errors are assumed to be independent, or simply uncorrelated, but their distributions are not assumed to be normal. As a result, the (joint) distribution of the data may not be fully specified up to a set of parameters. The following are some specific cases.

1. ANOVA model. Following Jiang (1996), a non-Gaussian (linear) mixed ANOVA model is defined by (1.1) and (1.2), where the components of α_i are i.i.d. with mean 0 and variance σ_i^2, $1 \le i \le s$; the components of ϵ are i.i.d. with mean 0 and variance τ^2; and $\alpha_1, \dots, \alpha_s, \epsilon$ are independent. All the other assumptions are the same as in the Gaussian case. Denote the common distribution of the components of α_i by F_i $(1 \le i \le s)$ and that of the components of ϵ by G. If the parametric forms of $F_1, \dots, F_s, G$ are not assumed, the distribution of y is not specified up to a set of parameters. In fact, even if the parametric forms of the F_is and G are known, as long as they are not normal, the (joint) distribution of y may not have an analytic expression. The vector θ of variance components is defined the same way as in the Gaussian case, having either the original or Hartley–Rao forms.

Example 1.5. Suppose that, in Example 1.1 the random effects $\alpha_1, \dots, \alpha_m$ are i.i.d. but their common distribution is t_3 instead of normal. Furthermore, the ϵ_{ij}s are i.i.d. but their common distribution is double exponential, rather than normal. It follows that the joint distribution of y_{ij}, $i = 1, \dots, m, j = 1, \dots, k_i$ does not have a closed-form expression.

2. Longitudinal model. The Gaussian longitudinal model may also be extended to the non-Gaussian case. The typical non-Gaussian case is such that $y_1, \dots, y_m$ are independent and (1.3) holds. Furthermore, for each i, α_i and ϵ_i are uncorrelated with $\mathrm{E}(\alpha_i) = 0$, $\mathrm{Var}(\alpha_i) = G_i$; $\mathrm{E}(\epsilon_i) = 0$, $\mathrm{Var}(\epsilon_i) = R_i$. Alternatively, the independence of y_i, $1 \le i \le m$ may be replaced by that of (α_i, ϵ_i), $1 \le i \le m$. All the other assumptions are the same as in the Gaussian case. Again, in this case, the distribution of y may not be fully specified up to a set of parameters, or, even if it is, may not have a closed-form expression.

Example 1.6. Consider Example 1.3. For simplicity, assume that the times $t_1, \dots, t_k$ are equally spaced, thus, without loss of generality, let $t_j = j$, $1 \le j \le k$. Let $\zeta_i = (\zeta_{ij})_{1 \le j \le k}$, $\epsilon_i = (\epsilon_{ij})_{1 \le j \le k}$. In the non-Gaussian case, it is assumed that $y_1, \ldots, y_m$ are independent, where $y_i = (y_{ij})_{1 \le j \le k}$, or, alternatively, $(\xi_i, \eta_i, \zeta_i, \epsilon_i)$, $1 \le i \le m$ are independent. Furthermore, assume that (ξ_i, η_i), ζ_i, and ϵ_i are mutually uncorrelated such that $\mathrm{E}(\xi_i) = \mu_1$, $\mathrm{E}(\eta_i) = \mu_2$, $\mathrm{E}(\zeta_i) = \mathrm{E}(\epsilon_i) = 0$; $\mathrm{var}(\xi_i) = \sigma_1^2$, $\mathrm{var}(\eta_i) = \sigma_2^2$, $\mathrm{cor}(\xi_i, \eta_i) = \rho$, $\mathrm{Var}(\zeta_i) = G_i$, $\mathrm{Var}(\epsilon_i) = \tau^2 I_k$, where G_i is the covariance matrix of ζ_i under the AR(1) model, whose (s, t) element is given by $\sigma_3^2 \phi^{-|s-t|}$ $(1 \le s, t \le k)$.

3. Marginal model. This is, perhaps, the most general model among all types. Under a marginal model, it is assumed that y, the vector of observations, satisfies $\mathrm{E}(y) = X\beta$ and $\mathrm{Var}(y) = V$, where V is specified up to a vector θ of variance components. A marginal model may arise by taking the mean and covariance matrix of a Gaussian mixed model (marginal or otherwise; see Section 1.2.1) and dropping the normality assumption. Similar to the Gaussian marginal model, the random effects are not present in this model. Therefore, the model has the disadvantage of not being suitable for inference about the random effects, if the latter are of interest. Also, because the model is so general that not many assumptions are made, it is often difficult to assess (asymptotic) properties of the estimators. See more discussion at the end of Section 1.4.1.

By not fully specifying the distribution, a non-Gaussian model may be more robust to violation of distributional assumptions. On the other hand, methods of inference that require specification of the parametric form of the distribution, such as maximum likelihood, may not apply to such a case. The inference about both Gaussian and non-Gaussian linear mixed models is discussed in the rest of this chapter.

1.3 Estimation in Gaussian Models

Standard methods of estimation in Gaussian mixed models are maximum likelihood (ML) and restricted maximum likelihood (REML). In this section we discuss these two methods. Historically, there have been other types of estimation that are either computationally simpler or more robust in a certain sense than the likelihood-based methods. These other methods are discussed in Section 1.5.

1.3.1 Maximum Likelihood

Although the ML method has a long and celebrated history going back to Fisher (1922), it had not been used in mixed model analysis until Hartley and Rao (1967). The main reason was that the estimation of the variance components in a linear mixed model was not easy to handle computationally in the old days, although the estimation of the fixed effects given the variance components is straightforward.

Point estimation. Under a Gaussian mixed model, the distribution of y is given by (1.4), which has a joint pdf

$$f(y) = \frac{1}{(2\pi)^{n/2}|V|^{1/2}} \exp\left\{-\frac{1}{2}(y - X\beta)'V^{-1}(y - X\beta)\right\},$$

where n is the dimension of y. Thus, the log-likelihood function is given by

$$l(\beta, \theta) = c - \frac{1}{2}\log(|V|) - \frac{1}{2}(y - X\beta)'V^{-1}(y - X\beta), \tag{1.6}$$

where θ represents the vector of all the variance components (involved in V), and c is a constant. By differentiating the log-likelihood with respect to the parameters (see Appendix B), we obtain the following,

$$\frac{\partial l}{\partial \beta} = X'V^{-1}y - X'V^{-1}X\beta, \tag{1.7}$$

$$\frac{\partial l}{\partial \theta_r} = \frac{1}{2}\left\{(y - X\beta)'V^{-1}\frac{\partial V}{\partial \theta_r}V^{-1}(y - X\beta) - \mathrm{tr}\left(V^{-1}\frac{\partial V}{\partial \theta_r}\right)\right\},$$
$$r = 1, \dots, q, \tag{1.8}$$

where θ_r is the rth component of θ, which has dimension q. The standard procedure of finding the ML estimator, or MLE, is to solve the ML equations $\partial l/\partial\beta = 0, \partial l/\partial\theta = 0$. However, finding the solution may not be the end of the story. In other words, the solution to (1.7) and (1.8) may or may not be the MLE. See Section 1.8 for further discussion. Let p be the dimension of β. For simplicity, we assume that X is of full (column) rank; that is, $\mathrm{rank}(X) = p$ (see Section 1.8). Let $(\hat\beta, \hat\theta)$ be the MLE. From (1.7) one obtains

$$\hat\beta = (X'\hat V^{-1}X)^{-1}X'\hat V^{-1}y, \tag{1.9}$$

where $\hat V = V(\hat\theta)$, that is, V with the variance components involved replaced by their MLE. Thus, once the MLE of θ is found, the MLE of β can be calculated by the "closed-form" expression (1.9). As for the MLE of θ, by (1.7) and (1.8) it is easy to show that it satisfies

$$y'P\frac{\partial V}{\partial \theta_r}Py = \mathrm{tr}\left(V^{-1}\frac{\partial V}{\partial \theta_r}\right), \quad r = 1, \dots, q, \tag{1.10}$$

where

$$P = V^{-1} - V^{-1}X(X'V^{-1}X)^{-1}X'V^{-1}. \tag{1.11}$$

Thus, one procedure is to first solve (1.10) for $\hat\theta$ and then compute $\hat\beta$ by (1.9). The computation of the MLE is discussed in Section 1.6.1.

In the special case of mixed ANOVA models (Section 1.2.1.1) with the original form of variance components, we have $V = \tau^2 I_n + \sum_{i=1}^s \sigma_i^2 Z_i Z_i'$, hence $\partial V/\partial\tau^2 = I_n$, $\partial V/\partial\sigma_i^2 = Z_i Z_i'$, $1 \le i \le s$. Similarly, with the Hartley–Rao form, we have $V = \lambda(I_n + \sum_{i=1}^s \gamma_i Z_i Z_i')$, hence $\partial V/\partial\lambda = V/\lambda$, $\partial V/\partial\gamma_i = \lambda Z_i Z_i'$, $1 \le i \le s$. With these expressions, the above equations may be further simplified. As for the longitudinal model (Section 1.2.1.2), the specification of (1.7)–(1.10) is left as an exercise (see Exercise 1.5).

For mixed ANOVA models (Section 1.2.1.1), asymptotic properties of the MLE, including consistency and asymptotic normality, were first studied by Hartley and Rao (1967). Also see Anderson (1969, 1971a), who studied asymptotic properties of the MLE under the marginal model (Section 1.2.1.3) with a linear covariance structure; and Miller (1977). In these papers the authors have assumed that the number of fixed effects (i.e., p) remains fixed or bounded as the sample size n increases. As it turns out, this assumption is critical to ensure consistency of the MLE. See Example 1.7 below. Jiang (1996) considered asymptotic behavior of the MLE when p increases with n, and compared it with that of the REML estimator. The results hold for non-Gaussian mixed ANOVA models (Section 1.2.2.1), which include the Gaussian case. See Section 1.8 for more details.

Asymptotic covariance matrix. Under suitable conditions (see Section 1.8 for discussion), the MLE is consistent and asymptotically normal with the asymptotic covariance matrix equal to the inverse of the Fisher information matrix. Let $\psi = (\beta', \theta')'$. Then, under regularity conditions, the Fisher information matrix has the following expressions,

$$\mathrm{Var}\left(\frac{\partial l}{\partial \psi}\right) = -\mathrm{E}\left(\frac{\partial^2 l}{\partial\psi\partial\psi'}\right). \tag{1.12}$$

By (1.7) and (1.8), further expressions can be obtained for the elements of (1.12). For example, assuming that V is twice continuously differentiable (with respect to the components of θ), then, using the results of Appendices B and C, it can be shown (Exercise 1.6) that

$$\mathrm{E}\left(\frac{\partial^2 l}{\partial\beta\partial\beta'}\right) = -X'V^{-1}X, \tag{1.13}$$

$$\mathrm{E}\left(\frac{\partial^2 l}{\partial\beta\partial\theta_r}\right) = 0, \qquad 1 \le r \le q, \tag{1.14}$$

$$\mathrm{E}\left(\frac{\partial^2 l}{\partial\theta_r\partial\theta_s}\right) = -\frac{1}{2}\mathrm{tr}\left(V^{-1}\frac{\partial V}{\partial\theta_r}V^{-1}\frac{\partial V}{\partial\theta_s}\right), \qquad 1 \le r, s \le q. \tag{1.15}$$

It follows that (1.12) does not depend on β, and therefore may be denoted by $I(\theta)$, as we do in the sequel.

We now consider some examples.

Example 1.1 (Continued). It can be shown (see Exercise 1.7) that, in this case, (1.6) has the following expression,

$$l(\mu, \sigma^2, \tau^2) = c - \frac{1}{2}(n-m)\log(\tau^2) - \frac{1}{2}\sum_{i=1}^{m}\log(\tau^2 + k_i\sigma^2)$$
$$- \frac{1}{2\tau^2}\sum_{i=1}^{m}\sum_{j=1}^{k_i}(y_{ij} - \mu)^2 + \frac{\sigma^2}{2\tau^2}\sum_{i=1}^{m}\frac{k_i^2}{\tau^2 + k_i\sigma^2}(\bar{y}_{i\cdot} - \mu)^2,$$

where c is a constant, $n = \sum_{i=1}^{m} k_i$, and $\bar{y}_{i\cdot} = k_i^{-1}\sum_{j=1}^{k_i} y_{ij}$. Furthermore, (1.7) and (1.8) become

$$\frac{\partial l}{\partial \mu} = \sum_{i=1}^{m}\frac{k_i}{\tau^2 + k_i\sigma^2}(\bar{y}_{i\cdot} - \mu),$$
$$\frac{\partial l}{\partial \tau^2} = -\frac{n-m}{2\tau^2} - \frac{1}{2}\sum_{i=1}^{m}\frac{1}{\tau^2 + k_i\sigma^2} + \frac{1}{2\tau^4}\sum_{i=1}^{m}\sum_{j=1}^{k_i}(y_{ij} - \mu)^2$$
$$- \frac{\sigma^2}{2\tau^2}\sum_{i=1}^{m}\left(\frac{1}{\tau^2} + \frac{1}{\tau^2 + k_i\sigma^2}\right)\frac{k_i^2}{\tau^2 + k_i\sigma^2}(\bar{y}_{i\cdot} - \mu)^2,$$
$$\frac{\partial l}{\partial \sigma^2} = -\frac{1}{2}\sum_{i=1}^{m}\frac{k_i}{\tau^2 + k_i\sigma^2} + \frac{1}{2}\sum_{i=1}^{m}\left(\frac{k_i}{\tau^2 + k_i\sigma^2}\right)^2(\bar{y}_{i\cdot} - \mu)^2.$$

The specification of the asymptotic covariance matrix in this case is left as an exercise (Exercise 1.7).

Example 1.7 (Neyman–Scott problem). Neyman and Scott (1948) gave the following example which shows that, when the number of parameters increases with the sample size, the MLE may not be consistent. Suppose that two observations are collected from m individuals. Each individual has its own (unknown) mean, say, μ_i for the ith individual. Suppose that the observations are independent and normally distributed with variance σ^2. The problem of interest is to estimate σ^2. The model may be expressed as the following, $y_{ij} = \mu_i + \epsilon_{ij}$, where ϵ_{ij}s are independent and distributed as $N(0, \sigma^2)$. Note that this may be viewed as a special case of the linear mixed model (1.1), in which $Z = 0$. However, it can be shown that the MLE of σ^2 is inconsistent (Exercise 1.8).

1.3.2 Restricted Maximum Likelihood

The problem in Example 1.7 is associated with the bias of the MLE. The MLE of the variance components are, in general, biased. Such a bias may not vanish as the sample size increases, if the number of the fixed effects is proportional to the sample size. In fact, in the latter case the MLE will be inconsistent (Jiang 1996).

Furthermore, in cases such as Example 1.7 the fixed effects are considered as nuisance parameters, and the main interest is the variance components.

However, with maximum likelihood one has to estimate all the parameters involved. It would be nice to have a method that can estimate the parameters of main interest without having to deal with the nuisance parameters. To introduce such a method, let us revisit Example 1.7.

Example 1.7 (Continued). In this case, there are $m+1$ parameters, of which the means $\mu_1, \dots, \mu_m$ are nuisance, while the parameter of main interest is σ^2. Clearly, the number of parameters is proportional to the sample size, which is $2m$. Now, instead of using the original data, consider the following simple transformation: $z_i = y_{i1} - y_{i2}$. It follows that $z_1, \dots, z_m$ are independent and distributed as $N(0, 2\sigma^2)$. What makes a difference is that the nuisance parameters are gone; they are not involved in the distribution of the zs. In fact, the MLE of σ^2 based on the new data $z_1, \dots, z_m$ is consistent (Exercise 1.8). Note that, after the transformation, one is in a situation with a single parameter, σ^2, and m observations.

The "trick" in the above example is no secret: apply a transformation to the data to eliminate the fixed effects, then use the transformed data to estimate the variance component. We now illustrate the method under a general setting.

Point Estimation. As before, we assume, w.l.o.g., that $\text{rank}(X) = p$. Let A be an $n \times (n-p)$ matrix such that

$$\text{rank}(A) = n - p, \quad A'X = 0. \tag{1.16}$$

Then, define $z = A'y$. It is easy to see that $z \sim N(0, A'VA)$. It follows that the joint pdf of z is given by

$$f_{\text{R}}(z) = \frac{1}{(2\pi)^{(n-p)/2}|A'VA|^{1/2}} \exp\left\{-\frac{1}{2}z'(A'VA)^{-1}z\right\},$$

where and hereafter the subscript R corresponds to "restricted". Thus, the log-likelihood based on z, which we call restricted log-likelihood, is given by

$$l_{\text{R}}(\theta) = c - \frac{1}{2}\log(|A'VA|) - \frac{1}{2}z'(A'VA)^{-1}z, \tag{1.17}$$

where c is a constant. By differentiating the restricted log-likelihood (see Appendix B), we obtain, expressed in terms of y,

$$\frac{\partial l_{\text{R}}}{\partial \theta_i} = \frac{1}{2}\left\{y'P\frac{\partial V}{\partial \theta_i}Py - \text{tr}\left(P\frac{\partial V}{\partial \theta_i}\right)\right\}, \qquad i = 1, \dots, q, \tag{1.18}$$

where

$$P = A(A'VA)^{-1}A' \quad = \quad \text{the right side of (1.11)} \tag{1.19}$$

(see Appendix B). The REML estimator of θ is defined as the maximizer of (1.17). As in the ML case, such a maximizer satisfies the REML equation $\partial l_{\text{R}}/\partial \theta = 0$. See Section 1.8 for further discussion.

Remark. Although the REML estimator is defined through a transforming matrix A, the REML estimator, in fact, does not depend on A. To see this, note that, by (1.18) and (1.19), the REML equations do not depend on A. A more thorough demonstration is left as an exercise (Exercise 1.9). This fact is important because, obviously, the choice of A is not unique, and one does not want the estimator to depend on the choice of the transformation.

Example 1.7 (Continued). It is easy to see that the transformation $z_i = y_{i1} - y_{i2}$, $i = 1, \dots, m$ corresponds to

$$A = \begin{pmatrix} 1 & -1 & 0 & 0 & \cdots & 0 & 0 \\ 0 & 0 & 1 & -1 & \cdots & 0 & 0 \\ \vdots & \vdots & \vdots & \vdots & & \vdots & \vdots \\ 0 & 0 & 0 & 0 & \cdots & 1 & -1 \end{pmatrix}'.$$

An alternative transforming matrix may be obtained as $B = AT$, where T is any $m \times m$ nonsingular matrix. But the resulting REML estimator of σ^2 is the same (Exercise 1.9).

Historical note. The REML method was first proposed by Thompson (1962) and was put on a broad basis by Patterson and Thompson (1971). The method is also known as residual maximum likelihood, although the abbreviation, REML, remains the same. There have been different derivations of REML. For example, Harville (1974) provided a Bayesian derivation of REML. He showed that the restricted likelihood can be derived as the marginal likelihood when β is integrated out under a noninformative, or flat, prior (Exercise 1.10). Also see Verbyla (1990). Barndorff-Nielsen (1983) derived the restricted likelihood as a modified profile likelihood. More recently, Jiang (1996) pointed out that the REML equations may be derived under the assumption of a multivariate t-distribution (instead of multivariate normal distribution). More generally, Heyde (1994) showed that the REML equations may be viewed as quasi-likelihood equations. See Heyde (1997) for further details. Surveys on REML can be found in Harville (1977), Khuri and Sahai (1985), Robinson (1987), and Speed (1997).

Note that the restricted log-likelihood (1.17) is a function of θ only. In other words, the REML method is a method of estimating θ (not β, because the latter is eliminated before the estimation). However, once the REML estimator of θ is obtained, β is usually estimated the same way as the ML, that is, by (1.9), where $V = V(\hat{\theta})$ with $\hat{\theta}$ being the REML estimator. Such an estimator is sometimes referred as the "REML estimator" of β.

Asymptotic covariance matrix. Under suitable conditions, the REML estimator is consistent and asymptotically normal (see Section 1.8 for discussion). The asymptotic covariance matrix is equal to the inverse of the restricted Fisher information matrix, which, under regularity conditions, has similar expressions as (1.12):

$$\mathrm{Var}\left(\frac{\partial l_{\mathrm{R}}}{\partial \theta}\right) = -\mathrm{E}\left(\frac{\partial^2 l_{\mathrm{R}}}{\partial \theta \partial \theta'}\right). \tag{1.20}$$

Further expressions may be obtained. For example, assuming, again, that V is twice continuously differentiable (with respect to the components of θ), then we have (Exercise 1.11)

$$\mathrm{E}\left(\frac{\partial^2 l_{\mathrm{R}}}{\partial \theta_i \partial \theta_j}\right) = -\frac{1}{2}\mathrm{tr}\left(P\frac{\partial V}{\partial \theta_i}P\frac{\partial V}{\partial \theta_j}\right), \qquad 1 \le i, j \le q. \tag{1.21}$$

Example 1.1 (Continued). For simplicity, consider the balanced case, that is, $k_i = k$, $1 \le i \le m$. It can be shown (Exercise 1.12) that, in this case, the REML equations $\partial l_{\mathrm{R}}/\partial \tau^2 = 0$ and $\partial l_{\mathrm{R}}/\partial \sigma^2 = 0$ are equivalent to the following,

$$\begin{cases} \tau^2 + k\sigma^2 = \mathrm{MSA}, \\ \tau^2 = \mathrm{MSE}, \end{cases} \tag{1.22}$$

where $\mathrm{MSA} = \mathrm{SSA}/(m-1)$, $\mathrm{SSA} = k\sum_{i=1}^m (\bar{y}_{i\cdot} - \bar{y}_{\cdot\cdot})^2$, $\bar{y}_{i\cdot} = k^{-1}\sum_{j=1}^k y_{ij}$, $\bar{y}_{\cdot\cdot} = (mk)^{-1}\sum_{i=1}^m\sum_{j=1}^k y_{ij}$; $\mathrm{MSE} = \mathrm{SSE}/m(k-1)$, $\mathrm{SSE} = \sum_{i=1}^m\sum_{j=1}^k (y_{ij} - \bar{y}_{i\cdot})^2$. The REML equations thus have an explicit solution: $\dot{\tau}^2 = \mathrm{MSE}$, $\dot{\sigma}^2 = k^{-1}(\mathrm{MSA} - \mathrm{MSE})$. Note that these are not necessarily the REML estimators, only the solution to the REML equations (although, in most cases, the two are identical). The derivation of the asymptotic covariance matrix is left as an exercise (Exercise 1.12).

1.4 Estimation in Non-Gaussian Models

The methods discussed in the previous section are based on the normality assumption. However, the normality assumption is likely to be violated in practice. For example, Lange and Ryan (1989) gave several examples showing that nonnormality of the random effects is, indeed, encountered in practice. The authors also developed a method for assessing normality of the random effects. Due to such concerns, linear mixed models without normality assumption, or non-Gaussian linear mixed models, have been considered. In this section, we focus on estimation in the two types of non-Gaussian linear mixed models described in Section 1.2.2, that is, the ANOVA model and longitudinal model. Sections 1.4.1 and 1.4.2 discuss estimation in ANOVA models, and Sections 1.4.3 and 1.4.4 deal with longitudinal models. It should be pointed out that, although the methods proposed here for the two types of models are different, it does not mean that one method cannot be applied to a different type of model. In fact, the two methods overlap in some special cases. See the discussion in Section 1.4.4.

1.4.1 Quasi-Likelihood Method

In this and the next sections we discuss estimation in non-Gaussian ANOVA models. Some remarks are made at the end of the next section on possible extension of the method to more general models.

First we must point out that, when normality is not assumed (fully) likelihood-based inference is difficult, or even impossible. To see this, first note that if the distributions of the random effects and errors are not specified, the likelihood function is simply not available. Furthermore, even if the (nonnormal) distributions of the random effects and errors are specified (up to some unknown parameters), the likelihood function is usually complicated. In particular, such a likelihood may not have an analytic expression. Finally, like normality, any other specific distributional assumptions may not hold in practice. These difficulties have led to consideration of methods other than maximum likelihood. One such method is Gaussian-likelihood, or, as we call it, quasi-likelihood.

The idea is to use normality-based estimators even if the data are not really normal. For the ANOVA models, the REML estimator of θ is defined as the solution to the (Gaussian) REML equations, provided that the solution belongs to the parameter space. See Section 1.8 for a discussion on how to handle cases where the solution is out of the parameter space. Similarly, the ML estimators of β and θ are defined as the solution to the (Gaussian) ML equations, provided that they stay in the parameter space. More specifically, under the ANOVA model with the original form of variance components, the REML equations are (Exercise 1.13)

$$\begin{cases} y'P^2y = \mathrm{tr}(P), & \\ y'PZ_iZ_i'Py = \mathrm{tr}(Z_i'PZ_i), & 1 \le i \le s. \end{cases} \tag{1.23}$$

With the same model and variance components, the ML equations are

$$\begin{cases} X'V^{-1}X\beta = X'V^{-1}y, & \\ y'P^2y = \mathrm{tr}(V^{-1}), & \\ y'PZ_iZ_i'Py = \mathrm{tr}(Z_i'V^{-1}Z_i), & 1 \le i \le s. \end{cases} \tag{1.24}$$

Similarly, the REML equations under the ANOVA model with the Hartley–Rao form of variance components are

$$\begin{cases} y'Py = n - p, & \\ y'PZ_iZ_i'Py = \mathrm{tr}(Z_i'PZ_i), & 1 \le i \le s. \end{cases} \tag{1.25}$$

The ML equations under ANOVA model and the Hartley–Rao form are

$$\begin{cases} X'V^{-1}X\beta = X'V^{-1}y, & \\ y'Py = n, & \\ y'PZ_iZ_i'Py = \mathrm{tr}(Z_i'V^{-1}Z_i), & 1 \le i \le s. \end{cases} \tag{1.26}$$

In order to justify such an approach, let us first point out that, although the REML estimator is defined as the solution to the REML equations, which are derived under normality, normal likelihood is not the only one that can lead to the REML equations. Jiang (1996) has pointed out that exactly the same equations will arise if one starts with a multivariate t-distribution, that is, $y \sim t_n(X\beta, V, d)$, which has a joint pdf

$$p(y) = \frac{\Gamma\{(n+d)/2\}}{(d\pi)^{n/2}\Gamma(d/2)|V|^{1/2}} \left\{1 + \frac{1}{d}(y - X\beta)'V^{-1}(y - X\beta)\right\}^{-(n+d)/2}$$

(Exercise 1.14). Here d is the degree of freedom of the multivariate t-distribution. More generally, Heyde (1994, 1997) showed that the REML equations can be derived from a quasi-likelihood. As it turns out, the likelihood under multivariate-t is a special case of Heyde's quasi-likelihood. For such a reason, the (Gaussian) REML estimation may be regarded as a method of quasi-likelihood. Similarly, the (Gaussian) ML estimation may be justified from a quasi-likelihood point of view. For simplicity, the corresponding estimators are still called REML or ML estimators.

Furthermore, it has been shown (Richardson and Welsh 1994; Jiang 1996, 1997a) that the REML estimator is consistent and asymptotically normal even if normality does not hold. Furthermore, Jiang (1996) showed that the ML estimator has similar asymptotic properties, provided that the number of fixed effects p remains bounded or increases at a slower rate than the sample size n. Again, the latter result does not require normality. See Section 1.8 for more details. Therefore, the quasi-likelihood approach is, at least, well-justified from an asymptotic point of view.

Although the method is justified for point estimation, there is a complication in assessing the variation of these estimators. Jiang (1996) derived the asymptotic covariance matrix of the REML estimator. As for the ML estimator, its asymptotic covariance matrix is the same as that obtained in Jiang (1998b), if, for example, p remains bounded. See Section 1.8 for more details. These results do not require normality. However, when normality does not hold, the asymptotic covariance matrix involves parameters other than the variance components, namely, the third and fourth moments of the random effects and errors. To see where exactly the problem occurs, note that, according to Jiang (1996), the asymptotic covariance matrix of the REML estimator is given by

$$\Sigma_{\mathrm{R}} = \left\{\mathrm{E}\left(\frac{\partial^2 l_{\mathrm{R}}}{\partial\theta\partial\theta'}\right)\right\}^{-1} \mathrm{Var}\left(\frac{\partial l_{\mathrm{R}}}{\partial\theta}\right) \left\{\mathrm{E}\left(\frac{\partial^2 l_{\mathrm{R}}}{\partial\theta\partial\theta'}\right)\right\}^{-1}. \tag{1.27}$$

If normality holds, then l_{R} is the true restricted log-likelihood, hence, under regularity conditions, we have $\mathcal{I}_1 = \mathrm{Var}(\partial l_{\mathrm{R}}/\partial\theta) = -\mathrm{E}(\partial^2 l_{\mathrm{R}}/\partial\theta\partial\theta') = -\mathcal{I}_2$, therefore (1.27) reduces to the inverse of (1.20). The factor $\mathcal{I}_2$ only depends on θ, whose estimator is already available. However, unlike $\mathcal{I}_2$, the factor $\mathcal{I}_1$ depends on, in addition to θ, the kurtoses of the random effects and errors.

Similarly, by Jiang's result (1998b), it can be shown that the asymptotic covariance matrix of the ML estimator of $\psi = (\beta', \theta')'$ is given by

$$\Sigma = \left\{ \mathrm{E}\left(\frac{\partial^2 l}{\partial\psi\partial\psi'} \right) \right\}^{-1} \mathrm{Var}\left(\frac{\partial l}{\partial\psi} \right) \left\{ \mathrm{E}\left(\frac{\partial^2 l}{\partial\psi\partial\psi'} \right) \right\}^{-1}. \tag{1.28}$$

Here $\mathcal{I}_2 = \mathrm{E}(\partial^2 l/\partial\psi\partial\psi')$ depends only on θ, but $\mathcal{I}_1 = \mathrm{Var}(\partial l/\partial\psi)$ depends on, in addition to θ, not only the kurtoses but also the third moments of random effects and errors.

Note that standard procedures, including ML, REML, and those discussed later in Section 1.5, do not produce estimators of these higher moments. Therefore, to make the quasi-likelihood method practical, it is necessary to develop a method of estimating the asymptotic covariance matrix of the REML (ML) estimator. In the next section we introduce such a method.

1.4.2 Partially Observed Information

It is clear that the key issue is how to estimate $\mathcal{I}_1$, which we call a quasi-information matrix (QUIM). Note that, when normality holds, QUIM is the (true) Fisher information matrix. Traditionally, there are two ways to estimate the Fisher information: (i) estimated information; and (ii) observed information. See, for example, Efron and Hinkley (1978) for a discussion and comparison of the two methods in the i.i.d. case. It is clear that (i) cannot be used to estimate $\mathcal{I}_1$, unless one finds some way to estimate the higher moments. Assuming that the random effects and errors are symmetrically distributed, in which case the third moments vanish, Jiang (2003b) proposed an empirical method of moments (EMM) to estimate the kurtoses of the random effects and errors. See Section 2.1.2.1 for more detail. The method has a limitation, because, like normality, symmetry may not hold in practice. When symmetry is not known to hold, the EMM does not provide estimates of the third moments, which are involved in the ML case. As for (ii), it is not all that clear how this should be defined in cases of correlated observations. For simplicity, let us consider ML estimation with a single unknown parameter, say, ϕ. Let l be the (true) log-likelihood. In the i.i.d. case, we have, under regularity conditions,

$$\mathcal{I}_1 = \mathrm{Var}\left(\frac{\partial l}{\partial\psi} \right) = \mathrm{E}\left\{ \sum_{i=1}^{n} \left(\frac{\partial l_i}{\partial\phi} \right)^2 \right\}, \tag{1.29}$$

where l_i is the log-likelihood based on y_i, the ith observation. Therefore, an observed information is $\tilde{\mathcal{I}}_1 = \sum_{i=1}^{n} (\partial l_i/\partial\phi|_{\hat{\phi}})^2$, where $\hat{\phi}$ is the ML estimator. This is a consistent estimator of $\mathcal{I}_1$ in the sense that $\tilde{\mathcal{I}}_1 - \mathcal{I}_1 = o_{\mathrm{P}}(\mathcal{I}_1)$ or, equivalently, $\tilde{\mathcal{I}}_1/\mathcal{I}_1 \to 1$ in probability. However, if the observations are correlated, (1.29) does not hold. In this case, because $\mathcal{I}_1 = \mathrm{E}\{(\partial l/\partial\psi)^2\}$, it might seem that an observed information would be $\tilde{\mathcal{I}}_1 = (\partial l/\partial\psi|_{\tilde{\psi}})^2$, which is 0 if $\tilde{\psi}$ is the MLE (i.e., the solution to the ML equation). Even if $\tilde{\psi}$ is not the MLE but a consistent estimator of ψ, the expression does not provide a consistent estimator for $\mathcal{I}_1$. For example, in the i.i.d. case, this is the same as $(\sum_{i=1}^n \partial l_i/\partial\psi|_{\tilde{\psi}})^2$, which, asymptotically, is equivalent to n times the square of a normal random variable. Therefore, it is not true that $\tilde{\mathcal{I}}_1 - \mathcal{I}_1 = o_{\mathrm{P}}(\mathcal{I}_1)$. The conclusion is that, in the case of correlated observations, (ii) does not work in general.

We now introduce a method that applies generally. Throughout the rest of this section, we consider the Hartley–Rao form of variance components: $\lambda = \tau^2$ and $\gamma_i = \sigma_i^2/\tau^2$, $1 \le i \le s$. Note that there is a simple transformation between the original form and the Hartley–Rao form of variance components:

$$\begin{pmatrix} \lambda \\ \gamma_1 \\ \vdots \\ \gamma_s \end{pmatrix} = \begin{pmatrix} 1 & 0 \\ 0 & I_s/\tau^2 \end{pmatrix} \begin{pmatrix} \tau^2 \\ \sigma_1^2 \\ \vdots \\ \sigma_s^2 \end{pmatrix}, \tag{1.30}$$

where 0 represents column or row vectors of zeros. By (1.30) and the results in Appendix C, it is easy to derive an estimator of the QUIM under one form of parameters given that under the other form.

To illustrate the basic idea, consider the following simple example.

Example 1.2 (Continued). Consider an element of the QUIM $\mathrm{Var}(\partial l_{\mathrm{R}}/\partial\theta)$ for REML estimation, say, $\mathrm{var}(\partial l_{\mathrm{R}}/\partial\lambda)$, where $\theta = (\lambda, \gamma_1, \gamma_2)'$. By the result of Jiang (2004, Example 2 in Section 5), it can be shown that $\partial l_{\mathrm{R}}/\partial\lambda = \{u'Bu - (mk-1)\lambda\}/2\lambda^2$, where $u = y - \mu 1_m \otimes 1_k$ with $y = (y_{ij})_{1\le i\le m, 1\le j\le k}$ (as a vector in which the components are ordered as $y_{11}, \ldots, y_{1k}, y_{21}, \ldots$), and

$$\begin{aligned} B &= I_m \otimes I_k - \frac{1}{k}\left(1 - \frac{1}{1+\gamma_1 k}\right) I_m \otimes J_k - \frac{1}{m}\left(1 - \frac{1}{1+\gamma_2 m}\right) J_m \otimes I_k \\ &\quad + \frac{1}{mk}\left(1 - \frac{1}{1+\gamma_1 k} - \frac{1}{1+\gamma_2 m}\right) J_m \otimes J_k \\ &= I_m \otimes I_k + \lambda_1 I_m \otimes J_k + \lambda_2 J_m \otimes I_k + \lambda_3 J_m \otimes J_k \end{aligned} \tag{1.31}$$

(see Appendix A for notation). Furthermore, it can be shown that (Exercise 1.15) $\mathrm{var}(\partial l_{\mathrm{R}}/\partial\lambda) = S_1 + S_2$, where

$$\begin{aligned} S_1 = \mathrm{E}\Bigg\{ &(a_0 + a_1 + a_2)\sum_{i,j} u_{ij}^4 - a_1 \sum_i \left(\sum_j u_{ij}\right)^4 \\ &- a_2 \sum_j \left(\sum_i u_{ij}\right)^4 \Bigg\}, \end{aligned} \tag{1.32}$$

whereas a_j, $j = 0, 1, 2$ and S_2 depend only on θ. Thus, S_2 can be estimated by replacing the variance components by their REML estimators, which are already available. As for S_1, it cannot be estimated in the same way for the reason given above. However, the form of S_1 [compare with (1.29)] suggests an "observed" estimator by taking out the expectation sign and replacing the parameters involved by their REML estimators. In fact, as $m, n \to \infty$, this observed S_1, say, $\hat{S}_1$, is consistent in the sense that $\hat{S}_1/S_1 \to 1$ in probability. It is interesting to note that S_2 cannot be consistently estimated by an observed form.

To summarize the basic idea, the elements of the QUIM can be expressed as $S_1 + S_2$, where S_1 cannot be estimated by an estimated form but can be estimated by an observed form; S_2 can be estimated by an estimated form (but not by an observed form). Thus, we have reached a balance. We propose to use such a method to estimate the QUIM. Because the estimator consists partially of an observed form and partially of an estimated one, it is called a partially observed quasi-information matrix, or POQUIM. The idea of POQUIM can be extended to a general non-Gaussian linear mixed model. See Section 1.8 for details.

Remark. The quasi-likelihood and POQUIM methods may be extended to the non-Gaussian marginal model. The ML and REML equations under the Gaussian marginal model are derived in Sections 1.3.1 and 1.3.2, respectively. Similar to the ANOVA case, the (quasi-) ML estimators of β and θ are defined as the solution to the ML equations; the (quasi-) REML estimator of θ is defined as the solution to the REML equation. However, because little assumption is made under the (non-Gaussian) marginal model, it is often difficult to study asymptotic properties of the estimators. In fact, some of the asymptotic results under the ANOVA model may not hold under the marginal model. For example, asymptotic normality often requires independence, to some extent, which may not exist at all under the marginal model. When asymptotic normality of the estimator does not hold, POQUIM may be meaningless because it is designed to estimate the asymptotic covariance matrix, which by definition is the covariance matrix of the asymptotic (multivariate) normal distribution. One exception is the non-Gaussian longitudinal model. See the discussion in Section 1.4.4.

1.4.3 Iterative Weighted Least Squares

In this and the next sections we discuss estimation in non-Gaussian longitudinal models. These models have been used in the analysis of longitudinal data (e.g., Diggle et al. 1996), where a traditional method of estimating the fixed effects is weighted least squares, or WLS. Suppose that the observations are collected from individuals over time. Let y denote the vector of observations, which may be correlated, and X a matrix of known covariates. Suppose that $\mathrm{E}(y) = X\beta$, where β is a vector of unknown regression coefficients. The WLS estimator of β is obtained by minimizing

$$(y - X\beta)'W(y - X\beta) , \tag{1.33}$$

where W is a known symmetric weighting matrix. As before, suppose, without loss of generality, that X is of full column rank p. Then, for any nonsingular W, the minimizer of (1.33) is given by

$$\hat{\beta}_W = (X'WX)^{-1}X'Wy \,. \tag{1.34}$$

As a special case, the ordinary least squares (OLS) estimator is obtained by choosing $W = I$, the identity matrix. This gives

$$\hat{\beta}_I = (X'X)^{-1}X'y \,. \tag{1.35}$$

On the other hand, the optimal choice of W in the sense of minimum variance is known to be $W = V^{-1}$, where $V = \mathrm{Var}(y)$. This is known as the best linear unbiased estimator, or BLUE, given by

$$\hat{\beta}_{\mathrm{BLUE}} = (X'V^{-1}X)^{-1}X'V^{-1}y \,. \tag{1.36}$$

However, inasmuch as V is typically unknown, the BLUE is not computable.

Let us turn our attention, for now, to a different problem which is related. The BLUE would be computable if V were known. On the other hand, it would be easier to estimate V if β were known. For example, an unbiased estimator of V is given by $\tilde{V} = (y - X\beta)(y - X\beta)'$. However, this is not a consistent estimator. In fact, if V is completely unknown, there are $n(n+1)/2$ unknown parameters in V, which is (far) more than the sample size n. Therefore, in such a case, one is not expected to have a consistent estimator of V no matter what. It is clear that some information about V must be available. For simplicity, let us first consider a special case.

Balanced case. Suppose that the observations are collected over a common set of times. Let y_{ij}, $j = 1, \ldots, k$ be the measures collected from the ith individual over times $t_1, \ldots, t_k$, respectively, and $y_i = (y_{ij})_{1 \leq j \leq k}$, $i = 1, \ldots, m$. Suppose that the vectors $y_1, \ldots, y_m$ are independent with

$$\mathrm{E}(y_i) = X_i\beta \quad \text{and} \quad \mathrm{Var}(y_i) = V_0, \tag{1.37}$$

where X_i is a matrix of known covariates, and $V_0 = (v_{qr})_{1 \leq q,r \leq k}$ is an unknown covariance matrix. It follows that $V = \mathrm{diag}(V_0, \ldots, V_0)$. Now the good thing is that V may be estimated consistently, if k is fixed. In fact, if β were known, a simple consistent estimator of V would be the following,

$$\begin{aligned} \hat{V} &= \mathrm{diag}(\hat{V}_0, \ldots, \hat{V}_0), \quad \text{where} \\ \hat{V}_0 &= \frac{1}{m}\sum_{i=1}^{m}(y_i - X_i\beta)(y_i - X_i\beta)' \,. \end{aligned} \tag{1.38}$$

To summarize the main idea, if V were known, one could use (1.36) to compute the BLUE of β; if β were known, one could use (1.38) to obtain a consistent estimator of V. It is clear that there is a cycle, which motivates the following algorithm when neither V nor β is assumed known: start with the OLS estimator (1.35) and compute $\hat{V}$ by (1.38) with β replaced by $\hat{\beta}_I$; then replace V on the right side of (1.36) by $\hat{V}$ just obtained to get the next step estimator of β; and repeat the process. We call such a procedure iterative weighted least squares, or I-WLS.

It can be shown that, if normality holds and the I-WLS converges, the limiting estimator is identical to the MLE. Also see Goldstein (1986). Therefore, the method may be regarded as quasi-likelihood. As shown later, such a property only holds in the balanced case. As for the convergence of the I-WLS algorithm, Jiang et al. (2006a) showed that, under mild conditions, the I-WLS converges exponentially with probability tending to one as the sample size increases. Such a result holds not only for the balanced case but also for the unbalanced case discussed below.

Unbalanced case. We now consider a more general case, in which the observations are not necessarily collected over a common set of times. This includes the cases: (i) the observations are supposed to be collected at a common set of times but there are missing observations; and (ii) the observations are not designed to be collected over a common set of times (e.g., some on Monday/Wednesday/Friday and some on Tuesday/Thursday). Let $T = \{t_1, \dots, t_k\}$ be the set of times at which at least one observation is collected. Then, (ii) may be viewed as a special case of (i), in which some observations are intentionally "missed." Therefore we may focus on case (i).

It is then more convenient to denote the observations as y_{ij}, $j \in J_i$, where J_i is a subset of $J = \{1, \dots, k\}$, $1 \le i \le m$, such that y_{ij} corresponds to the observation collected from the ith individual at time t_j. Write $y_i = (y_{ij})_{j \in J_i}$. Suppose that $y_1, \dots, y_m$ are independent with $\mathrm{E}(y_i) = X_i \beta$, where X_i is a matrix of known covariates whose jth row ($j \in J_i$) is x'_{ij} with $x_{ij} = (x_{ijk})_{1 \le k \le p}$. As for the covariance matrix, it follows that $V = \mathrm{diag}(V_1, \dots, V_m)$, where $V_i = \mathrm{Var}(y_i)$. If the V_is are completely unknown, there are still more unknown covariance parameters than the sample size. In such a case, again, consistent estimation of V is not possible. Therefore, one needs to further specify the V_is. For simplicity, we assume that for any $q, r \in J_i$, $\mathrm{cov}(y_{iq}, y_{ir})$ does not depend on i. This includes some important cases in practice. The following is an example.

Example 1.8. Suppose that the observation times are equally spaced. In such a case we may assume, without loss of generality, that $t_j = j$. Suppose that the observations y_{ij} satisfy

$$y_{ij} = x'_{ij}\beta + \xi_i + \zeta_{ij} + \epsilon_{ij} ,$$

$i = 1, \dots, m$, $j \in J_i \subset T = J = \{1, \dots, k\}$, where ξ_i, is an individual-specific random effect, ζ_{ij} corresponds to a serial correlation, and ϵ_{ij} represents a measurement error. It is assumed that the ξ_is are i.i.d. with mean 0 and variance σ_1^2, and the ϵ_{ij}s are i.i.d. with mean 0 and variance τ^2. Furthermore, the ζ_{ij}s satisfy the same AR(1) model as described in Example 1.3 except that the ω_{ij}s are i.i.d. (not necessarily normal) with mean 0 and variance $\sigma_2^2(1-\phi^2)$. Also, we assume that ξ, ζ, and ϵ are independent. It is easy to show that (e.g., Anderson 1971b, pp. 174)

$$\mathrm{cov}(y_{iq}, y_{ir}) = \sigma_1^2 + \sigma_2^2\phi^{|q-r|} + \tau^2\delta_{q,r} ,$$

where $\delta_{q,r} = 1$ if $q = r$ and 0 otherwise. Of course, in practice, the above covariance structure may not be known. Therefore, as a more robust approach one may have to estimate the covariances $\mathrm{cov}(y_{iq}, y_{ir})$, and, in this case, it is true that the latter do not depend on i.

Denote $\mathrm{cov}(y_{iq}, y_{ir})$ by v_{qr} for any i such that $q, r \in J_i$. Let $D = \{(q,r) : q, r \in J_i$ for some $1 \le i \le m$ and $q \le r\}$ and $d = |D|$, the cardinality of D. We assume that D does not change with m and neither does k (otherwise, the number of covariances changes with the sample size). Let $v = (v_{qr})_{(q,r)\in D}$ be the vector of different covariance parameters.

We now describe the I-WLS procedure. If v were known, the BLUE of β would be given by (1.36), which now has the expression

$$\hat{\beta}_{\mathrm{BLUE}} = \left(\sum_{i=1}^m X_i' V_i^{-1} X_i\right)^{-1} \sum_{i=1}^m X_i' V_i^{-1} y_i . \tag{1.39}$$

On the other hand, if β were known, a method of moments estimator of v would be given by $\hat{v} = (\hat{v}_{qr})_{(q,r)\in D}$, where

$$\hat{v}_{qr} = \frac{1}{m_{qr}} \sum_{i:q,r\in J_i} (y_{iq} - x_{iq}'\beta)(y_{ir} - x_{ir}'\beta) , \tag{1.40}$$

and $m_{qr} = |\{1 \le i \le m : q, r \in J_i\}|$. An estimator of V_i would then be $\hat{V}_i = (\hat{v}_{qr})_{q,r\in J_i}$, $1 \le i \le m$. Obviously, when there are no missing observations, this is the same as (1.38). On the other hand, when the data are unbalanced the $\hat{v}_{qr}$s cannot be derived from the quasi-likelihood. Nevertheless, under mild conditions $\hat{v}_{qr}$ is a consistent estimator of v_{qr}, if β is known and $m_{qr} \to \infty$. When both β and v are unknown, we iterate between (1.39) and (1.40), starting with the OLS estimator. This can be formulated as follows. Let $f(v)$ be the right side of (1.39), and $g(\beta) = \{g_{qr}(\beta)\}_{(q,r)\in D}$, where $g_{qr}(\beta)$ is given by the right side of (1.40). Then, similar to the balanced case, we have $\hat{v}^{(0)} = (\delta_{q,r})_{(q,r)\in D}$, where $\delta_{q,r}$ is defined in Example 1.8 above; $\hat{\beta}^{(1)} = f\{\hat{v}^{(0)}\}$, $\hat{v}^{(1)} = \hat{v}^{(0)}$; $\hat{\beta}^{(2)} = \hat{\beta}^{(1)}$, $\hat{v}^{(2)} = g\left\{\hat{\beta}^{(1)}\right\}$; .... In general, we have

$$\begin{aligned} \hat{\beta}^{(2h-1)} &= f\left\{\hat{v}^{(2h-2)}\right\}, & \hat{v}^{(2h-1)} &= \hat{v}^{(2h-2)}; \\ \hat{\beta}^{(2h)} &= \hat{\beta}^{(2h-1)}, & \hat{v}^{(2h)} &= g\left\{\hat{\beta}^{(2h-1)}\right\}; \end{aligned}$$

for $h = 1, 2, \ldots$. Similar to the balanced case, such a procedure is called iterative weighted least squares, or I-WLS.

Jiang et al. (2006a) showed that, under mild conditions, the I-WLS converges exponentially with probability tending to one as the sample size increases. Furthermore, the limiting estimator (i.e., the estimator obtained at convergence) is consistent and asymptotically as efficient as the BLUE.

1.4.4 Jackknife Method

Although, at convergence, the I-WLS leads to estimators of β and V, its main purpose is to produce an efficient estimator for β. One problem with I-WLS is that it only produces point estimators. A naive estimator of $\Sigma = \mathrm{Var}(\hat{\beta})$, where $\hat{\beta}$ is the I-WLS estimator of β, may be obtained by replacing V in the covariance matrix of the BLUE (1.36), which is $(X'V^{-1}X)^{-1}$, by $\hat{V}$, the I-WLS estimator of V. However, such an estimator of Σ is likely to be an underestimator of the true variation, because it does not take into account the additional variation caused by estimating V.

Furthermore, in some cases the covariance structure of the data is specified up to a set of parameters, that is, $V = V(\theta)$, where θ is a vector of unknown variance components. In such cases the problems of interest may include estimation of both β and θ. Note that the I-WLS is developed under a nonparametric covariance structure, and therefore does not apply directly to this case. On the other hand, a similar quasi-likelihood method to that discussed in Section 1.4.1 may apply to this case. In particular, the quasi-likelihood is obtained by first assuming that the longitudinal model is a Gaussian one. Note that, under the longitudinal model, the observations can be divided into independent blocks (i.e., $y_1, \dots, y_m$). Therefore, asymptotic results for quasi-likelihood estimators with independent observations may apply (see Heyde 1997). The asymptotic covariance matrix of the estimator may be estimated by the POQUIM method of Section 1.4.2.

Alternatively, the asymptotic covariance matrix may be estimated by the jackknife method. The jackknife method was first introduced by Quenouille (1949) and later developed by Tukey (1958). It has been used in estimating the bias and variation of estimators, mostly in the i.i.d. case. See Shao and Tu (1995). In the case of correlated observations with general M-estimators of parameters, the method was developed in the context of small area estimation (see Jiang, Lahiri, and Wan 2002). One advantage of the method is that it applies in the same way to different estimating procedures, including I-WLS and quasi-likelihood, and to generalized linear mixed models as well (see Section 3.6.2). In the following we describe such a method.

Consider the non-Gaussian longitudinal model defined in Section 1.2.2.2. Suppose that the vector $\psi = (\beta', \theta')$ is estimated by an M-estimating procedure, in which the estimator of ψ, $\hat{\psi}$, is a solution to the following equation,

$$\sum_{i=1}^{m} f_i(\psi, y_i) + a(\psi) = 0, \tag{1.41}$$

where $f_i(\cdot,\cdot)$ and $a(\cdot)$ are vector-valued with the same dimension as ψ such that $\mathrm{E}\{f_i(\psi, y_i)\} = 0$, $1 \le i \le m$. Such M-estimators include, for example, ML and REML estimators as well as the limiting I-WLS estimator. Similarly, the delete-i M-estimator of ψ, $\hat{\psi}_{-i}$, is a solution to the following equation:

$$\sum_{j \neq i} f_j(\psi, y_j) + a_{-i}(\psi) = 0, \tag{1.42}$$

where $a_{-i}(\cdot)$ has the same dimension as $a(\cdot)$. Let Σ be the asymptotic covariance matrix of $\hat{\psi}$. A jackknife estimator of Σ is then given by

$$\hat{\Sigma}_{\mathrm{Jack}} = \frac{m-1}{m} \sum_{i=1}^{m} (\hat{\psi}_{-i} - \hat{\psi})(\hat{\psi}_{-i} - \hat{\psi})'. \tag{1.43}$$

Jiang and Lahiri (2004) showed that, under suitable conditions, the jackknife estimator is consistent in the sense that $\hat{\Sigma}_{\mathrm{Jack}} = \Sigma + O_{\mathrm{P}}(m^{-1-\delta})$ for some $\delta > 0$. As is shown, the same jackknife estimator (1.43) also applies to longitudinal generalized linear mixed models such that the same asymptotic property holds (see Section 3.6.2).

1.5 Other Methods of Estimation

The ML and REML methods require maximization of multivariate nonlinear functions, namely, the likelihood or restricted likelihood functions, or (at least) simultaneously solving systems of nonlinear equations. Such tasks were quite challenging computationally many years ago, when computer technology was not as advanced. On the other hand, methods that are computationally simpler were developed, mostly before the ML and REML methods became popularized. Among these are analysis of variance (ANOVA) estimation, proposed by C. R. Henderson for unbalanced data and minimum norm quadratic unbiased estimation, or MINQUE, proposed by C. R. Rao. A common feature of these methods is that they do not require normality of the data. In this section we discuss these two methods. The discussion is restricted to the mixed ANOVA model (1.1) and (1.2).

1.5.1 Analysis of Variance Estimation

The basic idea of ANOVA estimation came from the method of moments. Suppose that there are q variance components involved in a linear mixed model. Let Q be a q-dimensional vector whose components are quadratic functions of the data. The ANOVA estimators of the variance components are obtained by solving the system of equations $\mathrm{E}(Q) = Q$. The only thing not clear at this point is how to choose Q. For balanced data the choice is straightforward; whereas for unbalanced data it is less obvious. We first consider the balanced case.

1. Balanced data. For balanced data, the components of Q are determined by the ANOVA tables (e.g., Scheffé 1959). We illustrate the method by a simple example. The general description of the rules can be found in Henderson (1969).

Example 1.1 (continued). Consider the balanced case of Example 1.1, that is, $k_i = k$, $1 \le i \le m$. If the α_is are (fixed) treatment effects, the ANOVA table for analyzing the treatment effects is given below, where $\mathrm{SSA} = k\sum_{i=1}^m (\bar{y}_{i\cdot} - \bar{y}_{\cdot\cdot})^2$, $\mathrm{SSE} = \sum_{i=1}^m \sum_{j=1}^k (y_{ij} - \bar{y}_{i\cdot})^2$, and $\mathrm{SS}_{\mathrm{total}} = \sum_{i=1}^m \sum_{j=1}^k (y_{ij} - \bar{y}_{\cdot\cdot})^2$. Because there are two variance components, σ^2 and τ^2, the components of

Table 1.1. ANOVA table

Source	SS	df	MS	F
Treatment	SSA	$m-1$	$\mathrm{MSA} = \mathrm{SSA}/(m-1)$	MSA/MSE
Error	SSE	$m(k-1)$	$\mathrm{MSE} = \mathrm{SSE}/m(k-1)$	
Total	$\mathrm{SS}_{\mathrm{total}}$	$mk-1$		

Q consist of SSA and SSE, and we have $\mathrm{E(SSA)} = (m-1)k\sigma^2 + (m-1)\tau^2$, $\mathrm{E(SSE)} = m(k-1)\tau^2$. Thus, the ANOVA estimating equations are

$$\begin{cases} (m-1)k\sigma^2 + (m-1)\tau^2 = \mathrm{SSA}, \\ m(k-1)\tau^2 \qquad\qquad\qquad = \mathrm{SSE}. \end{cases}$$

The resulting ANOVA estimators are therefore $\hat{\sigma}^2 = (\mathrm{MSA} - \mathrm{MSE})/k$, $\hat{\tau}^2 = \mathrm{MSE}$ (Exercise 1.16).

Note. It is seen that, unlike ML and REML, ANOVA estimators of the variance components may not belong to the parameter space. For example, in the above example, $\hat{\sigma}^2$ will be negative if MSA $<$ MSE. This is one of the drawbacks of the ANOVA method.

For balanced data, there is a remarkable relationship between the AVOVA estimator and REML estimator discussed earlier. It is known that, under a balanced mixed ANOVA model, the ANOVA estimator of $\theta = (\tau^2, \sigma_i^2, 1 \le i \le s)'$ is identical to the solution of the REML equations (e.g., Anderson 1979). Of course, the solution to the REML equations is not necessarily the REML estimator, because, by definition, the latter has to be in the parameter space. On the other hand, when the solution does belong to the parameter space, the REML and ANOVA estimators are identical. This result holds regardless of the normality assumption (see Section 1.4.1), but it does require being balanced (Exercise 1.16).

Unbalanced data. Henderson (1953) proposed three methods, known as Henderson's Methods I, II and III, for ANOVA estimation with unbalanced data. Here we introduce the third method, which applies most broadly to unbalanced cases. To determine the ANOVA equations, all one has to do is to determine the quadratic forms that are the components of Q. Henderson's Method III proposes to do so by decomposing the (regression) sum of squares of residuals (SSR). Again, we illustrate the method by an example and for more details refer the reader to Henderson's paper.

Example 1.9. Consider a special case of the mixed ANOVA model (1.1) and (1.2): $y = X\beta + Z_1\alpha_1 + Z_2\alpha_2 + \epsilon$, where the terms are defined as in Section 1.2.1.1 except that normality is not required.

First write the model as $y = W\gamma + \epsilon$, where $W = (X, Z)$ with $Z = (Z_1, Z_2)$, and $\gamma = (\beta', \alpha_1', \alpha_2)'$. If this were a fixed effects model (i.e., linear regression), one would have $\text{SSR} = \text{SSR}(\alpha, \beta) = |P_W y|^2 = y'P_W y$, where P_W is the projection matrix (see Appendix B). On the other hand, if there were no random effects, one would have $y = X\beta + \epsilon$, and the corresponding SSR is $\text{SSR}(\beta) = |P_X y|^2 = y'P_X y$. Thus, the difference in SSR is $\text{SSR}(\alpha|\beta) = \text{SSR}(\alpha, \beta) - \text{SSR}(\beta) = y'P_{Z\ominus X}y$, where $Z \ominus X = P_{X^\perp}Z$ with $P_{X^\perp} = I - P_X$. Here we use the facts that $P_W = P_{X,Z} = P_X + P_{Z\ominus X}$ and the last two projections are orthogonal to each other (Exercise 1.17). Thus, the first quadratic form for ANOVA estimation is $\text{SSR}(\alpha|\beta)$.

Next, we have, similarly, $\text{SSR}(\alpha_2|\beta, \alpha_1) = \text{SSR}(\alpha, \beta) - \text{SSR}(\alpha_1, \beta) = y'P_{Z_2\ominus(X,Z_1)}y$, where $Z_2 \ominus (X, Z_1) = P_{(X,Z_1)^\perp}Z_2$ and $P_{Z_2\ominus(X,Z_1)} = P_W - P_{(X,Z_1)}$ (Exercise 1.17). This is the second quadratic form for ANOVA estimation.

Finally, the last quadratic form for ANOVA estimation is $\text{SSE} = y'P_{W^\perp}y$.

In conclusion, the three quadratic forms for estimating σ_1^2, σ_2^2, and τ^2 are $\text{SSR}(\alpha|\beta)$, $\text{SSR}(\alpha_2|\beta, \alpha_1)$, and SSE, which are the components of Q.

In order to determine $\text{E}(Q)$, note that the expected value of each of the above quadratic forms is a linear function of σ_1^2, σ_2^2, and τ^2. Thus, all one has to do is to determine the coefficients in those linear functions. The following lemma may be helpful in this regard.

Lemma 1.1. Let A be any symmetric matrix such that $X'AX = 0$. Under the mixed ANOVA model (1.1) and (1.2) without the normality assumption, the coefficient of σ_i^2 in $\text{E}(y'Ay)$ is $\text{tr}(AZ_iZ_i')$, $0 \le i \le s$, where $\sigma_0^2 = \tau^2$ and $Z_0 = I_n$.

Proof. By Appendix C we have $\text{E}(y'Ay) = \text{tr}(AV) + \beta'X'AX\beta = \text{tr}(AV)$, where $V = \sum_{i=0}^s \sigma_i^2 Z_i Z_i'$. Thus, $\text{E}(y'Ay) = \sum_{i=0}^s \sigma_i^2 \text{tr}(AZ_iZ_i')$. ∎

In $\text{SSR}(\alpha|\beta)$ of Example 1.9, the coefficient of σ_i^2 is $\text{tr}(P_{X^\perp}Z_iZ_i')$, $i = 1, 2$ and the coefficient of τ^2 is $\text{tr}(P_{Z\ominus X}) = \text{rank}(W) - \text{rank}(X)$. Here we use the identity $P_{Z\ominus X}Z_i = P_{X^\perp}Z_i$, $i = 1, 2$. Similarly, in $\text{SSR}(\alpha_2|\beta, \alpha_1)$ and SSE, the coefficients for σ_1^2, σ_2^2, τ^2 are 0, $\text{tr}\{P_{(X,Z_1)^\perp}Z_2Z_2'\}$, $\text{rank}(W) - \text{rank}\{(X, Z_1)\}$, and 0, 0, $n - \text{rank}(W)$, respectively, where n is the sample size (i.e., dimension of y; Exercise 1.17).

1.5.2 Minimum Norm Quadratic Unbiased Estimation

This method, known as MINQUE, was proposed by C. R. Rao in a series of papers (1970, 1971, 1972). The form of the estimator in MINQUE is similar to that of ANOVA, that is, obtained by solving a system of equations $\mathrm{E}(Q) = Q$, where Q is a vector of quadratic forms. Again, the question is: what Q?

Write $\theta = (\sigma_i^2)_{0 \le i \le s}$, where, as usual, $\sigma_0^2 = \tau^2$. Consider estimation of a linear function of θ, say, $\eta = b'\theta$, where $b = (b_i)_{0 \le i \le s}$. Suppose that the estimator is a quadratic form in y, say, $\hat{\eta} = y'Ay$. If we can determine the A here, we can subsequently determine Q. We assume that A is symmetric such that $A'X = 0$, and the estimator $\hat{\eta}$ is unbiased. By Lemma 1.1, the latter assumption implies that

$$b_i = \mathrm{tr}(AZ_iZ_i'), \qquad 0 \le i \le s. \tag{1.44}$$

Furthermore, we have, by (1.1), $\hat{\eta} = \alpha' Z_*' A Z_* \alpha$, where $Z_* = (Z_0, Z_1, \dots, Z_s)$, $\alpha = (\alpha_0', \alpha_1', \dots, \alpha_s')'$ with $\alpha_0 = \epsilon$. On the other hand, if α_i were observable, a method of moment estimator of σ_i^2 would be $m_i^{-1} \sum_{j=1}^{m_i} \alpha_{ij}^2 = m_i^{-1} |\alpha_i|^2$, $0 \le i \le s$, where m_i is the dimension of α_i with $m_0 = n$. Thus, an unbiased estimator of η would be $\tilde{\eta} = \sum_{i=0}^{s} b_i m_i^{-1} |\alpha_i|^2 = \alpha' B \alpha$, where $B = \mathrm{diag}(b_i m_i^{-1} I_{m_i}, 0 \le i \le s)$. The reality is: $\hat{\eta}$ is the actual estimator. By Lemma 1.2 below it can be shown that, under normality, the mean squared difference $\mathrm{E}(|\hat{\eta} - \tilde{\eta}|^2)$ is equal to $2\mathrm{tr}[\{(Z_*AZ_* - B)D\}^2]$, where $D = \mathrm{diag}(\sigma_i^2 I_{m_i}, 0 \le i \le s)$. Note that (1.44) implies that $\mathrm{E}(\hat{\eta} - \tilde{\eta}) = 0$ (Exercise 1.18). Thus, under normality, A may be chosen by minimizing the mean squared difference between the actual estimator $\hat{\eta}$ and "would be" estimator $\tilde{\eta}$. Without normality, the matrix A may still be chosen this way, with the interpretation that it minimizes a weighted Euclidean norm $\mathrm{tr}[\{(Z_*AZ_* - B)D\}^2] = \|D^{1/2}(Z_*AZ_* - B)D^{1/2}\|_2^2$. The resulting estimator $\hat{\eta}$ is called the minimum norm quadratic unbiased estimator, or MINQUE, of η. By suitably choosing b one obtains MINQUE for σ_i^2, $0 \le i \le s$.

However, before the minimization is performed one needs to know D or, equivalently, σ_i^2, $0 \le i \le s$, which are exactly the variance components we wish to estimate. It is suggested that the σ_i^2s be replaced by some initial values σ_{0i}^2, $0 \le i \le s$ in order to compute the MINQUE. It follows that the estimator depends on the initial values. On the other hand, the fact that MINQUE depends on the initial values motivates an iterative procedure in which the (current) MINQUE is used as initial values to repeat the calculation, and repeat again. This procedure is called iterative MINQUE, or I-MINQUE. Unlike MINQUE, I-MINQUE, if it converges, is not affected by initial values. This is because the limiting I-MINQUE satisfies the REML equations. In other words, I-MINQUE is identical to the REML estimator (see Section 1.4.1) if the restriction that the latter belong to the parameter space is not imposed (e.g., Searle et al. 1992, §11.3). However, neither MINQUE nor I-MINQUE is guaranteed to lie in the parameter space. Brown (1976) showed that, under suitable conditions, I-MINQUE is consistent and asymptotically normal. For more about MINQUE and related methods, see Rao and Kleffe (1988).

Lemma 1.2. Let A_1, A_2 be symmetric matrices and $\xi \sim N(0, \Sigma)$. Then, $\mathrm{E}[\{\xi' A_1 \xi - \mathrm{E}(\xi' A_1 \xi)\}\{\xi' A_2 \xi - \mathrm{E}(\xi' A_2 \xi)\}] = 2\mathrm{tr}(A_1 \Sigma A_2 \Sigma)$.

1.6 Notes on Computation and Software

1.6.1 Notes on Computation

1. Computation of the ML and REML estimators. From a computational standpoint, the more challenging part of the analysis of linear mixed models is the computation of maximum likelihood and restricted maximum likelihood estimators. Because the likelihood or restricted likelihood functions under a Gaussian mixed model can be expressed in closed-forms, the maximization of these functions can be done, in principle, by standard numerical procedures, such as Newton–Raphson. However, a more efficient algorithm may be developed based on the nature of the Gaussian mixed model.

To see this, let us first consider ML estimation under a Gaussian mixed model. Note that a general Gaussian mixed model can be expressed as (1.1), where $\epsilon \sim N(0, R)$. In many cases, the covariance matrix R is equal to $\tau^2 I$, where τ^2 is an unknown positive variance and I the $(n \times n)$ identity matrix. As for $G = \mathrm{Var}(\alpha)$, there is a decomposition such that $G = \tau^2 UU'$, although the decomposition is not unique. One that is frequently used is Cholesky's decomposition, in which U is lower-triangular. Another well-known decomposition is the eigenvalue decomposition, in which $U = T\mathrm{diag}(\sqrt{\lambda_1}/\tau, \ldots, \sqrt{\lambda_m}/\tau)$, T is an orthogonal matrix, and $\lambda_1, \ldots, \lambda_m$ are the eigenvalues of G (m is the dimension of α). See Appendix B for more about these decompositions. Suppose that G is specified up to a vector ψ of dispersion parameters; that is, $G = G(\psi)$. Then, $U = U(\psi)$. Furthermore, if R is nonsingular, so is U. Denote the conditional density function of y given α by $f(y|\alpha)$, and the density function of α by $f(\alpha)$. Then, we have

$$\begin{aligned} f(y|\alpha)f(\alpha) &= \frac{1}{(2\pi\tau^2)^{n/2}} \exp\left(-\frac{1}{2\tau^2}|y - X\beta - Z\alpha|^2\right) \\ &\quad \times \frac{1}{(2\pi\tau^2)^{m/2}|U|} \exp\left(-\frac{1}{2\tau^2}|U^{-1}\alpha|^2\right) \\ &= \frac{1}{(2\pi\tau^2)^{(m+n)/2}|U|} \exp\left(-\frac{1}{2\tau^2}|\tilde{y} - \tilde{X}\beta - \tilde{Z}\alpha|^2\right), \end{aligned}$$

where

$$\tilde{y} = \begin{pmatrix} y \\ 0 \end{pmatrix}, \qquad \tilde{X} = \begin{pmatrix} X \\ 0 \end{pmatrix}, \qquad \tilde{Z} = \begin{pmatrix} Z \\ U^{-1} \end{pmatrix}.$$

For a given β, write $\tilde{u} = \tilde{y} - \tilde{X}\beta$. According to the geometry of the least squares, we have the orthogonal decomposition:

$$
\begin{aligned}
|\tilde{u} - \tilde{Z}\alpha|^2 &= |\tilde{u} - \tilde{Z}\tilde{\alpha}|^2 + |\tilde{Z}(\tilde{\alpha} - \alpha)|^2 \\
&= |\tilde{u} - \tilde{Z}\tilde{\alpha}|^2 + (\alpha - \tilde{\alpha})'\tilde{Z}'\tilde{Z}(\alpha - \tilde{\alpha}),
\end{aligned}
$$

where $\tilde{\alpha} = (\tilde{Z}'\tilde{Z})^{-1}\tilde{Z}'\tilde{u}$ so that $\tilde{Z}\tilde{\alpha} = P_{\tilde{Z}}\tilde{u}$, the projection of $\tilde{u}$ to $\mathcal{L}(\tilde{Z})$, the linear space spanned by the columns of $\tilde{Z}$. Note that

$$
\int \exp\left\{-\frac{1}{2\tau^2}(\alpha - \tilde{\alpha})'\tilde{Z}'\tilde{Z}(\alpha - \tilde{\alpha})\right\} d\alpha = \frac{(2\pi\tau^2)^{m/2}}{|\tilde{Z}'\tilde{Z}|^{1/2}}.
$$

Thus, we have

$$
\int f(y|\alpha)f(\alpha)d\alpha = \frac{1}{(2\pi\tau^2)^{n/2}|U|\cdot|\tilde{Z}'\tilde{Z}|^{1/2}} \exp\left(-\frac{1}{2\tau^2}|\tilde{u} - \tilde{Z}\tilde{\alpha}|^2\right).
$$

Also note that $|U|\cdot|\tilde{Z}'\tilde{Z}|^{1/2} = |UU'|^{1/2}|Z'Z+(UU')^{-1}|^{1/2} = |I_m+UU'Z'Z|^{1/2}$, and $\tilde{u} - \tilde{Z}\tilde{\alpha} = y^* - X^*\beta$, where $y^* = P_{\tilde{Z}^\perp}\tilde{y}$, $X^* = P_{\tilde{Z}^\perp}\tilde{X}$ with $P_{A^\perp} = I - P_A$, the projection to the linear space orthogonal to $\mathcal{L}(A)$. It follows that the log-likelihood function can be expressed as

$$
\begin{aligned}
l = c &- \frac{n}{2}\log(\tau^2) - \frac{1}{2}\log(|I_m + UU'Z'Z|) \\
&- \frac{1}{2\tau^2}|y^* - X^*\beta|^2,
\end{aligned}
\tag{1.45}
$$

where c is a constant. It is seen that, given ψ, the maximization of (1.45) is equivalent to fitting the linear regression $y^* = X^*\beta + \epsilon$, where the components of ϵ are i.i.d. and distributed as $N(0, \tau^2)$. Thus, the maximizer of (1.45) given ψ is given by

$$
\begin{aligned}
\tilde{\beta} &= (X^{*'}X^*)^{-1}X^{*'}y^* \\
&= (\tilde{X}'P_{\tilde{Z}^\perp}\tilde{X})^{-1}\tilde{X}'P_{\tilde{Z}^\perp}\tilde{y} \\
&= P(X)^{-1}P(y),
\end{aligned}
\tag{1.46}
$$

here for any matrix or vector A, $P(A) = X'A - X'Z\{Z'Z + (UU')^{-1}\}^{-1}Z'A$ (see Appendix B for properties of projection matrices), and

$$
\begin{aligned}
\tilde{\tau}^2 &= \frac{1}{n}|y^* - X^*\tilde{\beta}|^2 \\
&= \frac{1}{n}|P_{\tilde{Z}^\perp}(\tilde{y} - \tilde{X}\tilde{\beta})|^2 \\
&= \frac{1}{n}[|y - X\tilde{\beta}|^2 \\
&\quad -(y - X\tilde{\beta})'Z\{Z'Z + (UU')^{-1}\}^{-1}Z'(y - X\tilde{\beta})].
\end{aligned}
\tag{1.47}
$$

This leads to the following algorithm. Express the log-likelihood function as a profile log-likelihood by plugging (1.46) and (1.47) into (1.45); that is,

$$l_{\rm p}(\psi) = c - \frac{n}{2}\log(\tilde{\tau}^2) - \frac{1}{2}\log(|I_m + UU'Z'Z|), \tag{1.48}$$

where c is another constant; then maximize (1.48) with respect to ψ to find the MLE for ψ, say, $\hat{\psi}$. The MLE for β and τ^2, say, $\hat{\beta}$ and $\hat{\tau}^2$, are thus computed by (1.46) and (1.47) with ψ replaced by $\hat{\psi}$. We use a simple example to illustrate the algorithm.

Example 1.1 (Continued). Consider a special case of Example 1.1 with $k_i = k$, $1 \leq i \leq m$. The model can be written in the standard form (1.1), with $X = 1_m \otimes 1_k$, $\beta = \mu$, $Z = I_m \otimes 1_k$, $R = \tau^2 I_m \otimes I_k$, and $G = \sigma^2 I_m$. Furthermore, we have $G = \tau^2 UU'$ with $U = \sqrt{\psi} I_m$ and $\psi = \sigma^2/\tau^2$. Thus, it is easy to show that $P(X) = mk/(1+k\psi)$, $P(y) = y_{\cdot\cdot}/(1+k\psi)$. Thus, $\tilde{\beta} = \tilde{\mu} = y_{\cdot\cdot}/mk = \bar{y}_{\cdot\cdot}$. Furthermore, it can be shown (Exercise 1.19) that

$$\tilde{\tau}^2 = \frac{1}{n}\left(\mathrm{SSW} + \frac{\mathrm{SSB}}{1+k\psi}\right),$$

where $\mathrm{SSB} = k\sum_{i=1}^m (\bar{y}_{i\cdot} - \bar{y}_{\cdot\cdot})^2$, $\mathrm{SSW} = \sum_{i=1}^m \sum_{j=1}^k (y_{ij} - \bar{y}_{i\cdot})^2$ with $\bar{y}_{i\cdot} = k^{-1}\sum_{j=1}^k y_{ij}$. It follows that the profile log-likelihood has the form

$$l_{\rm p}(\psi) = c - \frac{n}{2}\log(\mathrm{SSB} + \lambda\mathrm{SSW}) + \frac{n-m}{2}\log\lambda,$$

where $\lambda = 1 + k\psi$. The maximum of $l_{\rm p}$ is given by

$$\hat{\lambda} = \left(1 - \frac{1}{m}\right)\frac{\mathrm{MSB}}{\mathrm{MSW}},$$

where $\mathrm{MSB} = \mathrm{SSB}/(m-1)$ and $\mathrm{MSW} = \mathrm{SSW}/(n-m)$, or

$$\hat{\psi} = \frac{1}{k}\left\{\left(1 - \frac{1}{m}\right)\frac{\mathrm{MSB}}{\mathrm{MSW}} - 1\right\}.$$

Thus, the MLE for τ^2 is given by $\hat{\tau}^2 = \mathrm{MSW}$.

Now consider REML estimation under a Gaussian mixed model. Again, assume that $R = \tau^2 I_n$. Let A be the matrix (of full rank) corresponding to the orthogonal transformation in REML (see Section 1.3.2). Because the REML estimator is not affected by the choice of A, one may choose A such that, in addition to (1.16),

$$A'A = I_{n-p}, \tag{1.49}$$

where $p = \mathrm{rank}(X)$, so that

$$\begin{aligned} AA' &= I_n - X(X'X)^{-1}X' \\ &\equiv P \end{aligned} \tag{1.50}$$

(see Appendix B). Then, we have $z = A'y = H\alpha + \zeta$, where $H = A'Z$ and $\zeta \sim N(0, \tau^2 I_{n-p})$. Thus, by similar arguments, we have

$$f(z|\alpha)f(\alpha) = \frac{1}{(2\pi\tau^2)^{(n-p+m)/2}|U|} \exp\left(-\frac{1}{2\tau^2}|\tilde{z} - \tilde{H}\alpha|^2\right),$$

where

$$\tilde{z} = \begin{pmatrix} z \\ 0 \end{pmatrix}, \qquad \tilde{H} = \begin{pmatrix} H \\ U^{-1} \end{pmatrix},$$

so that the restricted log-likelihood can be expressed as

$$\begin{aligned} l_{\mathrm{R}} = c &- \frac{n-p}{2}\log(\tau^2) - \frac{1}{2}\log(|I_m + UU'Z'PZ|) \\ &- \frac{1}{2\tau^2}|P_{\tilde{H}^\perp}\tilde{z}|^2, \end{aligned} \tag{1.51}$$

where $\tilde{\alpha} = (\tilde{H}'\tilde{H})^{-1}\tilde{H}'\tilde{z}$ so that $\tilde{H}\tilde{\alpha} = P_{\tilde{H}}\tilde{z}$. Given ψ, the maximizer of (1.51) is given by

$$\begin{aligned} \tilde{\tau}^2 &= \frac{1}{n-p}|P_{\tilde{H}^\perp}\tilde{z}|^2 \\ &= \frac{1}{n-p}[y'Py - y'PZ\{Z'PZ + (UU')^{-1}\}^{-1}Z'Py], \end{aligned} \tag{1.52}$$

hence the profile restricted log-likelihood has the form

$$l_{\mathrm{R,p}}(\psi) = c - \frac{n-p}{2}\log(\tilde{\tau}^2) - \frac{1}{2}\log(|I_m + UU'Z'PZ|), \tag{1.53}$$

where c is another constant. The maximizer of (1.53), say, $\hat{\psi}$, will be the REML estimator of ψ; the REML estimator of τ^2 is then given by (1.52) with ψ replaced by $\hat{\psi}$. Again, we consider a simple example.

Example 1.1 (Continued). In this case, $P = I_n - n^{-1}J_n$, so that

$$\tilde{\tau}^2 = \frac{1}{n-1}\left(\mathrm{SSW} + \frac{\mathrm{SSB}}{1+k\psi}\right).$$

It follows that the restricted profile log-likelihood has the form

$$l_{\mathrm{R,p}}(\psi) = c - \frac{n-1}{2}\log(\mathrm{SSB} + \lambda\mathrm{SSW}) + \frac{n-m}{2}\log\lambda,$$

where λ is defined earlier. This is the same as l_{p} except that n is replaced by $n-1$ in the term that involves SSB and SSW. The maximizer of $l_{\mathrm{R,p}}$ is given by

$$\hat{\lambda} = \frac{\mathrm{MSB}}{\mathrm{MSW}},$$

or

$$\hat{\psi} = \frac{1}{k}\left(\frac{\text{MSB}}{\text{MSW}} - 1\right).$$

The REML estimator of τ^2 remains the same as $\tau^2 = \text{MSW}$.

Note that the maximizer of the (restricted) profile log-likelihood may fall outside the parameter space of ψ, say, Ψ. For example, in Example 1.1 (continued), $\Psi = [0, \infty)$, but the maximizer of ψ obtained as the solution to the (restricted) profile likelihood equation may be negative. (The likelihood equations are obtained by setting the derivatives equal to zero.) In such a situation, the maximizer within Ψ lies on the boundary of Ψ, and therefore cannot be obtained by solving the (restricted) profile likelihood equation. However, the maximizer within the parameter space may still be obtained by searching for the maximum along the boundary of Ψ. See, for example, Searle et al. (1992, §8.1) for more details.

2. The EM algorithm. According to the discussion above, the key component in the computation of ML and REML estimators is the maximization of the profile log-likelihood, or restricted profile log-likelihood, with respect to ψ. Note that this is a nonlinear maximization problem. Although standard numerical procedures, such as Newton–Raphson, are available, the procedure is often sensitive to initial values.

Alternatively, one may use the EM algorithm (Dempster et al. 1977) to compute the MLE. The idea is to treat the random effects as "missing data." See Section 4.1.1 for more details. The EM algorithm is known to converge slower than the Newton–Raphson procedure. For example, Thisted (1988, pp. 242) gave an example, in which the first iteration of EM was comparable to four iterations of Newton–Raphson in terms of convergence speed; however, after the first iteration, the EM flattened and eventually converged in more than five times as many iterations as Newton–Raphson. On the other hand, the EM is more robust to initial values than the Newton–Raphson. The two procedures may be combined to utilize the advantages of both. For example, one could start with the EM, which is more capable of converging with poor initial values, and then switch to Newton–Raphson (with the simplifications given above) after a few iterations.

1.6.2 Notes on Software

Standard routines for Gaussian mixed model analysis are available in several major statistical packages including SAS, S-Plus, SPSS, and Stata. Here we briefly summarize the software available in SAS and S-Plus.

The main procedure for linear mixed model analysis in SAS is PROC MIXED, although in some cases, a similar analysis may be carried out by PROC GLM. Note that here GLM refers to general linear models (rather than generalized linear models). In fact, PROC GLM was the procedure of fitting linear mixed models prior to the advent of PROC MIXED, and the latter has advantages over the former on various occasions. For example, in obtaining estimates for the fixed effects, PROC GLM computes the OLS estimator, whereas PROC MIXED gives the empirical (or estimated) BLUE, or EBLUE, which is (asymptotically) more efficient than the OLS estimator. See Sections 1.4.3 and 2.3.1.2. In addition, PROC GLM does not provide a valid estimate for the standard error of the OLS estimator, because the correlations among the observations are ignored. In contrast, the standard error estimate in PROC MIXED for the EBLUE is more accurate, using the method of Kackar and Harville (1984; see Section 2.3.1.3).

It should be pointed out that these analyses are for Gaussian mixed models, although some of the results do not require normality. For example, the standard method of variance component estimation in PROC MIXED is REML, and the standard error calculations in REML are all based on the asymptotic covariance matrix (ACM) of the REML estimator under the normality assumption. As we know, it is inappropriate to use the ACM under normality for non-Gaussian linear mixed models. However, new methods of estimating the ACM for non-Gaussian linear mixed models, such as POQUIM (see Section 1.4.2) have not been developed in SAS. On the other hand, the (point) REML estimates, for example, are the same whether or not normality is assumed. See Section 1.4.1. Note that the REML estimators in non-Gaussian linear mixed models are defined as quasi-likelihood estimators, which are solutions to the Gaussian REML equations. For further details about PROC MIXED, see, for example, Littell et al. (1996).

On the other hand, S-Plus has established itself as a powerful tool for application, research, and education. The main function in S-Plus that deals with linear mixed effects models is **lme** (linear mixed effects). One of the main advantages with S-Plus is that it offers a variety of graphic displays of the data as well as the fitted models. Once again, we illustrate the practice of **lme** using a real-life data example. For further details, see, for example, Pinheiro and Bates (2000).

The use of both SAS PROC MIXED and S-Plus **lme** for analysis of linear mixed models is illustrated in the next section.

1.7 Real-Life Data Examples

As mentioned in the preface, there is a vast literature on the applications of linear mixed models. In this section, we consider two examples of such applications as case studies. The main goal is to illustrate the situations in which these models may be useful, the procedures of modeling and data analysis under the assumed models, and the interpretation of the results with respect to the real-life problems.

1.7.1 Analysis of Birth Weights of Lambs

Harville and Fenech (1985) presented a dataset of birth weights of lambs and used it to illustrate the analysis of linear mixed models. The observations consist of birth weights of 62 single-birth male lambs. These lambs were progenies of 23 rams, so that each lamb had a different dam. The ages of the dams were recorded as a covariate. A second covariate was the (distinct) population lines. There were two control lines and three selection lines.

We record the data in Table 1.2 in a way different from Harville and Fenech (1985), so that it better matches the linear mixed model introduced below. In this model, the sire (ram) effects are considered random effects. These random effects are nested within lines, and are denoted by s_{ij}, $1 \leq i \leq 5$, $j = 1, \ldots, n_i$, where $n_1 = n_2 = n_3 = 4$, $n_4 = 3$, and $n_5 = 8$. The s_{ij}s are assumed independent and normally distributed with mean 0 and variance σ_s^2. The age of the dam, which is a categorical variable with three categories numbered 1 (1–2 years), 2 (2–3 years), 3 (over 3 years), is considered a fixed covariate. Let $x_{ijk,1} = 1$ if the age of the kth dam corresponding to line i and sire j is in category 1, and $x_{ijk,1} = 0$ otherwise; similarly, let $x_{ijk,2} = 1$ if the age of the kth dam corresponding to line i and sire j is in category 2, and $x_{ijk,2} = 0$ otherwise. Another fixed effect is the line effect, denoted by l_i, $i = 1, \ldots, 5$. Finally, the random errors e_{ijk}, $1 \leq i \leq 5$, $1 \leq j \leq n_i$, $k = 1, \ldots, n_{ij}$ are added to the model to represent the variation due to the environment and other unexplained factors. The e_{ijk}s are assumed independent and normally distributed with mean 0 and variance σ_e^2, and independent of the s_{ij}s. The last assumption may be interpreted as that the sire effects are orthogonal to the environmental effects. Here n_{ij} is the number of measures in the (i, j) cell. For example, $n_{11} = 1$, $n_{13} = 6$, and $n_{42} = 2$. A linear mixed model can be expressed as

$$y_{ijk} = l_i + a_1 x_{ijk,1} + a_2 x_{ijk,2} + s_{ij} + e_{ijk},$$

$i = 1, \ldots, 5$, $j = 1, \ldots, n_i$, and $k = 1, \ldots, n_{ij}$. It can be formulated in the standard form (1.1); that is,

$$y = X\beta + Zs + e,$$

where $y = (y_{111}, y_{121}, \ldots, y_{585})'$ is the vector of all the observations, $\beta = (l_1, \ldots, l_5, a_1, a_2)'$ is the vector of all the fixed effects, X is the matrix of covariates corresponding to β, $s = (s_{11}, s_{12}, \ldots, s_{58})'$ is the vector of sire effects, Z is the design matrix corresponding to s, and $e = (e_{111}, e_{121}, \ldots, e_{585})'$ is the vector of errors. For example, verify that the first row of X is $(1, 0, 0, 0, 0, 1, 0)$; and the 13th row of X is $(0, 1, 0, 0, 0, 0, 0)$. Note that X is of full rank. Also note that Z is a standard design matrix in that it consists of zeros and ones; there is exactly one 1 in each row, and at least one 1 in each column.

The model is fitted using SAS PROC MIXED with the option *REML*. The REML estimates of the two variance components, σ_s^2 and σ_e^2, are obtained as $\hat{\sigma}_s^2 = 0.511$ and $\hat{\sigma}_e^2 = 2.996$.

Table 1.2. Lamb birth weights

Obs.	6.2	13.0	9.5	10.1	11.4	11.8	12.9	13.1	10.4
Sire	11	12	13	13	13	13	13	13	14
Line	1	1	1	1	1	1	1	1	1
Age	1	1	1	1	1	2	3	3	1
Obs.	8.5	13.5	10.1	11.0	14.0	15.5	12.0	11.5	10.8
Sire	14	21	22	22	22	22	23	24	24
Line	1	2	2	2	2	2	2	2	2
Age	2	3	2	3	3	3	1	1	3
Obs.	9.0	9.0	12.6	11.0	10.1	11.7	8.5	8.8	9.9
Sire	31	31	31	32	32	32	32	32	32
Line	3	3	3	3	3	3	3	3	3
Age	2	3	3	1	2	2	3	3	3
Obs.	10.9	11.0	13.9	11.6	13.0	12.0	9.2	10.6	10.6
Sire	32	32	32	33	33	34	41	41	41
Line	3	3	3	3	3	3	4	4	4
Age	3	3	3	1	3	2	1	1	1
Obs.	7.7	10.0	11.2	10.2	10.9	11.7	9.9	11.7	12.6
Sire	41	41	41	42	42	43	43	51	51
Line	4	4	4	4	4	4	4	5	5
Age	3	3	3	1	1	1	3	1	1
Obs.	9.0	11.0	9.0	12.0	9.9	13.5	10.9	5.9	10.0
Sire	52	52	53	53	54	55	56	56	57
Line	5	5	5	5	5	5	5	5	5
Age	1	3	3	3	3	2	2	3	2
Obs.	12.7	13.2	13.3	10.7	11.0	12.5	9.0	10.2	
Sire	57	57	57	58	58	58	58	58	
Line	5	5	5	5	5	5	5	5	
Age	2	3	3	1	1	1	3	3	

In order to obtain estimates of the fixed effects, it is important to fit the model without intercept, because otherwise the line and age effects are not all identifiable. The estimates of the fixed effects without intercept are given by Table 1.3 below. It is seen that all the line effects are significantly different from zero (in fact, positive), whereas all the age effects are not significant. Here the level of significance is understood as 0.01. The results suggest that whereas the (average) birth weights of lambs appear different from line to line, there seem to be no such differences among the age groups for the dams. This example is revisited in the next chapter, where estimates of the random effects are obtained.

SAS PROC MIXED also provides alternative methods for fitting a linear mixed model. For example, if the *REML* option is replaced by *MIVQUE0*, which is a special case of MINQUE (see Section 1.5.2), the estimates of the variance components are $\hat{\sigma}_s^2 = 0.323$ and $\hat{\sigma}_e^2 = 3.116$. The estimates of the fixed effects are given by Table 1.4 below. It is observed that the estimates of

Table 1.3. Estimates of the fixed effects (REML)

Effect	Line	Age	Est.	S.E.	t-value	Pr > $\|t\|$
Line	1		10.5008	0.8070	13.01	$< .0001$
Line	2		12.2998	0.7569	16.25	$< .0001$
Line	3		11.0425	0.6562	16.83	$< .0001$
Line	4		10.2864	0.7882	13.05	$< .0001$
Line	5		10.9625	0.5438	20.16	$< .0001$
Age		1	-0.0097	0.5481	-0.02	0.9861
Age		2	-0.1651	0.6435	-0.26	0.7989

the fixed effects using *MIVQUE0* are very close to those using *REML*, even though the estimates of σ_s^2 using the two methods are quite different.

Table 1.4. Estimates of the fixed effects (MIVQUE0)

Effect	Line	Age	Est.	S.E.	t-value	Pr > $\|t\|$
Line	1		10.5637	0.7730	13.67	$< .0001$
Line	2		12.3028	0.7277	16.91	$< .0001$
Line	3		10.9962	0.6134	17.93	$< .0001$
Line	4		10.2640	0.7484	13.72	$< .0001$
Line	5		10.9650	0.5255	20.86	$< .0001$
Age		1	-0.0109	0.5509	-0.02	0.9844
Age		2	-0.1393	0.6495	-0.21	0.8313

However, PROC MIXED does not provide an option for ANOVA estimation of the variance components. If one wishes to obtain ANOVA estimates of the variance components, one may use the GLM procedure in SAS, PROC GLM. For example, the ANOVA estimates of the variance components using Henderson's method III (see Section 1.5.1, which gives the same result as Henderson's method I in this case) are $\hat{\sigma}_s^2 = 0.764$ and $\hat{\sigma}_e^2 = 2.796$. Again, the estimate of σ_s^2 is quite different from the ones using *REML* and *MIVQUE0* in PROC MIXED.

1.7.2 Analysis of Hip Replacements Data

In this section, we use a data set presented by Hand and Crowder (1996) regarding hip replacements to illustrate the iterative WLS method of longitudinal data analysis introduced in Section 1.4.3. Thirty patients were involved in this study. Each patient was measured four times, once before the operation and three times after, for hematocrit, TPP, vitamin E, vitamin A, urinary zinc, plasma zinc, hydroxyprolene (in milligrams), hydroxyprolene (index), ascorbic acid, carotine, calcium, and plasma phosphate (12 variables). One important feature of the data is that there is considerable amount of missing observations. In fact, most of the patients have at least one missing observation for all 12 measured variables. In other words, the longitudinal data are (seriously) unbalanced.

We consider two of the measured variables: hematocrit and calcium. The first variable was considered by Hand and Crowder (1996) who used the data to assess age, sex, and time differences. The authors assumed an equicorrelated model and obtained Gaussian estimates of regression coefficients and variance components (i.e., MLE under normality). Here we take a robust approach without assuming a specific covariance structure. The covariates consist of the same variables as suggested by Hand and Crowder. The variables include an intercept, sex, occasion dummy variables (three), sex by occasion interaction dummy variables (three), age, and age by sex interaction. For the hematocrit data the I-WLS algorithm converged in seven iterations. The results are shown in Table 1.5. The first row is I-WLS estimates corresponding to, from left to right, intercept, sex, occasions (three), sex by occasion interaction (three), age, and age by sex interaction; the second row is estimated standard errors corresponding to the I-WLS estimates; the third row is the Gaussian maximum likelihood estimates obtained by Hand and Crowder (1996, pp. 106) included for comparison.

Table 1.5. Estimates for hematocrit.

Coef.	β_1	β_2	β_3	β_4	β_5
I-WLS	3.19	0.08	0.65	-0.34	-0.21
s.e.	0.39	0.14	0.06	0.06	0.07
Gaussian	3.28	0.21	0.65	-0.34	-0.21
Coef.	β_6	β_7	β_8	β_9	β_{10}
I-WLS	0.12	-0.051	-0.051	0.033	-0.001
s.e.	0.06	0.061	0.066	0.058	0.021
Gaussian	0.12	-0.050	-0.048	0.019	-0.020

It is seen that the I-WLS estimates are similar to the Gaussian estimates, especially for the parameters that are found significant. This is, of course, not surprising, because the Gaussian and I-WLS estimators should both be close to BLUE, provided that the covariance model suggested by Hand and Crowder is correct (the authors believed that their method was valid in this case). Taking into account the estimated standard errors, we found the coefficients β_1, β_3, β_4, β_5, and β_6 to be significant and the rest of the coefficients to be insignificant, where the β_1, β_2, ... are the coefficients corresponding to the variables described in the first paragraph of this section in that order. This suggests that, for example, the recovery of hematocrit improves over time at least for the period of measurement times. The findings are consistent with those of Hand and Crowder with the only exception of β_6. Hand and Crowder considered jointly testing the hypothesis that $\beta_6 = \beta_7 = \beta_8 = 0$ and found an insignificant result. In our case, the coefficients are considered separately, and we found β_7 and β_8 to be insignificant and β_6 to be barely significant at the 5% level. However, because Hand and Crowder did not publish the individual standard errors, this does not necessarily imply a difference. The interpretation of the significance of β_6, which corresponds to the interaction between sex and the first occasion, appears to be less straightforward (Exercise 1.21).

Next, we consider the calcium data. We use the same covariate variables to assess age, sex, and time differences. In this case, our algorithm converged in 13 iterations. The results are given in Table 1.6. The first row is I-WLS estimates corresponding to, from left to right, intercept, sex, occasions (three), sex by occasion interaction (three), age, and age by sex interaction; the second row is estimated standard errors corresponding to the estimates. It is seen that, except for β_1, β_3, and β_4, all the coefficients are not significant (at the 5% level). In particular, there seems to be no difference in terms of sex and age. Also, the recovery of calcium after the operation seems to be a little quicker than that of hematocrit, because β_5 is no longer significant. Hand and Crowder (1996) did not analyze this dataset.

Table 1.6. Estimates for calcium.

Coef.	β_1	β_2	β_3	β_4	β_5
I-WLS	20.1	0.93	1.32	-1.89	-0.13
s.e.	1.3	0.57	0.16	0.13	0.16
Coef.	β_6	β_7	β_8	β_9	β_{10}
I-WLS	0.09	0.17	-0.15	0.19	-0.12
s.e.	0.16	0.13	0.16	0.19	0.09

1.8 Further Results and Technical Notes

1. A note on finding the MLE. By definition, the MLE is the global maximum of the (log-)likelihood function. A method that, in principle, ensures finding the MLE is called the "fine grid." The idea is to divide the parameter space into (many) small subspaces, or grids, and compute the value of the log-likelihood at a given point within each grid. Then, by comparing the values of the log-likelihood at those points, one obtains an approximate global maximum of the log-likelihood. As the grid becomes finer, the approximation becomes more accurate. However, this method is impractical for the calculation of MLE under linear mixed models, especially if the parameter is multidimensional.

Alternatively, one may look for stationary points that are solutions to the ML equations. Several cases may occur as a result, then: (i) the solution is a global maximum; (ii) the solution is a local maximum; and (iii) the solution is something else (e.g., local minimum, saddle point). In case (i) the goal of finding the MLE is achieved. In cases (ii) and (iii), the goal is not achieved, but there is a way to separate these two cases by computing the Hessian matrix of the log-likelihood, which is supposed to be positive definite at a local maximum. Because one is unsure whether the solution found is a global maximum, a complete and successful implementation of this method may require finding all the solutions to the ML equations, comparing the values of the log-likelihood among the solutions and with the values on the boundary of the parameter space to identify the global maximum. Sometimes such a procedure is also quite expensive, even impractical, especially if the number of solutions is unknown. Some methods have been discussed in Section 1.6.1 regarding computation of the MLE. All of these methods lead to, at least, a local maximum of the log-likelihood function. The same discussion also applies to REML estimation.

Alternatively, Gan and Jiang (1999) proposed a statistical method for identifying whether a given root is the global maximizer of the likelihood function. Their method consists of a test for the global maximum or, more precisely, a test for the asymptotic efficiency of a root to the ML equation. Unfortunately, the Gan–Jiang test applies only to the case of i.i.d. observations with a single (one-dimensional) parameter.

2. Note on X not of full rank. If X is not of full rank, the matrix $X'V^{-1}X$ will be singular. However, most of the results in this chapter still hold with $(X'V^{-1}X)^{-1}$ replaced by $(X'V^{-1}X)^{-}$, where M^{-} represents the generalized inverse of matrix M (see Appendix B). For example, $(X'V^{-1}X)^{-1}$ in the definitions of $\hat{\beta}$ in (1.9) and P in (1.11) can be replaced by $(X'V^{-1}X)^{-}$.

3. Asymptotic behavior of ML and REML estimators in non-Gaussian mixed ANOVA models. Asymptotic properties of ML estimators under the normality assumption have been considered by Hartley and Rao (1967), Anderson (1969, 1971a), and Miller (1977), among others. Asymptotic behavior of REML estimators has been studied by Das (1979) and Cressie and Lahiri (1993) under the normality assumption, and by Richardson and Welsh (1994) and Jiang (1996, 1997a) without the normality assumption. All but Jiang (1996, 1997a) have assumed that the rank p of the design matrix X is fixed or bounded, which turns out to be a critical assumption. This is because, under such an assumption, the ML and REML estimators are asymptotically equivalent. On the other hand, earlier examples, including the famous Neyman–Scott problem (Neyman and Scott 1948), showed apparent asymptotic superiority of REML over ML in cases where the number of fixed effects increases with the sample size. In other words, to uncover the true superiority of REML one has to look at the case where the number of fixed effects grows with the sample size.

For simplicity, here we state some results for balanced mixed ANOVA models. More general results can be found in Jiang (1996, 1997a). Here we consider the Hartley–Rao form of variance components λ and γ_i, $1 \le i \le s$ (see Section 1.2.1.1).

First introduce some notation for a general linear mixed model (not necessarily balanced). The model is called unconfounded if (i) the fixed effects are not confounded with the random effects and errors; that is, $\text{rank}(X, Z_i) > p$, $\forall i$ and $X \neq I$; and (ii) the random effects and errors are not confounded; that is, the matrices I and $Z_i Z_i'$, $1 \le i \le s$ are linearly independent (e.g., Miller 1977). The model is called nondegenerate if $\text{var}(\alpha_{i1}^2)$, $0 \le i \le s$ are bounded away from zero, where α_{i1} is the first component of α_i. Note that if $\text{var}(\alpha_{i1}^2) = 0$, $\alpha_{i1} = -c$ or c with probability one for some constant c. A sequence of estimators $\hat{\lambda}_n, \hat{\gamma}_{1,n}, \dots, \hat{\gamma}_{s,n}$ is asymptotically normal if there are sequences of positive constants $p_{i,n} \to \infty$, $0 \le i \le s$ and a sequence of matrices $\mathcal{M}_n$ such that $\limsup(\|\mathcal{M}_n^{-1}\| \vee \|\mathcal{M}_n\|) < \infty$ such that

$$\mathcal{M}_n(p_{0,n}(\hat{\lambda}_n - \lambda), p_{1,n}(\hat{\gamma}_{1,n} - \gamma_1), \dots, p_{s,n}(\hat{\gamma}_{s,n} - \gamma_s))' \xrightarrow{\mathcal{D}} N(0, I_{s+1}).$$

The subscript n is often suppressed, as we do in the sequel, to simplify the notation. Let A be as in (1.16). Define $V(A, \gamma) = A'A + \sum_{i=1}^s \gamma_i A' Z_i Z_i' A$, and $V(\gamma) = AV(A, \gamma)^{-1} A'$. Note that $V(\gamma)$ does not depend on the choice of A. Let $V_0(\gamma) = b(\gamma) V(\gamma) b(\gamma)'$, where $b(\gamma) = (I, \sqrt{\gamma_1} Z_1, \dots, \sqrt{\gamma_s} Z_s)'$, and $V_i(\gamma) = b(\gamma) V(\gamma) Z_i Z_i' V(\gamma) b(\gamma)'$, $1 \le i \le s$. Furthermore, we define $V_0 = I_{n-p}/\lambda$, $V_i = V(A, \gamma)^{-1/2} A' Z_i Z_i' A V(A, \gamma)^{-1/2}$, $1 \le i \le s$. Let $\mathcal{I}$ be the matrix whose (i, j) element is $\text{tr}(V_i V_j)/p_i p_j$, $0 \le i, j \le s$, where p_i, $0 \le i \le s$ are given in the following theorem, and $\mathcal{K}$ the matrix whose (i, j) element is

$$\sum_{l=1}^{m+n} (\text{E}\omega_l^4 - 3) V_{i,ll}(\gamma) V_{j,ll}(\gamma)/p_i p_j \lambda^{1_{(i=0)} + 1_{(j=0)}},$$

where $m = \sum_{i=1}^s m_i$ and

$$\omega_l = \begin{cases} \epsilon_l/\sqrt{\lambda}, & 1 \le l \le n, \\ \alpha_{i,l-n-\sum_{j<i} m_j}/\sqrt{\lambda\gamma_i}, & n + \sum_{j<i} m_j + 1 \le l \le n + \sum_{j \le i} m_j, \\ & 1 \le i \le s, \end{cases}$$

where $V_{i,kl}(\gamma)$ is the (k, l) element of $V_i(\gamma)$, $0 \le i \le s$, and m_i the dimension of α_i, $1 \le i \le s$.

In the balanced case, it is more convenient to use the multiple indices $a = (a_1, \dots, a_{w+1})$, and $b_i = (b_{i,1}, \dots, b_{i,w+1}) \in S_{w+1}$ introduced above Section 1.2.1.1. Let $u, v \in S_{w+1}$. Define $u \vee v = (u_1 \vee v_1, \dots, u_{w+1} \vee v_{w+1})$, $S_u = \{v \in S : v \le u\}$, $m_u = \prod_{l:u_l=0} n_l$, $m_{u,S} = \min_{v \in S_u} m_v$ if $S_u \neq \emptyset$, and $m_{u,S} = 1$ if $S_u = \emptyset$. The following theorems are due to Jiang (1996).

Theorem 1.1. Let the balanced mixed ANOVA model be unconfounded, and the variance components be positive. As $n \to \infty$ and $m_i \to \infty$, $1 \le i \le s$, the following hold.

(i) There exist with probability tending to one REML estimators $\hat{\lambda}$ and $\hat{\gamma}_i$, $1 \le i \le s$ which are consistent, and the sequence $(\sqrt{n-p}(\hat{\lambda}-\lambda), \sqrt{m_1}(\hat{\gamma}_1 - \gamma_1), \dots, \sqrt{m_s}(\hat{\gamma}_s - \gamma_s))'$ is bounded in probability.

(ii) If, moreover, the model is nondegenerate, the REML estimators in (i) are asymptotically normal with $p_0 = \sqrt{n-p}$, $p_i = \sqrt{m_i}$, $1 \le i \le s$, and $\mathcal{M} = \mathcal{J}^{-1/2}\mathcal{I}$, where $\mathcal{J} = 2\mathcal{I} + \mathcal{K}$.

Theorem 1.2. Let the balanced mixed ANOVA model be unconfounded, and the variance components be positive. As $n \to \infty$ and $m_i \to \infty$, $1 \le i \le s$, the following hold.

(i) There exist with probability tending to one MLE that are consistent if and only if

$$\frac{p}{n} \to 0, \qquad \frac{m_{b_i \vee a} m_{b_i \vee a, S}}{m_{b_i}^2} \to 0, \qquad 1 \le i \le s.$$

(ii) If, moreover, the model is nondegenerate, there exist with probability tending to one MLE that are asymptotically normal if and only if

$$p_0 \sim \sqrt{n-p}, \qquad p_i \sim \sqrt{m_i}, \qquad 1 \le i \le s,$$

and

$$\frac{p}{\sqrt{n}} \to 0, \qquad \frac{m_{b_i \vee a} m_{b_i \vee a, S}}{m_{b_i}^{3/2}} \to 0, \qquad 1 \le i \le s.$$

When these conditions are satisfied, the MLE are asymptotically normal with the same p_i, $0 \le i \le s$, and M as for the REML estimators.

The consistency of the REML (and ML) estimators in the above theorems are of Cramér type (Cramér 1946), in which the existence of a sequence of consistent roots to the REML (ML) equations is ensured with no indication of which root is consistent when the roots are not unique. On the other hand, Hartley and Rao (1967) proved Wald consistency (Wald 1949) of the MLE under the normality assumption. By Wald consistency it means that the global maximizer of the likelihood function is consistent, therefore there is no uncertainty regarding the consistent root. Jiang (1997a) considered Wald consistency of REML estimators without assuming normality. The following is the result for the balanced case.

Theorem 1.3. Let the balanced mixed ANOVA model be unconfounded. As $n \to \infty$ and $m_i \to \infty$, $1 \le i \le s$, the following hold.

(i) The global maximizer of the Gaussian restricted log-likelihood $\hat{\theta}$ is consistent, and the sequence $(\sqrt{n-p}(\hat{\lambda}-\lambda), \sqrt{m_1}(\hat{\gamma}_1-\gamma_1), \dots, \sqrt{m_s}(\hat{\gamma}_s-\gamma_s))'$ is bounded in probability.

(ii) If, moreover, the variance components are positive and the model is nondegenerate, the global maximizer $\hat{\theta}$ in (i) is asymptotically normal with $p_0 = \sqrt{n-p}$, $p_i = \sqrt{m_i}$, $1 \le i \le s$, and $\mathcal{M} = \mathcal{J}^{-1/2}\mathcal{I}$, where $\mathcal{J} = 2\mathcal{I} + \mathcal{K}$.

Note. Unlike Theorem 1.1 and Theorem 1.2, part (i) of Theorem 1.3 does not require positiveness of the variance components.

4. Truncated estimator. For non-Gaussian linear mixed models, the REML estimator is defined as the solution to the (Gaussian) REML equations, if the solution lies within the parameter space. If the solution is out of the parameter space, it is customary to truncate the solution at the boundary of the parameter space. For example, for ANOVA models, let $\dot{\theta} = (\dot{\tau}^2, \dot{\sigma}_1^2, \dots, \dot{\sigma}_s^2)'$ be the solution. Suppose that $\dot{\tau}^2 > 0$, $\dot{\sigma}_1^2 < 0$, and $\dot{\sigma}_i^2 \geq 0$, $2 \leq i \leq s$. Then, the truncated REML estimator is $(\dot{\tau}^2, 0, \dot{\sigma}_2^2, \dots, \dot{\sigma}_s^2)'$.

5. POQUIM in general. We consider REML estimation. Similar results for ML can be found in Jiang (2005a). This case is relatively simpler (compared to ML) because only estimation of the variance components is involved. Furthermore, as shown, the QUIM in this case does not involve the third moments of the random effects and errors.

Under the ANOVA model and normality, we have (1.18), which can be further expressed as $\partial l_{\mathrm{R}}/\partial\theta_j = u'B_j u - b_j$, $0 \leq j \leq s$, where $\theta_0 = \lambda$, $\theta_j = \gamma_j$, $1 \leq j \leq s$; $u = y - X\beta$; $B_0 = (2\lambda)^{-1}P$, $B_j = (\lambda/2)PZ_jZ_j'P$, $1 \leq j \leq s$; $b_0 = (n-p)/2\lambda$, and $b_j = (\lambda/2)\mathrm{tr}(PZ_jZ_j')$, $1 \leq j \leq s$. Note that $b_j = \mathrm{E}(u'B_ju)$, $0 \leq j \leq s$.

Let $u_i = y_i - x_i'\beta$ be the ith component of u, where x_i' is the ith row of X. The kurtoses of the random effects and errors are defined as $\kappa_t = \mathrm{E}(\alpha_{t1}^4) - 3\sigma_t^4 = \mathrm{E}(\alpha_{t1}^4) - 3(\lambda\gamma_t)^2$, $0 \leq t \leq s$, where $\alpha_0 = \epsilon$ and $\gamma_0 = 1$. Also, with a slight abuse of the notation, let z_{it}' and z_{tl} be the ith row and lth column of Z_t, respectively, $0 \leq t \leq s$, where $Z_0 = I$. Define $\Gamma(i_1, i_2) = \sum_{t=0}^s \gamma_t(z_{i_1t} \cdot z_{i_2t})$. Here, the dot product of vectors $a_1, \dots, a_k$ of the same dimension is defined as $a_1 \cdot a_2 \cdots a_k = \sum_l a_{1l}a_{2l}\cdots a_{kl}$. Also, let m_t be the dimension of α_t, $0 \leq t \leq s$ (so that $m_0 = n$). We begin with an expression for $\mathrm{cov}(u_{i_1}u_{i_2}, u_{i_3}u_{i_4})$ as well as one for $\mathrm{cov}(\partial l_{\mathrm{R}}/\partial\theta_j, \partial l_{\mathrm{R}}/\partial\theta_k)$, the (j,k) element of of $\mathcal{I}_1$.

Lemma 1.3. We have

$$\mathrm{cov}(u_{i_1}u_{i_2}, u_{i_3}u_{i_4}) = \lambda^2\{\Gamma(i_1,i_3)\Gamma(i_2,i_4) + \Gamma(i_1,i_4)\Gamma(i_2,i_3)\} + \sum_{t=0}^s \kappa_t z_{i_1t}\cdots z_{i_4t}, \tag{1.54}$$

where $z_{i_1t}\cdots z_{i_4t} = z_{i_1t}\cdot z_{i_2t}\cdot z_{i_3t}\cdot z_{i_4t}$. Furthermore, we have

$$\mathrm{cov}\left(\frac{\partial l_{\mathrm{R}}}{\partial\theta_j}, \frac{\partial l_{\mathrm{R}}}{\partial\theta_k}\right) = 2\mathrm{tr}(B_jVB_kV) + \sum_{t=0}^s \kappa_t \sum_{l=1}^{m_t}(z_{tl}'B_jz_{tl})(z_{tl}'B_kz_{tl}). \tag{1.55}$$

Let $f_1, \dots, f_L$ be the different nonzero functional values of

$$f(i_1, \dots, i_4) = \sum_{t=0}^{s} \kappa_t z_{i_1 t} \cdots z_{i_4 t}. \tag{1.56}$$

Note that this is the second term on the right side of (1.54). Here by functional value it means $f(i_1, \dots, i_4)$ as a function of $\kappa = (\kappa_t)_{0 \le t \le s}$. For example, $\kappa_0 + \kappa_1$ and $\kappa_2 + \kappa_3$ are different functions (even if their values may be the same for some κ). Also, let 0 denote the zero function (of κ). Then, without using (1.55), we have

$$\begin{aligned}
\mathrm{cov}\left(\frac{\partial l_{\mathrm{R}}}{\partial \theta_j}, \frac{\partial l_{\mathrm{R}}}{\partial \theta_k}\right) &= \sum_{i_1, \dots, i_4} B_{j,i_1,i_2} B_{k,i_3,i_4} \mathrm{cov}(u_{i_1} u_{i_2}, u_{i_3} u_{i_4}) \\
&= \sum_{f(i_1,\dots,i_4)=0} B_{j,i_1,i_2} B_{k,i_3,i_4} \mathrm{cov}(u_{i_1} u_{i_2}, u_{i_3} u_{i_4}) \\
&\quad + \sum_{l=1}^{L} \sum_{f(i_1,\dots,i_4)=f_l} B_{j,i_1,i_2} B_{k,i_3,i_4} \mathrm{cov}(u_{i_1} u_{i_2}, u_{i_3} u_{i_4}) \\
&= \sum_{l=0}^{L} S_l
\end{aligned} \tag{1.57}$$

with S_l, $0 \le l \le L$ defined in obvious ways. According to Lemma 1.3, the left side of (1.57) depends on the higher moments only through κ. By (1.54) and (1.56), we have

$$S_0 = 2\lambda^2 \sum_{f(i_1,\dots,i_4)=0} B_{j,i_1,i_2} B_{k,i_3,i_4} \Gamma(i_1, i_3) \Gamma(i_2, i_4), \tag{1.58}$$

which depends only on θ. Furthermore, for $1 \le l \le L$, write

$$\begin{aligned}
S_l &= c_l \sum_{f(i_1,\dots,i_4)=f_l} \mathrm{cov}(u_{i_1} u_{i_2}, u_{i_3} u_{i_4}) \\
&\quad + \sum_{f(i_1,\dots,i_4)=f_l} (B_{j,i_1,i_2} B_{k,i_3,i_4} - c_l) \mathrm{cov}(u_{i_1} u_{i_2}, u_{i_3} u_{i_4}) \\
&= S_{l,1} + S_{l,2},
\end{aligned}$$

where c_l is a constant to be determined later on. By (1.54), we have

$$\begin{aligned}
S_{l,2} &= \sum_{f(i_1,\dots,i_4)=f_l} (B_{j,i_1,i_2} B_{k,i_3,i_4} - c_l)[f_l + \lambda^2\{\cdots\}] \\
&= f_l \sum_{f(i_1,\dots,i_4)=f_l} (B_{j,i_1,i_2} B_{k,i_3,i_4} - c_l) + \cdots,
\end{aligned}$$

where $\cdots$ depends only on θ. If we let the coefficient of f_l in the above equal zero, we have

$$c_l = \frac{1}{|\{f(i_1,\dots,i_4)=f_l\}|}\sum_{f(i_1,\dots,i_4)=f_l} B_{j,i_1,i_2}B_{k,i_3,i_4}, \tag{1.59}$$

where $|\cdot|$ denotes cardinality. With this choice of c_l, we have

$$\begin{aligned}
S_{l,2} &= \lambda^2 \sum_{f(i_1,\dots,i_4)=f_l} (B_{j,i_1,i_2}B_{k,i_3,i_4} - c_l)\{\Gamma(i_1,i_3)\Gamma(i_2,i_4) \\
&\quad +\Gamma(i_1,i_4)\Gamma(i_2,i_3)\} \\
&= 2\lambda^2 \sum_{f(i_1,\dots,i_4)=f_l} (B_{j,i_1,i_2}B_{k,i_3,i_4} - c_l)\Gamma(i_1,i_3)\Gamma(i_2,i_4),
\end{aligned}$$

which depends only on θ. Note that c_l depends only on θ. On the other hand, note that $u_i = \sum_{t=0}^{s} u_{it}$ with $u_{it} = \sum_{l=1}^{m_t} z_{itl}\alpha_{tl}$, hence $\mathrm{E}(u_{i_1}u_{i_2}) = \sum_{t=0}^{s}\mathrm{E}(u_{i_1t}u_{i_2t}) = \sum_{t=0}^{s}\sigma_t^2 z'_{i_1t}z_{i_2t} = \lambda\Gamma(i_1,i_2)$. Thus, we have

$$\begin{aligned}
S_{l,1} &= c_l \sum_{f(i_1,\dots,i_4)=f_l} \{\mathrm{E}(u_{i_1}\cdots u_{i_4}) - \lambda^2\Gamma(i_1,i_2)\Gamma(i_3,i_4)\} \\
&= \mathrm{E}\left(c_l \sum_{f(i_1,\dots,i_4)=f_l} u_{i_1}\cdots u_{i_4}\right) - \lambda^2 c_l \sum_{f(i_1,\dots,i_4)=f_l} \Gamma(i_1,i_3)\Gamma(i_2,i_4).
\end{aligned}$$

Note that $\sum_{f(i_1,\dots,i_4)=f_l}\Gamma(i_1,i_2)\Gamma(i_3,i_4) = \sum_{f(i_1,\dots,i_4)=f_l}\Gamma(i_1,i_3)\Gamma(i_2,i_4)$, because $f(i_1,\dots,i_4)$ is symmetric in $i_1,\dots,i_4$. Therefore, we have by combining the above

$$\begin{aligned}
S_l &= \mathrm{E}\left(c_l \sum_{f(i_1,\dots,i_4)=f_l} u_{i_1}\cdots u_{i_4}\right) \\
&\quad +2\lambda^2 \sum_{f(i_1,\dots,i_4)=f_l} B_{j,i_1,i_2}B_{k,i_3,i_4}\Gamma(i_1,i_3)\Gamma(i_2,i_4) \\
&\quad -3\lambda^2 c_l \sum_{f(i_1,\dots,i_4)=f_l} \Gamma(i_1,i_3)\Gamma(i_2,i_4).
\end{aligned} \tag{1.60}$$

Note that c_l defined by (1.59) depends on j and k; that is, $c_l = c_{j,k,l}$. If we define $c_{j,k}(i_1,\dots,i_4) = c_{j,k,l}$, if $f(i_1,\dots,i_4) = f_l$, $1 \le l \le L$, then, by (1.57), (1.58), and (1.60), it can be shown that

$$\begin{aligned}
\mathrm{cov}\left(\frac{\partial l_{\mathrm{R}}}{\partial\theta_j}, \frac{\partial l_{\mathrm{R}}}{\partial\theta_k}\right) &= \mathrm{E}\left\{\sum_{f(i_1,\dots,i_4)\neq 0} c_{j,k}(i_1,\dots,i_4)u_{i_1}\cdots u_{i_4}\right\} \\
&\quad +2\mathrm{tr}(B_jVB_kV) \\
&\quad -3\lambda^2 \sum_{f(i_1,\dots,i_4)\neq 0} c_{j,k}(i_1,\dots,i_4)\Gamma(i_1,i_3)\Gamma(i_2,i_4).
\end{aligned}$$

We summarize the result in terms of a theorem. Write

$$\mathcal{I}_{1,jk} = \text{cov}\left(\frac{\partial l_{\text{R}}}{\partial \theta_j}, \frac{\partial l_{\text{R}}}{\partial \theta_k}\right),$$

which is the j, k element of the QUIM $\mathcal{I}_1 = \text{Var}(\partial l_{\text{R}}/\partial \theta)$.

Theorem 1.4. For any non-Gaussian mixed ANOVA model, we have

$$\begin{aligned}
\mathcal{I}_{1,jk} &= 2\text{tr}(B_j V B_k V) + \sum_{t=0}^{s} \kappa_t \sum_{l=1}^{m_t} (z'_{tl} B_j z_{tl})(z'_{tl} B_k z_{tl}) \\
&= \text{E}\left\{ \sum_{f(i_1,\ldots,i_4)\neq 0} c_{j,k}(i_1,\ldots,i_4) u_{i_1} \cdots u_{i_4} \right\} \\
&\quad + \left\{ 2\text{tr}(B_j V B_k V) - 3\lambda^2 \sum_{f(i_1,\ldots,i_4)\neq 0} c_{j,k}(i_1,\ldots,i_4) \Gamma(i_1,i_3)\Gamma(i_2,i_4) \right\} \\
&= \mathcal{I}_{1,1,jk} + \mathcal{I}_{1,2,jk}, \qquad (1.61)
\end{aligned}$$

$0 \leq j,k \leq s$, where $c_{j,k}(i_1,\ldots,i_4) = c_{j,k,l}$, if $f(i_1,\ldots,i_4) = f_l$, $1 \leq l \leq L$ with

$$c_{j,k,l} = \frac{1}{|\{f(i_1,\ldots,i_4) = f_l\}|} \sum_{f(i_1,\ldots,i_4)=f_l} B_{j,i_1,i_2} B_{k,i_3,i_4}. \qquad (1.62)$$

Of course, (1.61) can be verified directly, but the derivation above also explains where the idea comes from, which is, after all, quite natural. Note that $2\text{tr}(B_j V B_k V)$ is the Gaussian covariance between $\partial l_{\text{R}}/\partial \theta_j$ and $\partial l_{\text{R}}/\partial \theta_k$. This means that, under normality, $\mathcal{I}_{1,1,jk}$ is identical to the second term in $\mathcal{I}_{1,2,jk}$ with the negative sign removed. Of course, this can be easily verified using (1.54). On the other hand, without normality, $\mathcal{I}_{1,1,jk}$ may involve higher moments of the random effects and errors, and this is why the expectation is not taken inside the summation. Instead, we propose to estimate $\mathcal{I}_{1,1,jk}$ by taking out the expectation sign, and replacing any parameter involved by its REML estimator; that is, $\hat{\mathcal{I}}_{1,1,jk} = \sum_{f(i_1,\ldots,i_4)\neq 0} \hat{c}_{j,k}(i_1,\ldots,i_4)\hat{u}_{i_1}\cdots\hat{u}_{i_4}$, where $\hat{c}_{j,k}(i_1,\ldots,i_4)$ is defined in the same way as $c_{j,k}(i_1,\ldots,i_4)$ except with θ replaced by $\hat{\theta}$, and $\hat{u}_i = y_i - x'_i\hat{\beta}$. Here $\hat{\theta}$ is the REML estimator of θ, and $\hat{\beta}$ is given by (1.9) with $V = \hat{V}$, which is V with θ replaced by $\hat{\theta}$. Note that the set $\{(i_1,\ldots,i_4) : f(i_1,\ldots,i_4) = f_l\}$ does not depend on θ. It follows that $\hat{c}_{j,k}(i_1,\ldots,i_4) = \hat{c}_{j,k,l}$, if $f(i_1,\ldots,i_4) = f_l$, $1 \leq l \leq L$, where $\hat{c}_{j,k,l} = |\{f(i_1,\ldots,i_4) = f_l\}|^{-1} \sum_{f(i_1,\ldots,i_4)=f_l} \hat{B}_{j,i_1,i_2}\hat{B}_{k,i_3,i_4}$, and $\hat{B}_{j,i_1,i_2}$ is B_{j,i_1,i_2} with θ replaced by $\hat{\theta}$, and so on. This is the observed part. On the other hand, $\mathcal{I}_{1,2,jk}$ only depends on θ, and therefore can be estimated by replacing θ by $\hat{\theta}$. The result, denoted $\hat{\mathcal{I}}_{1,2,jk}$, is the estimated part. An estimator of $\mathcal{I}_{1,jk}$ is then $\hat{\mathcal{I}}_{1,1,jk} + \hat{\mathcal{I}}_{1,2,jk}$, hence an estimator of $\mathcal{I}_1$ is given by

$\hat{\mathcal{I}}_1 = \hat{\mathcal{I}}_{1,1} + \hat{\mathcal{I}}_{1,2}$, where $\hat{\mathcal{I}}_{1,r} = (\hat{\mathcal{I}}_{1,r,jk})_{0 \le j,k \le s}$, $r = 1, 2$. Because the estimator consists partially of an observed form and partially of an estimated one, it is called partially observed quasi information matrix, or POQUIM. It can be shown that, under some regularity conditions, the POQUIM estimator of $\mathcal{I}_1$ and the resulting estimator of Σ_{R} are consistent. See Jiang (2005a) for details. We now use another simple example to illustrate the POQUIM decomposition.

Example 1.1 (Continued). Consider the special case of Example 1.1 with $k_i = k$, $1 \le i \le m$. It is easy to show that $f(i_1 j_1, \dots, i_4 j_4) = 0$, if not $i_1 = \cdots = i_4$; κ_1, if $i_1 = \cdots = i_4$ but not $j_1 = \cdots = j_4$; and $\kappa_0 + \kappa_1$, if $i_1 = \cdots = i_4$ and $j_1 = \cdots j_4$. Thus, $L = 2$ [note that L is the number of different functional values of $f(i_1 j_1, \dots, i_4 j_4)$]. Define the following functions of θ, where $\theta = (\lambda, \gamma_1)'$: $t_0 = 1 - \gamma_1/(1 + \gamma_1 k) - 1/\{(1 + \gamma_1 k)mk\}$, $t_1 = (m-1)k/\{m(1 + \gamma_1 k)\}$, and $t_3 = \{k(1 + \gamma_1 k)^2 - (1 + \gamma_1)^2\}/(k^3 - 1)$. Then, the POQUIM is given by $\hat{\mathcal{I}}_{1,st} = \hat{\mathcal{I}}_{1,1,st} + \hat{\mathcal{I}}_{1,2,st}$, $s, t = 0, 1$, where

$$
\begin{aligned}
\hat{\mathcal{I}}_{1,1,00} &= \frac{\hat{t}_1^2 - \hat{t}_0^2 k}{4\hat{\lambda}^4 k(k^3 - 1)} \left\{ \sum_i \left(\sum_j \hat{u}_{ij} \right)^4 - \sum_{i,j} \hat{u}_{ij}^4 \right\} + \frac{\hat{t}_0^2}{4\hat{\lambda}^4} \sum_{i,j} \hat{u}_{ij}^4, \\
\hat{\mathcal{I}}_{1,1,01} &= \frac{(m-1)(\hat{t}_1 k - \hat{t}_0)}{4\hat{\lambda}^3 (1 + \hat{\gamma}_1 k)^2 m (k^3 - 1)} \left\{ \sum_i \left(\sum_j \hat{u}_{ij} \right)^4 - \sum_{i,j} \hat{u}_{ij}^4 \right\}, \\
&\quad + \frac{(m-1)\hat{t}_0}{4\hat{\lambda}^3 (1 + \hat{\gamma}_1 k)^2 m} \sum_{i,j} \hat{u}_{ij}^4, \\
\hat{\mathcal{I}}_{1,1,11} &= \frac{(m-1)^2}{4\hat{\lambda}^2 (1 + \hat{\gamma}_1 k)^4 m^2} \sum_i \left(\sum_j \hat{u}_{ij} \right)^4; \\
\hat{\mathcal{I}}_{1,2,00} &= \frac{1}{2\hat{\lambda}^2} \left[mk - 1 - \frac{3}{2} mk \hat{t}_0^2 \{(1 + \hat{\gamma}_1)^2 - \hat{t}_3\} - \frac{3}{2} m \hat{t}_1^2 \hat{t}_3 \right], \\
\hat{\mathcal{I}}_{1,2,01} &= \frac{(m-1)k}{2\hat{\lambda}(1 + \hat{\gamma}_1 k)} \left\{ 1 - \left(\frac{3}{2} \right) \frac{(\hat{t}_1 k - \hat{t}_0)\hat{t}_3 + (1 + \hat{\gamma}_1)^2 \hat{t}_0}{1 + \hat{\gamma}_1 k} \right\}, \\
\hat{\mathcal{I}}_{1,2,11} &= -\frac{(m-1)(m-3)k^2}{4m(1 + \hat{\gamma}_1 k)^2},
\end{aligned}
$$

$\hat{u}_{ij} = y_{ij} - \bar{y}_{\cdot\cdot}$, and the $\hat{t}$s are the ts with θ replaced by $\hat{\theta}$, the REML estimator.

The following outlines a numerical algorithm for POQUIM.

1. Determine the sets of indices $\mathcal{S}_l = \{(i_1, \dots, i_4) : f(i_1, \dots, i_4) = f_l\}$, $1 \le l \le L$. Then, for each (j, k), $0 \le j \le k \le s$, do the following.
2. Compute the constants $c_{j,k,l}$ given by (1.62), $1 \le l \le L$. Note that the denominator is $|\mathcal{S}_l|$.
3. Compute

$$\hat{\mathcal{I}}_{1,1,jk} = \sum_{f(i_1,\dots,i_4)\neq 0} \hat{c}_{j,k}(i_1,\dots,i_4)\hat{u}_{i_1}\cdots\hat{u}_{i_4},$$

where $\hat{c}_{j,k}(i_1,\dots,i_4)$ is defined the same way as $c_{j,k}(i_1,\dots,i_4)$ above (1.62) with θ replaced by $\hat{\theta}$, and $\hat{u}_i = y_i - x_i'\hat{\beta}$. Note that $\sum_{f((i_1,\dots,i_4)\neq 0} = \sum_{\mathcal{S}_1} + \cdots + \sum_{\mathcal{S}_L}$.
4. Compute $\hat{\mathcal{I}}_{1,2,jk}$ which is $\mathcal{I}_{1,2,jk}$ with θ replaced by $\hat{\theta}$. See step 3 for the summation.
5. Let $\hat{\mathcal{I}}_{1,jk} = \hat{\mathcal{I}}_{1,1,jk} + \hat{\mathcal{I}}_{1,2,jk}$.

All except step 1 are fairly straightforward. As for step 1, the sets may be determined as follows. First, the index $(1,1,1,1)$ belongs to $\mathcal{S}_1$. Also compute the vector $v_{1,1,1,1} = (z_{1t} \cdot z_{1t} \cdot z_{1t} \cdot z_{1t})_{0\le t\le s}$. Then, compute the vector $v_{1,1,1,2} = (z_{1t} \cdot z_{1t} \cdot z_{1t} \cdot z_{2t})_{0\le t\le s}$. If $v_{1,1,1,2} = v_{1,1,1,1}$, the index $(1,1,1,2)$ belongs to $\mathcal{S}_1$; otherwise, it belongs to $\mathcal{S}_2$, and so on.

1.9 Exercises

1.1. Show that the one-way random effects model in Example 1.1 can be expressed as (1.1), where y is given in Example 1.1. What are X and Z in this case?

1.2. Show that the two-way random effects model in Example 1.2 can be expressed as (1.1). Also, show that this model is a special case of the balanced mixed ANOVA model defined in Section 1.2.1.1.

1.3. Show that the growth curve model of Example 1.3 can be expressed as the standard form (1.1). Note that in (1.1) the random effects are assumed to have mean zero, therefore you may need to define some new random effects that satisfy the basic assumptions for (1.1). Furthermore, show that Example 1.3 is a special case of the general longitudinal model (1.3).

1.4. Show that the first two stages of Example 1.4, (i) and (ii), are equivalent to the (classical) linear mixed model of Example 1.1 with μ replaced by $x_{ij}'\beta$.

1.5. Specify Equations (1.7)–(1.10) for the longitudinal model Section 1.2.1.2.

1.6. Verify Equations (1.13)–(1.15).

1.7. Verify the expressions in Example 1.1 (Continued) in Section 1.3.1 for the log-likelihood and its derivatives with respect to μ, σ^2, and τ^2. Also, obtain expressions for $I(\theta)$ in this particular case.

1.8. Show that in the Neyman–Scott example (Example 1.7), the MLE is inconsistent as the number of individuals increases. Furthermore, show that the MLE based on $z_i = y_{i1} - y_{i2}$, $i = 1,\dots,m$ [see Example 1.7 (Continued) in Section 1.3.2], that is, the REML estimator of σ^2, is consistent as m increases.

1.9. Show that the REML estimators do not depend on the choice of A; that is, if A is replaced by $B = AT$, where T is any $(n-p)\times(n-p)$ nonsingular matrix, the REML estimators will not change.

1.10. Suppose that, under the marginal model (1.4), we have a prior for β which is noninformative (i.e., the "density function" for β is a positive constant, say, 1, everywhere). Note that this is an improper prior. Show that the marginal likelihood for the variance components involved in V, obtained by integrating out β with respect to its prior, is identical to the restricted likelihood.

1.11. Verify Equation (1.21). Also show the following.

$$\mathrm{cov}\left(\frac{\partial l_{\mathrm{R}}}{\partial \theta_i}, \frac{\partial l_{\mathrm{R}}}{\partial \theta_j}\right) = \frac{1}{2}\mathrm{tr}\left(P\frac{\partial V}{\partial \theta_i}P\frac{\partial V}{\partial \theta_j}\right), 1 \le i, j \le q.$$

1.12. Show that, under the balanced one-way random effects model (i.e., Example 1.1 with $k_i = k$, $1 \le i \le m$), the REML equations for estimating σ^2 and τ^2 are equivalent to (1.22). Obtain the solution to these equations. Also derive the asymptotic covariance matrix of the REML estimators.

1.13. Show that, under ANOVA models with the original form of variance components $\tau^2, \sigma_1^2, \dots, \sigma_s^2$, the REML and ML equations are given by (1.23) and (1.24), respectively; under the Hartley–Rao form of variance components $\lambda, \gamma_1, \dots, \gamma_s$ (see Section 1.2.1.1), the REML and ML equations are given by (1.25) and (1.26), respectively.

1.14. Show that the REML equations derived under the multivariate t-distribution (see Section 1.4.1) are equivalent to those derived under the multivariate normal distribution.

*1.15**. Consider Example 1.2 (Continued) in Section 1.4.2.

a. Verify the expression (1.31).

b. Verify that $\mathrm{var}(\partial l_{\mathrm{R}}/\partial \lambda)$ can be expressed as $S_1 + S_2$, where S_1 can be expressed as (1.32) with the coefficients a_j, $j = 0, 1, 2$ given below. First define $t_0 = 1 + \lambda_1 + \lambda_2 + \lambda_3$, $t_1 = \{(m-1)n\}/\{m(1+\gamma_1 n)\}$, $t_2 = m(n-1)/\{n(1 + \gamma_2 m)\}$; $m_0 = mn$, $m_1 = m$, and $m_2 = n$. Then,

$$a_0 = \frac{t_0^2}{4\lambda^4},$$

$$a_1 = \frac{nt_0^2 - t_1^2}{4\lambda^4 n(n^3-1)},$$

$$a_2 = \frac{mt_0^2 - t_2^2}{4\lambda^4 m(m^3-1)}.$$

Furthermore, define $t_3 = \{n(1+\gamma_2+\gamma_1 n)^2 - (1+\gamma_1+\gamma_2)^2\}/(n^3-1)$ and $t_4 = \{m(1+\gamma_1+\gamma_2 m)^2 - (1+\gamma_1+\gamma_2)^2\}/(m^3-1)$. We have

$$S_2 = \frac{mn-1}{2\lambda^2} - \frac{3mnt_0^2}{4\lambda^2}\{(1+\gamma_1+\gamma_2)^2 - (t_3+t_4)\} - \frac{3(t_1^2 t_3 m + t_2^2 t_4 n)}{4\lambda^2}.$$

Hence S_2 depends only on θ.

1.16. Show that, in the balanced one-way random effects model (i.e., Example 1.1 with $k_i = k$, $1 \le i \le m$), the ANOVA estimators of σ^2 and τ^2

are $\hat{\sigma}^2 = (\text{MSA} - \text{MSE})/k$ and $\hat{\tau}^2 = \text{MSE}$. Are these estimators identical to the solution to the REML equations in this particular case? To answer the latter question you should not refer to the general result mentioned in Section 1.5.1.1 but, instead, derive the REML equations and see if the solution is the same as the ANOVA estimators. When will the ANOVA estimators be identical to the REML estimators? Furthermore, suppose that the true parameters are $\mu = 0.5$ and $\sigma^2 = \tau^2 = 1.0$, and the observations are normally distributed. Evaluate empirically the probability of negative estimator (for σ^2) and note how the probability changes with the sample size. The following sample sizes may be considered: $m = 20, 40, 100$, and 200, and $k = 5$ in all cases.

1.17. Show that, in Example 1.9, we have $P_W = P_X + P_{Z\ominus X}$, where the two matrices on the right side of the equation are projections orthogonal to each other. Also show that $Z_i' P_{Z\ominus X} Z_j = Z_i' P_{X^\perp} Z_j$, $i, j = 1, 2$. [In fact, using the result of Searle et al. (1992, Theorem M.1 on page 449), it can be shown that $P_{Z\ominus X} Z_j = P_{X^\perp} Z_j, j = 1, 2$]. Finally, verify that the coefficients for σ_1^2, σ_2^2, and τ^2 in $\text{SSR}(\alpha_2|\beta, \alpha_1)$ are 0, $\text{tr}\{P_{(X,Z_1)^\perp} Z_2 Z_2'\}$ and $\text{rank}(W) - \text{rank}\{(X, Z_1)\}$, respectively; the corresponding coefficients are 0, 0 and $n - \text{rank}(W)$ in SSE.

1.18. Refer to Section 1.5.2. Show that, by Lemma 1.2, $\text{E}(\hat{\eta} - \tilde{\eta}|^2) = 2\text{tr}[\{(Z_* A Z_* - B)D\}^2]$, where $D = \text{diag}(\sigma_i^2 I_{m_i}, 0 \le i \le s)$. Also show that $\text{E}(\hat{\eta} - \tilde{\eta}) = 0$.

1.19. Verify the expressions for $\tilde{\beta}$, $\tilde{\tau}^2$, $\hat{\lambda}$, $\hat{\psi}$, and $\hat{\tau}^2$ in Example 1.1 (Continued) in Section 1.6.1.

1.20. Verify the expression for the restricted log-likelihood (1.51).

1.21. Interpret the result of the data analysis summarized by Table 1.5 in terms of the medical research problems considered there (see Hand and Crowder 1996). In particular, how would you explain the significance of the coefficient β_6 to a researcher of the medical research problem?

2

Linear Mixed Models: Part II

2.1 Tests in Linear Mixed Models

The previous section dealt with point estimation and related problems in linear mixed models. In this section, we consider a different type of inference, namely, tests in linear mixed models. Section 2.1.1 discusses statistical tests in Gaussian mixed models. As shown, exact F-tests can often be derived under Gaussian ANOVA models. Furthermore, in some special cases, optimal tests such as uniformly most powerful unbiased (UMPU) tests exist and coincide with the exact F-tests. Section 2.1.2 considers tests in non-Gaussian linear mixed models. In such cases, exact/optimal tests typically do not exist. Therefore, statistical tests are usually developed based on asymptotic theory.

2.1.1 Tests in Gaussian Mixed Models

1. Exact tests. For ANOVA models, exact F-tests can often be derived using the following method. The original idea was due to Wald (1947). Consider the mixed ANOVA model (1.1) and (1.2). Suppose that one wishes to test the hypothesis H_0: $\sigma_1^2 = 0$. Note that the model can be written as

$$y = X\beta + Z_1\alpha_1 + Z_{-1}\alpha_{-1} + \epsilon, \tag{2.1}$$

where $\alpha_{-1} = (\alpha_2', \dots, \alpha_s')'$ and $Z_{-1} = (Z_2, \dots, Z_s)$. Consider the quadratic form $q_1 = \tau^{-2}y'P_{Z_1\ominus(X,Z_{-1})}y = y'\{P_{Z_1\ominus(X,Z_{-1})}/\tau^2\}y$, where $\ominus$ is introduced in Example 1.9. Note that, under the null hypothesis, we have $y \sim N(X\beta, V_0)$, where $V_0 = \tau^2 I + \sum_{i=2}^s \sigma_i^2 Z_i Z_i'$. Furthermore, we have

$$\begin{aligned}\left(\frac{P_{Z_1\ominus(X,Z_{-1})}}{\tau^2}\right)V_0 &= P_{Z_1\ominus(X,Z_{-1})} + \sum_{i=2}^{2}\left(\frac{\sigma_i^2}{\tau^2}\right)P_{Z_1\ominus(X,Z_{-1})}Z_iZ_i' \\ &= P_{Z_1\ominus(X,Z_{-1})},\end{aligned}$$

which is idempotent. Therefore, by Theorem C.1 in Appendix C, we have $q_1 \sim \chi^2_{r_1}$, where $r_1 = \text{rank}\{P_{Z_1\ominus(X,Z_{-1})}\} = \text{rank}\{(X, Z)\} - \text{rank}\{(X, Z_{-1})\}$.

Note that $P_{Z_1 \ominus (X,Z_{-1})} X = 0$ and $P_{(X,Z)} = P_{(X,Z_{-1})} + P_{Z_1 \ominus (X,Z_{-1})}$, where the two projections on the right side are orthogonal to each other (see Example 1.9 and Exercise 1.17).

On the other hand, consider the quadratic form $q_2 = \tau^{-2} y' P_{(X,Z)^\perp} y = y'\{P_{(X,Z)^\perp}/\tau^2\} y$. Note that $y \sim N(X\beta, V)$, where $V = \tau^2 I + \sum_{i=1}^s \sigma_i^2 Z_i Z_i'$. Thus, we have

$$\begin{aligned} \left(\frac{P_{(X,Z)^\perp}}{\tau^2}\right) V &= P_{(X,Z)^\perp} + \sum_{i=1}^{2} \left(\frac{\sigma_i^2}{\tau^2}\right) P_{(X,Z)^\perp} Z_i Z_i' \\ &= P_{(X,Z)^\perp}, \end{aligned} \tag{2.2}$$

which is idempotent. Therefore, by the same theorem, we have $q_2 \sim \chi^2_{r_2}$, where $r_2 = \text{rank}\{P_{(X,Z)^\perp}\} = n - \text{rank}\{(X,Z)\}$. Note that $P_{(X,Z)^\perp} X = 0$. Also note that, unlike q_1, the distribution of q_2 is unaffected by the null hypothesis.

Finally, because $P_{(X,Z)^\perp} V P_{Z_1 \ominus (X,Z_{-1})} = \tau^2 P_{(X,Z)^\perp} P_{Z_1 \ominus (X,Z_{-1})} = 0$ by (2.2), the two quadratic forms q_1 and q_2 are independent (again, this fact does not depend on the null hypothesis; see Appendix C). It follows that

$$\begin{aligned} F_1 &= \frac{y' P_{Z_1 \ominus (X,Z_{-1})} y / r_1}{y' P_{(X,Z)^\perp} y / r_2} \\ &= \frac{q_1/r_1}{q_2/r_2} \quad \sim \quad F_{r_1, r_2}. \end{aligned} \tag{2.3}$$

In words, F_1 has an exact (central) F-distribution with degrees of freedom r_1 and r_2 for testing the hypothesis H_0: $\sigma_1^2 = 0$.

It should be pointed out that, for the above test to be effective one must have $Z_1 \ominus (X, Z_{-1}) \neq \emptyset$. For example, if $\mathcal{L}(Z_1) \subset \mathcal{L}(Z_{-1})$, then the test will not work. We now consider an example.

Example 2.1 (Balanced two-way random effects model). First consider the case where there is no interaction between the random effect factors. The model can be expressed as

$$y_{ijk} = \mu + u_i + v_j + e_{ijk},$$

$i - 1, \ldots, a$, $j = 1, \ldots, b$, $k = 1, \ldots, c$, where u_is and v_js are random effects and e_{ijk}s are errors such that u_is are independent $N(0, \sigma_1^2)$, v_js are independent $N(0, \sigma_2^2)$, e_{ijk}s are independent $N(0, \tau^2)$, and u, v, e are independent. Using matrix expressions, we have

$$y = X\mu + Z_1 u + Z_2 v + e,$$

where $X = 1_a \otimes 1_b \otimes 1_c$, $Z_1 = I_a \otimes 1_b \otimes 1_c$, and $Z_2 = 1_a \otimes I_b \otimes 1_c$. Clearly, $Z_1 \ominus (X, Z_2) \neq \emptyset$, thus (2.3) may be applied for testing H_0: $\sigma_1^2 = 0$. In this case, we have $r_1 = (a+b-1) - b = a - 1$ and $r_2 = n - (a+b-1) = abc - a - b + 1$.

Next, we consider the case where there is interaction between u and v. In this case, the model can be expressed as

$$y_{ijk} = \mu + u_i + v_j + w_{ij} + e_{ijk},$$

$i = 1, \dots, a$, $j = 1, \dots, b$, $k = 1, \dots, c$, where, in addition, the interactions w_{ij}s are independent $N(0, \sigma_3^2)$, and u, v, w, e are independent. Similarly, the model may be written as

$$y = X\mu + Z_1 u + Z_2 v + Z_3 w + e,$$

where $Z_3 = I_a \otimes I_b \otimes 1_c$. However, neither $\sigma_1^2 = 0$ nor $\sigma_2^2 = 0$ can be tested using the exact F-test derived above, because $\mathcal{L}(Z_j) \subset \mathcal{L}(Z_3)$, $j = 1, 2$. Nevertheless, the hypothesis H_0: $\sigma_3^2 = 0$ can be tested using (2.3). In this case, $r_1 = ab - (a + b - 1) = (a - 1)(b - 1)$ and $r_2 = n - ab = ab(c - 1)$ (Exercise 2.1).

Further results on exact tests in Gaussian mixed models can be found in Khuri et al. (1998).

2. Optimal tests. It is known that optimal tests, such as UMPU and uniformly most powerful invariant unbiased tests (UMPIU), exist in some special cases of the mixed ANOVA models, assuming that normality holds. For example, Mathew and Sinha (1988) considered a balanced mixed ANOVA model, which can be expressed as

$$y = X_1\beta_1 + \cdots + X_t\beta_t + Z_1\alpha_1 + \cdots + Z_s\alpha_s + \epsilon, \tag{2.4}$$

where the βs and αs are, respectively, vectors of fixed and random effects in the analysis of variance; that is, they are main effects, interactions, nested effects, and the like. (e.g., Scheffé 1959), and ϵ is a vector of errors. Furthermore, assume that the random effects and errors are independent such that the components of α_i are distributed as $N(0, \sigma_i^2)$, and the components of ϵ are distributed as $N(0, \tau^2)$. The design matrices $X_1, \dots, X_t$ and $Z_1, \dots, Z_s$ are assumed known with $X_1 = 1_n$. Let P_i, $i = 1, \dots, t$ and Q_i, $i = 1, \dots, s$ be projection matrices such that $P_1 = n^{-1}J_n$, where $J_n = 1_n 1_n'$, $y'P_i y$ the sum of squares due to β_i, $2 \le i \le t$, and $y'Q_i y$ the sum of squares due to α_i (as in a fixed effects model), $1 \le i \le s$ (Searle 1971, §9.6). Note that each P_i (Q_i) is a Kronecker product of matrices of the form I_a, $a^{-1}J_a$ or $I_a - a^{-1}J_a$, so that P_i, $i = 1, \dots, t$ and Q_i, $i = 1, \dots, s + 1$ are orthogonal to each other, where $Q_{s+1} = I_n - \sum_{i=1}^t P_i - \sum_{i=1}^s Q_i$. With these notations, the likelihood function can be expressed as

$$f(y) = c(\theta) \times \exp\left[-\frac{1}{2}\left\{\sum_{i=1}^{s+1} \xi_i y' Q_i y + \sum_{i=1}^{t} \eta_i (S_i' y - \lambda_i)'(S_i' y - \lambda_i)\right\}\right], \tag{2.5}$$

where $c(\theta)$ depends only on the variance components, $\theta = (\sigma_1^2, \dots, \sigma_s^2, \tau^2)'$; ξ_i and η_i are linear functions of the variance components; $S_i S_i' = P_i$ and

$\lambda_i = S_i' X\beta$, $1 \le i \le t$. Here $X\beta$ is as in (1.1) when (2.4) is written in this way. By (2.5), it can be shown that $S_i'y$, $i = 1, \ldots, t$ and $y'Q_i y$, $i = 1, \ldots, s+1$ are complete sufficient statistics for the parameters ξ_is, η_is and λ_is. Furthermore, standard theory for the multiparameter exponential family (e.g., Lehmann and Casella 1998, §1) may be applied to derive UMPU and other optimal tests. For example, Mathew and Sinha (1988) obtained the following results.

1. Suppose that the hypothesis of interest is H_0: $\lambda_i = 0$ versus H_1: $\lambda_i \ne 0$. If η_i equals some ξ_j, say, ξ_1, an exact F-test is based on $y'P_i y/y'Q_1 y$; if λ_i is a scalar, then this test is UMPU; if λ_i is multidimensional, a UMPU test does not exist, however, the above F-test is UMPIU.
2. Suppose that the hypothesis of interest is H_0: $\xi_1 = \xi_2$ versus H_1: $\xi_2 > \xi_1$. The F-test based on $y'Q_2 y/y'Q_1 y$ is UMPU and UMPIU.

Note that, in some cases, a hypothesis such as $\sigma_i^2 = 0$ is equivalent to the equality of two ξ_is. We consider some examples.

Example 2.2 (Balanced one-way random effects model). Consider a special case of the one-way random effects model of Example 1.1 with $k_i = k$, $1 \le i \le m$. In this case, $y'Q_1 y$ is equal to the treatment sum of squares and $y'Q_2 y$ error sum of squares, that is, $y'Q_1 y = \text{SSA} = k\sum_{i=1}^m (\bar{y}_{i\cdot} - \bar{y}_{\cdot\cdot})^2$, $y'Q_2 y = \text{SSE} = \sum_{i=1}^m \sum_{j=1}^k (y_{ij} - \bar{y}_{i\cdot})^2$, and $S_1'y = \sqrt{mk}\bar{y}_{\cdot\cdot}$. Furthermore, we have $\xi_1^{-1} = \tau^2 + k\sigma^2$, $\xi_2^{-1} = \tau^2$, $\eta_1^{-1} = \tau^2 + k\sigma^2$, and $\lambda_1 = \sqrt{mk}\mu$.

Consider the hypothesis $\mu = 0$. Because $\eta_1 = \xi_1$ and λ_1 is a scalar, by the first result above, the F-test based on $\bar{y}_{\cdot\cdot}^2/\text{SSA}$ is UMPU and UMPIU. As for the hypothesis $\sigma^2 = 0$, because it is equivalent to $\xi_1 = \xi_2$, the F-test based on SSA/SSE is UMPU and UMPIU.

Example 2.1 (Continued). Consider the case without interaction and that $k = 1$. In this case, the model can simply be expressed as

$$y_{ij} = \mu + u_i + v_j + e_{ij},$$

$i = 1, \ldots, a$, $j = 1, \ldots, b$. In this case, we have $y'Q_1 y = b\sum_{i=1}^a (\bar{y}_{i\cdot} - \bar{y}_{\cdot\cdot})^2 \equiv \text{SSA}$, $y'Q_2 y = a\sum_{j=1}^b (\bar{y}_{\cdot j} - \bar{y}_{\cdot\cdot})^2 \equiv \text{SSB}$, and $y'Q_3 y = \sum_{i=1}^a \sum_{j=1}^b (y_{ij} - \bar{y}_{i\cdot} - \bar{y}_{\cdot j} + \bar{y}_{\cdot\cdot})^2 \equiv \text{SSE}$, which correspond to $\xi_1^{-1} = \tau^2 + b\sigma_1^2$, $\xi_2^{-1} = \tau^2 + a\sigma_2^2$, and $\xi_3^{-1} = \tau^2$, respectively. Furthermore, we have $S_1'y = \sqrt{ab}\bar{y}_{\cdot\cdot}$ with $\eta_1^{-1} = \tau^2 + a\sigma_2^2 + b\sigma_1^2$ and $\lambda_1 = \sqrt{ab}\mu$.

The hypotheses $\sigma_1^2 = 0$ and $\sigma_2^2 = 0$ correspond to $\xi_1 = \xi_3$ and $\xi_2 = \xi_3$, respectively. Thus, the F-tests based on SSA/SSE and SSB/SSE are, respectively, optimal (i.e., UMPU and UMPIU) for testing these hypotheses. However, unlike the previous example, no exact optimal test (in the same sense) exists for testing $\mu = 0$, because η_1 is not equal to any of the ξs.

These examples show that the results of Mathew and Sinha (1988) may be useful in some cases to obtain optimal tests, but there are cases where these results do not yield optimal tests (see Exercise 2.2 for an additional example). For more discussion on optimal tests, see Khuri et al. (1998).

3. Likelihood-ratio tests. The likelihood-ratio is a well-known method of constructing statistical tests. The theory of likelihood-ratio tests is fully developed in the i.i.d. case (e.g., Lehmann 1999, §7.7). However, the literature on likelihood-ratio tests in the context of linear mixed models is much less extensive, from a theoretical point of view. Hartley and Rao (1967) was the first paper that addressed the issue. Let $\psi = (\beta', \theta')'$ be the vector of all the unknown parameters involved in a Gaussian mixed model, where θ represents the vector of variance components. Many of the hypotheses are concerned with testing whether a subvector of θ, say, $\theta^{(1)}$, is identical to a known vector, $\theta_0^{(1)}$. Let $\theta^{(2)}$ denote the subvector of θ complementary to $\theta^{(1)}$. Then, the likelihood function may be expressed as $L(\theta) = L(\theta^{(1)}, \theta^{(2)})$. [Note that $L(\theta)$ depends on y and therefore should be properly denoted by $L(\theta|y)$, but we suppress y for notational simplicity.] Let $\hat{\theta}$ be the (global) maximizer of $L(\theta|y)$ over $\theta \in \Theta$, where Θ is the parameter space, and $\hat{\theta}^{(2)}$ be the (global) maximizer of $L(\theta_0^{(1)}, \theta^{(2)})$ over $\theta^{(2)} \in \Theta^{(2)}$, where $\Theta^{(2)}$ is the parameter space for $\theta^{(2)}$. Then, the likelihood ratio is given by

$$\mathcal{R} = \frac{L(\theta_0^{(1)}, \hat{\theta}^{(2)})}{L(\hat{\theta})} . \tag{2.6}$$

Hartley and Rao (1967) stated without giving a proof that the asymptotic null distribution of $-2\log\mathcal{R}$ is a central χ^2 with r degrees of freedom, where r is the dimension of $\theta^{(1)}$. See Jiang (2005c) for a rigorous proof of this result, which also applies to non-Gaussian linear mixed models (see Section 2.1.2.4). We consider a simple example.

Example 2.3 (One-way random effects model). Consider the one-way random effects model of Example 1.1 with normality assumption. It was shown in Section 1.3.1 (see Example 1.1 (Continued)) that the log-likelihood function is given by

$$\begin{aligned} l(\mu, \sigma^2, \tau^2) &= c - \frac{1}{2}(n-m)\log(\tau^2) - \frac{1}{2}\sum_{i=1}^m \log(\tau^2 + k_i\sigma^2) \\ &\quad - \frac{1}{2\tau^2}\sum_{i=1}^m\sum_{j=1}^{k_i}(y_{ij}-\mu)^2 + \frac{\sigma^2}{2\tau^2}\sum_{i=1}^m \frac{k_i^2}{\tau^2 + k_i\sigma^2}(\bar{y}_{i\cdot}-\mu)^2, \end{aligned}$$

where c is a constant, $n = \sum_{i=1}^m k_i$, and $\bar{y}_{i\cdot} = k_i^{-1}\sum_{j=1}^{k_i} y_{ij}$. Let $\hat{\mu}$, $\hat{\sigma}^2$ and $\hat{\tau}^2$ be the MLE of μ, σ^2, and τ^2. Suppose that one is interested in testing the hypothesis $\sigma^2 = 0$. Under the null hypothesis, we have

$$l(\mu, 0, \tau^2) = c - \frac{n}{2}\log(\tau^2) - \frac{1}{2\tau^2}\sum_{i=1}^m\sum_{j=1}^{k_i}(y_{ij}-\mu)^2.$$

The MLE under the null are $\tilde{\mu} = \bar{y}_{\cdot\cdot}$ and $\tilde{\sigma}^2 = n^{-1}\sum_{i=1}^m \sum_{j=1}^{k_i}(y_{ij} - \bar{y}_{\cdot\cdot})^2$, where $\bar{y}_{\cdot\cdot} = n^{-1}\sum_{i=1}^m \sum_{j=1}^{k_i} y_{ij}$. Thus, an expression for $-2\log\mathcal{R}$ can be easily derived (Exercise 2.3).

2.1.2 Tests in Non-Gaussian Linear Mixed Models

For non-Gaussian linear mixed models, exact or optimal tests typically do not exist. This is because under a non-Gaussian model, the distribution of y is not fully specified, therefore it is (usually) not possible either to derive the exact distribution of a test statistic or to study the power function of the test. In such cases, statistical tests are usually based on asymptotic theory. In this section, we consider asymptotic tests in non-Gaussian linear mixed models. Please note that the results of this section also apply to Gaussian mixed models, especially in cases where exact/optimal tests do not exist.

A basic idea of deriving an asymptotic test is the following. Consider a non-Gaussian linear mixed model (1.1). Let $\psi = (\beta', \theta')'$, where θ represents the vector of variance components involved. Then ψ is the vector of all the unknown parameters involved in the model. Suppose that an estimator of ψ, say, $\hat{\psi}$, can be obtained, which is asymptotically normal, that is, there exists a sequence of positive definite matrices, $\Sigma = \Sigma_n$, such that

$$\Sigma^{-1/2}(\hat{\psi} - \psi) \longrightarrow N(0, I), \quad \text{in distribution}, \tag{2.7}$$

where I is the $(p+q)$-dimensional identity matrix with $p = \dim(\beta)$ and $q = \dim(\theta)$. Σ is called the asymptotic covariance matrix of $\hat{\psi}$. Suppose that one wishes to test a linear hypothesis of the form

$$H_0: \quad K'\psi = c, \tag{2.8}$$

where K is a known matrix of full (column) rank, say, r, and c is a known vector. Under (2.8), (2.7) implies that

$$(K'\hat{\psi} - c)'(K'\Sigma K)^{-1}(K'\hat{\psi} - c) \longrightarrow \chi_r^2, \quad \text{in distribution}. \tag{2.9}$$

Thus, (2.9) can be used to test the hypothesis (2.8).

Typically, the asymptotic covariance matrix depends not only on θ but also on some additional parameters. For example, under the mixed ANOVA model Section 1.2.2.1, the asymptotic covariance matrix of the REML estimator of $\theta = (\tau^2, \sigma_1^2, \dots, \sigma_s^2)'$ depends not only on θ but also on the kurtoses of the random effects and errors; the asymptotic covariance matrix of the ML estimator of ψ depends not only on θ but also on the kurtoses as well as the third moments of the random effects and errors. (See Section 2.2.2 for more details; note that, under normality both the third moments and the kurtoses vanish, so there is no such problem for Gaussian mixed models.) Therefore, for the asymptotic test (2.9) to be applicable, one has to find some way to consistently estimate Σ, because standard procedures in mixed model analysis such as ML and REML do not produce estimators of higher (i.e., third and fourth) moments of the random effects and errors. In the following we discuss several methods of estimating Σ. Typically, when Σ in (2.9) is replaced by a consistent estimator, say $\hat{\Sigma}$, the asymptotic distribution on the right side does not change. The test therefore rejects if

$$(K'\hat{\psi} - c)'(K'\hat{\Sigma}K)^{-1}(K'\hat{\psi} - c) > \chi^2_{r,\rho}, \tag{2.10}$$

where ρ is the significance level.

1. Empirical method of moments. Consider the case of the mixed ANOVA model (1.1) and (1.2). As mentioned, the asymptotic covariance matrix of the REML (ML) estimator involves higher moments, thus, a natural approach would be to find consistent estimators of those higher moments. Jiang (2003) proposed an empirical method of moments and gave a number of applications, including estimation of the kurtoses in mixed ANOVA models. The basic idea is the following. Let θ be a vector of parameters. Suppose that a consistent estimator of θ, $\hat{\theta}$, is available. Let ϕ be a vector of additional parameters about which knowledge is needed. Let $\vartheta = (\theta' \ \phi')'$, and $M(\vartheta, y) = M(\theta, \phi, y)$ be a vector-valued function of the same dimension as ϕ that depends on ϑ and y, a vector of observations. Suppose that $\mathrm{E}\{M(\vartheta, y)\} = 0$ when ϑ is the true vector of parameters. Then, if θ were known, a method of moments estimator of ϕ would be obtained by solving

$$M(\theta, \phi, y) = 0 \tag{2.11}$$

for ϕ. Note that this is more general than the classical method of moments, in which the function M is a vector of sample moments minus their expected values. In econometric literature, this is referred to as the generalized method of moments (e.g., Hansen 1982, Newey 1985). Because θ is unknown, we replace it in (2.11) by $\hat{\theta}$. The result is called an empirical method of moments (EMM) estimator of ϕ, denoted by $\hat{\phi}$, which is obtained by solving

$$M(\hat{\theta}, \phi, y) = 0\,. \tag{2.12}$$

Note that here we use the words "an EMM estimator" instead of "the EMM estimator", because sometimes there may be more than one consistent estimator of θ, and each may result in a different EMM estimator of ϕ. In general, ML estimators may be viewed as a special kind of EMM estimator (Exercises 2.4 and 2.5). To see this, let $l(\vartheta; y) = l(\theta, \phi; y)$ be the log-likelihood function. Then, the ML estimator, $\hat{\vartheta} = (\hat{\theta}' \ \hat{\phi}')'$ satisfies $\partial l / \partial \vartheta = 0$, and hence $\hat{\phi}$, the ML estimator of ϕ, satisfies

$$\frac{\partial}{\partial \vartheta} l(\hat{\theta}, \phi; y) = 0\,. \tag{2.13}$$

On the other hand, (2.13) is a special case of (2.12), in which $M(\theta, \phi, y) = \partial l / \partial \vartheta$. Note that $\mathrm{E}(\partial l / \partial \vartheta) = 0$ when ϑ is the true vector of parameters. Jiang (2003) showed that, under mild conditions, $\hat{\phi}$ is consistent.

To apply EMM to non-Gaussian mixed ANOVA models, let θ be the vector of variance components. It is clear that a consistent estimator of θ, $\hat{\theta}$, exists. For example, $\hat{\theta}$ can be the REML or ML estimator (e.g., Jiang 1996). Furthermore, assume that the third moments of the random effects and errors vanish; that is,

$$\mathrm{E}(\epsilon_1^3) = 0 \quad \text{and} \quad \mathrm{E}(\alpha_{i1}^3) = 0, \qquad 1 \le i \le s, \tag{2.14}$$

where α_{i1} is the first component of α_i and ϵ_1 the first component of ϵ. Then, the asymptotic covariance matrix of the REML (ML) estimator involves only the kurtoses, in addition to the variance components [in fact, the asymptotic covariance matrix of REML estimator does not involve the third moments regardless of (2.14)]. For notational convenience, write $\sigma_0^2 = \tau^2$. Then, the (unscaled) kurtoses are defined by $\kappa_0 = \mathrm{E}(\epsilon_1^4) - 3\sigma_0^4$, $\kappa_i = \mathrm{E}(\alpha_{i1}^4) - 3\sigma_i^4$, $1 \le i \le s$. For any matrix $A = (a_{ij})$, we define $\|A\|_4 = (\sum_{i,j} a_{ij}^4)^{1/4}$. Similarly, if $a = (a_i)$ is a vector, then $\|a\|_4 = (\sum_i a_i^4)^{1/4}$. Let L be a linear space, then $L^\perp$ represents the linear space $\{a : a'b = 0, \forall b \in L\}$. If L_1, L_2 are linear spaces such that $L_1 \subset L_2$, then $L_2 \ominus L_1$ represents the linear space $\{a : a \in L_2, a'b = 0, \forall b \in L_1\}$ (note that the notation is consistent with that in Example 1.9). If $M_1, \ldots, M_k$ are matrices with the same number of rows, then $\mathcal{L}(M_1, \ldots, M_k)$ represents the linear space spanned by the columns of $M_1, \ldots, M_k$. Let the matrices $Z_1, \ldots, Z_s$ be suitably ordered such that

$$L_i \ne \{0\}, \qquad 0 \le i \le s\,, \tag{2.15}$$

where $L_0 = \mathcal{L}(Z_1, \ldots, Z_s)^\perp$, $L_i = \mathcal{L}(Z_i, \ldots, Z_s) \ominus \mathcal{L}(Z_{i+1}, \ldots, Z_s)$, $1 \le i \le s-1$, and $L_s = \mathcal{L}(Z_s)$. Let C_i be a matrix whose columns constitute a base of L_i, $0 \le i \le s$. We define $a_{ij} = \|Z_j' C_i\|_4^4$, $0 \le j \le i \le s$, where $Z_0 = I$, the identity matrix. It is easy to see that, under (2.15), $a_{ii} > 0$, $0 \le i \le s$. Let n_i be the number of columns of C_i, and c_{ik} the kth column of C_i, $1 \le k \le n_i$, $0 \le i \le s$. Define

$$b_i(\sigma^2) = 3 \sum_{k=1}^{n_i} \left(\sum_{j=0}^{i} |Z_j' c_{ik}|^2 \sigma_j^2 \right)^2, \qquad 0 \le i \le s.$$

where $\sigma^2 = (\sigma_j^2)_{0 \le j \le s}$. Let $\kappa = (\kappa_j)_{0 \le j \le s}$, and $M(\beta, \sigma^2, \kappa, y)$ be the vector whose ith component is

$$M_i(\beta, \sigma^2, \kappa, y) = \|C_i'(y - X\beta)\|_4^4 - \sum_{j=0}^{i} a_{ij}\kappa_j - b_i(\sigma^2), \qquad 0 \le i \le s.$$

Then, by the following lemma and the definition of the C_is, it can be shown that $\mathrm{E}\{M(\beta, \sigma^2, \kappa, y)\} = 0$ when β, σ^2, κ correspond to the true parameters (Exercise 2.6). Thus, a set of EMM estimators can be easily obtained by solving $M(\hat{\beta}, \hat{\sigma}^2, \kappa, y) = 0$, where $\hat{\beta}$ and $\hat{\sigma}^2$ are the REML or ML estimators. Furthermore, the EMM estimators can be computed recursively as follows.

$$\hat{\kappa}_0 = a_{00}^{-1}\hat{d}_0 ,$$

$$\hat{\kappa}_i = a_{ii}^{-1}\hat{d}_i - \sum_{j=0}^{i-1}\left(\frac{a_{ij}}{a_{ii}}\right)\hat{\kappa}_j, \qquad 1 \le i \le s, \tag{2.16}$$

where $\hat{d}_i = \|C_i'(y - X\hat{\beta})\|_4^4 - b_i(\hat{\sigma}^2)$, $0 \le i \le s$.

Lemma 2.1. Let $\xi_1, \ldots, \xi_n$ be independent random variables such that $\mathrm{E}\xi_i = 0$ and $\mathrm{E}\xi_i^4 < \infty$, and $\lambda_1, \ldots, \lambda_n$ be constants. Then,

$$\mathrm{E}\left(\sum_{i=1}^n \lambda_i \xi_i\right)^4 = 3\left[\sum_{i=1}^n \lambda_i^2 \mathrm{var}(\xi_i)\right]^2 + \sum_{i=1}^n \lambda_i^4\{\mathrm{E}\xi_i^4 - 3[\mathrm{var}(\xi_i)]^2\} .$$

Example 2.2 (Continued). Here we have $\kappa_0 = \mathrm{E}(\epsilon_{11}^4) - 3\tau^4$ and $\kappa_1 = \mathrm{E}(\alpha_1^4) - 3\sigma^4$. The model can be written as $y = X\mu + Z\alpha + \epsilon$, where $X = 1_m \otimes 1_k$, and $Z = I_m \otimes 1_k$. Let

$$D_k = \begin{pmatrix} 1 & \cdots & 1 \\ -1 & \cdots & 0 \\ \vdots & \ddots & \vdots \\ 0 & \cdots & -1 \end{pmatrix}_{k \times (k-1)} .$$

Then, it is easy to show that $C_0 = I_m \otimes D_k$, $C_1 = Z = I_m \otimes 1_k$. It follows from (2.16) that, in closed form,

$$\hat{\kappa}_0 = \frac{1}{2m(k-1)}\sum_{i=1}^m\sum_{j=2}^k (y_{i1} - y_{ij})^4 - 6\hat{\tau}^4 ,$$

$$\begin{aligned}\hat{\kappa}_1 = {} & \frac{1}{mk^4}\sum_{i=1}^m (y_{i\cdot} - k\hat{\mu})^4 - \frac{1}{2mk^3(k-1)}\sum_{i=1}^m\sum_{j=2}^k (y_{i1} - y_{ij})^4 \\ & -\frac{3}{k^2}\left(1 - \frac{2}{k}\right)\hat{\tau}^4 - \frac{6}{k}\hat{\tau}^2\hat{\sigma}^2 - 3\hat{\sigma}^4 ,\end{aligned}$$

where $y_{i\cdot} = \sum_{j=1}^k y_{ij}$, $\hat{\mu} = \bar{y}_{\cdot\cdot}$, and $\hat{\tau}^2$, $\hat{\sigma}^2$ are the REML or ML estimators. It can be shown (Exercise 2.7) that the EMM estimators are consistent provided that $m \to \infty$ and $k \ge 2$.

2. Partially observed information. One important assumption that we have made in the application of EMM is (2.14). This assumption holds, for example, if the random effects and errors are symmetrically distributed. However, from a practical point of view, such an assumption is not very pleasant because, like normality, symmetry may not hold in practice. On the other hand, a method called partially observed information was proposed in Section 1.4.2 for estimating the asymptotic covariance matrices of REML or ML estimators. This method applies to a general non-Gaussian mixed ANOVA model regardless of (2.14). We consider an example.

Example 2.2 (Continued). Suppose that one wishes to test the hypothesis H_0: $\gamma_1 = 1$; that is, the variance contribution due to the random effects is the same as that due to the errors. Note that in this case $\theta = (\lambda, \gamma_1)'$, so the null hypothesis corresponds to (2.8) with $K = (0,1)'$ and $c = 1$. Furthermore, we have $K'\Sigma_{\mathrm{R}}K = \Sigma_{\mathrm{R},11}$, which is the asymptotic variance of $\hat{\gamma}_1$, the REML estimator of γ_1. Thus, the test statistic is $\hat{\chi}^2 = (\hat{\gamma}_1 - 1)^2/\hat{\Sigma}_{\mathrm{R},11}$, where $\hat{\Sigma}_{\mathrm{R},11}$ is the POQUIM estimator of $\Sigma_{\mathrm{R},11}$ (see Section 1.8.5) given by

$$\hat{\Sigma}_{\mathrm{R},11} = \frac{\hat{\mathcal{I}}_{1,11}\hat{\mathcal{I}}_{2,00}^2 - 2\hat{\mathcal{I}}_{1,01}\hat{\mathcal{I}}_{2,00}\hat{\mathcal{I}}_{2,01} + \hat{\mathcal{I}}_{1,00}\hat{\mathcal{I}}_{2,01}^2}{(\hat{\mathcal{I}}_{2,00}\hat{\mathcal{I}}_{2,11} - \hat{\mathcal{I}}_{2,01}^2)^2},$$

where $\hat{\mathcal{I}}_{1,st} = \hat{\mathcal{I}}_{1,1,st} + \hat{\mathcal{I}}_{1,2,st}$, $s,t = 0,1$, and $\hat{\mathcal{I}}_{1,r,st}$, $r = 1,2$ are given in Example 1.1 (continued) in Section 1.8.5 but with $\hat{\gamma}_1$ replaced by 1, its value under H_0; furthermore, we have

$$\begin{aligned}
\hat{\mathcal{I}}_{2,00} &= -\frac{(mk-1)}{2\hat{\lambda}^2}, \\
\hat{\mathcal{I}}_{2,01} &= -\frac{(m-1)k}{2\hat{\lambda}(1+\hat{\gamma}_1 k)}, \\
\hat{\mathcal{I}}_{2,11} &= -\frac{(m-1)k^2}{2(1+\hat{\gamma}_1 k)^2},
\end{aligned}$$

again with $\hat{\gamma}_1$ replaced by 1, where $\hat{\lambda}$ is the REML estimator of λ (Exercise 2.8). The asymptotic null distribution of the test is χ_1^2.

3. Jackknife method. For non-Gaussian longitudinal models, the asymptotic covariance matrix of the REML (ML) estimator may be estimated using the jackknife method discussed in Section 1.4.4. One advantage of the jackknife method is that it is one-formula-works-for-all. In fact, the same jackknife estimator not only applies to longitudinal linear mixed models, it also applies to longitudinal generalized linear mixed models, which we discuss in Chapters 4 and 5. Let ψ be the vector of all the parameters involved in a non-Gaussian longitudinal model, which includes fixed effects and variance components. Let $\hat{\psi}$ be the REML or ML estimator of ψ. Then, the jackknife estimator of the asymptotic covariance matrix of $\hat{\psi}$ is given by (1.43). Jiang and Lahiri (2004) showed that, under suitable conditions, the jackknife estimator is consistent in the sense that $\hat{\Sigma}_{\mathrm{Jack}} = \Sigma + O_{\mathrm{P}}(m^{-1-\delta})$ for some $\delta > 0$. Therefore, one may use $\hat{\Sigma} = \hat{\Sigma}_{\mathrm{Jack}}$ on the left side of (2.10) for the asymptotic test. We consider a simple example.

Example 2.4 (The James–Stein estimator). Let y_i, $i = 1,\dots,m$ be independent such that $y_i \sim N(\theta_i, 1)$. In the context of simultaneous estimation of

$\theta = (\theta_1, \dots, \theta_m)'$, it is well known that for $m \geq 3$, the James–Stein estimator dominates the maximum likelihood estimator, given by $y = (y_1, \dots, y_m)'$ in terms of the frequentist risk under the sum of squared error loss function (e.g., Lehmann and Casella 1998, pp. 272-273). Efron and Morris (1973) provided an empirical Bayes justification of the James–Stein estimator. Their Bayesian model can be equivalently written as the following simple random effects model: $y_i = \alpha_i + \epsilon_i$, $i = 1, \dots, m$, where the sampling errors $\{\epsilon_i\}$ and the random effects $\{\alpha_i\}$ are independently distributed with $\alpha_i \sim N(0, \psi)$ and $\epsilon_i \sim N(0, 1)$, and ϵ and α are independent.

Now we drop the normality assumption. Instead, we assume that y_i, $1 \leq i \leq m$ $(m > 1)$ are i.i.d. with $\mathrm{E}(y_1) = 0$, $\mathrm{var}(y_1) = \psi + 1$ and $\mathrm{E}(|y_1|^d) < \infty$ $(d > 4)$. Then, an M-estimator for ψ, which is the solution to the ML equation, is given by $\hat{\psi} = m^{-1}\sum_{i=1}^m y_i^2 - 1$. The delete-$i$ M-estimator is $\hat{\psi}_{-i} = (m-1)^{-1}\sum_{j \neq i} y_j^2 - 1$. The jackknife estimator of the asymptotic variance of $\hat{\psi}$ is given by

$$\hat{\sigma}^2_{\text{jack}} = \frac{m-1}{m}\sum_{i=1}^m (\hat{\psi}_{-i} - \hat{\psi})^2.$$

4. Robust versions of classical tests. Robust testing procedures have been studied extensively in the literature. In particular, robust versions of the classical tests, that is, the Wald, score, and likelihood-ratio tests (e.g., Lehmann 1999, §7) have been considered. In the case of i.i.d. observations, see, Foutz and Srivastava (1977), Kent (1982), Hampel et. al. (1986), and Heritier and Ronchetti (1994), among others. In the case of independent but not identically distributed observations, see, for example, Schrader and Hettmansperger (1980), Chen (1985), Silvapulle (1992), and Kim and Cai (1993). In contrast to the independent cases, the literature on robust testing with dependent observations is not extensive. In particular, in the case of linear mixed models, such tests as the likelihood-ratio test were studied only under the normality assumption (e.g., Hartley and Rao 1967). Because the normality assumption is likely to be violated in practice, it would be interesting to know if the classical tests developed under normality are robust against departure from such a distributional assumption.

Jiang (2005c) considered robust versions of the Wald, score, and likelihood-ratio tests in the case of dependent observations, which he called *W*-, *S*- and *L*-tests, and applied the results to non-Gaussian linear mixed models. The approach is briefly described as follows with more details given in Section 2.7. Let $y = (y_k)_{1 \leq k \leq n}$ be a vector of observations not necessarily independent. Let ψ be a vector of unknown parameters that are associated with the joint distribution of y, but the entire distribution of y may not be known given ψ (and possibly other parameters). We are interested in testing the hypothesis:

$$H_0 : \psi \in \Psi_0 \tag{2.17}$$

versus H_1: $\psi \notin \Psi_0$, where $\Psi_0 \subset \Psi$, and Ψ is the parameter space. Suppose that there is a new parameterization ϕ such that, under the null hypothesis (2. 17), $\psi = \psi(\phi)$ for some ϕ. Here $\psi(\cdot)$ is a map from Φ, the parameter space of ϕ, to Ψ. Note that such a reparameterization is almost always possible, but the key is to try to make ϕ unrestricted (unless completely specified, such as in Example 2.5 below). The following are some examples.

Example 2.5. Suppose that, under the null hypothesis, ψ is completely specified; that is, H_0: $\psi = \psi_0$. Then, under H_0, $\psi = \phi = \psi_0$.

Example 2.6. Let $\psi = (\psi_1, \ldots, \psi_p, \psi_{p+1}, \ldots, \psi_q)'$, and suppose that one wishes to test the hypothesis H_0: $\psi_j = \psi_{0j}$, $p+1 \leq j \leq q$, where ψ_{0j}, $p+1 \leq j \leq q$ are known constants. Then, under the null hypothesis, $\psi_j = \phi_j$, $1 \leq j \leq p$, and $\psi_j = \psi_{0j}$, $p+1 \leq j \leq q$ for some (unrestricted) $\phi = (\phi_j)_{1 \leq j \leq p}$.

Example 2.7. Suppose that the null hypothesis includes inequality constraints: H_0: $\psi_j > \psi_{0j}$, $p_1+1 \leq j \leq p$, and $\psi_j = \psi_{0j}$, $p+1 \leq j \leq q$, where $p_1 < p < q$. Then, under the null hypothesis, $\psi_j = \phi_j$, $1 \leq j \leq p_1$, $\psi_j = \psi_{0j} + e^{\phi_j}$, $p_1+1 \leq j \leq p$, and $\psi_j = \psi_{0j}$, $p+1 \leq j \leq q$ for some (unrestricted) $\phi = (\phi_j)_{1 \leq j \leq p}$.

Let $L(\psi, y)$ be a function of ψ and y that takes positive values, and $l(\psi, y) = \log L(\psi, y)$. Let $L_0(\phi, y) = L(\psi(\phi), y)$, and $l_0(\phi, y) = \log L_0(\phi, y)$. Let q and p be the dimensions of θ and ϕ, respectively. Let $\hat{\psi}$ be an estimator of ψ, and $\hat{\phi}$ an estimator of ϕ. Note that here we do not require that $\hat{\psi}$ and $\hat{\phi}$ be the (global) maximizers of $l(\psi, y)$ and $l_0(\phi, y)$, respectively. But we require that $\hat{\psi}$ be a solution to $\partial l/\partial\psi = 0$, and $\hat{\phi}$ a solution to $\partial l_0/\partial\phi = 0$.

We now loosely define matrices A, B, C, and Σ with the exact definitions given in section 2.7: A is the limit of the matrix of second derivatives of l with respect to θ; B is the limit of the matrix of second derivatives of l_0 with respect to ϕ; C is the limit of the matrix of first derivatives of θ with respect to ϕ; and Σ is the asymptotic covariance matrix of $\partial l/\partial\theta$, all subject to suitable normalizations. As shown in Section 2.7, the normalizations are associated with sequences of nonsingular symmetric matrices G and H. The W-test is closely related to the following quantity.

$$\mathcal{W} = [\hat{\psi} - \psi(\hat{\phi})]' G Q_w^- G [\hat{\psi} - \psi(\hat{\phi})] , \tag{2.18}$$

where Q_w^- is the unique Moore–Penrose inverse (see Appendix B) of

$$Q_w = [A^{-1} - C(C'AC)^{-1}C']\Sigma[A^{-1} - C(C'AC)^{-1}C'] .$$

Let $\hat{Q}_w^-$ be a consistent estimator of Q_w^- in the sense that $\|\hat{Q}_w^- - Q_w^-\| \to 0$ in probability. The W-test statistic, $\hat{\mathcal{W}}$, is defined by (2.18) with Q_w^- replaced by $\hat{Q}_w^-$. Similarly, we define the following:

$$\mathcal{S} = \left(\left.\frac{\partial l}{\partial \psi}\right|_{\psi(\hat{\phi})} \right)' G^{-1} A^{-1/2} Q_s^- A^{-1/2} G^{-1} \left(\left.\frac{\partial l}{\partial \psi}\right|_{\psi(\hat{\phi})} \right) , \tag{2.19}$$

where Q_s^- is the unique Moore-Penrose inverse of

$$Q_s = (I - P)A^{-1/2}\Sigma A^{-1/2}(I - P) ,$$

and $P = A^{1/2}C(C'AC)^{-1}C'A^{1/2}$. Let $\hat{A}$ and $\hat{Q}_s^-$ be consistent estimators of A and Q_s^-, respectively, in the same sense as above. Note that, quite often, A only depends on ψ, of which a consistent estimator; that is, $\hat{\psi}$, is available. The S-test statistic, $\hat{\mathcal{S}}$, is defined by (2.19) with A and Q_s^- replaced by $\hat{A}$ and $\hat{Q}_s^-$, respectively. Finally, the L-ratio for testing (2.17) is defined as

$$\mathcal{R} = \frac{L_0(\hat{\phi}, y)}{L(\hat{\psi}, y)} .$$

Note that the L-ratio is the same as the likelihood ratio when $L(\psi, y)$ is a likelihood function. The L-test statistic is then $-2\log\mathcal{R}$.

Jiang (2005c) showed that, under some regularity conditions, both the W- and S-tests have an asymptotic χ_r^2 distribution, where the degrees of freedom $r = \text{rank}\{\Sigma^{1/2}A^{-1/2}(I-P)\}$ with P given below (2.19). As for the L-test, the asymptotic distribution of $-2\log\mathcal{R}$ is the same as $\lambda_1\xi_1^2 + \cdots + \lambda_r\xi_r^2$, where r is the same as before, $\lambda_1, \ldots, \lambda_r$ are the positive eigenvalues of

$$Q_l = [A^{-1} - C(C'AC)^{-1}C']^{1/2}\Sigma[A^{-1} - C(C'AC)^{-1}C']^{1/2} , \quad (2.20)$$

and $\xi_1, \ldots, \xi_r$ are independent $N(0,1)$ random variables. In particular, if Σ is nonsingular, then $r = q - p$. These general results apply, in particular, to non-Gaussian linear mixed models. See Section 2.7 for more details.

We now consider application of the robust versions of classical tests to non-Gaussian mixed ANOVA models. The models are defined in Section 1.2.2 and the estimation problems discussed in Section 1.4. Consider the Hartley–Rao variance components: $\lambda = \sigma_0^2$, $\gamma_i = \sigma_i^2/\sigma_0^2$, $1 \le i \le s$. Let $\gamma = (\gamma_i)_{1\le i\le s}$, and $\psi = (\beta' \ \lambda \ \gamma')'$. Then, ψ is a vector of parameters, which alone may not completely determine the distribution of y. Nevertheless, in many cases, people are interested in testing hypotheses of the form (2.17), where $\Psi_0 \subset \Psi = \{\psi : \lambda > 0, \gamma_i \ge 0, 1 \le i \le s\}$, versus H_1: $\psi \notin \Psi_0$. We assume that there is a new parameterization ϕ such that, under the null hypothesis, $\psi = \psi(\phi)$ for some $\phi = (\phi_k)_{1\le k\le d}$. Here $\psi(\cdot)$ is a map from Φ, the parameter space of ϕ, to Ψ. More specifically, let $q = p + s + 1$, which is the dimension of ψ. We assume that there is a subset of indices $1 \le i_1 < \cdots < i_d \le q$ such that

$$\begin{cases} \psi_{i_k}(\phi) \text{ is a function of } \phi_k, & 1 \le k \le d, \quad \text{and} \\ \psi_i(\phi) \text{ is a constant}, & i \in \{1, \ldots, q\} \setminus \{i_1, \ldots, i_d\}. \end{cases} \quad (2.21)$$

Intuitively, the null hypothesis imposes constraints on ψ, therefore there are less free parameters under H_0, and ϕ represents the vector of free parameters after some changes of variables. Note that such a reparameterization almost always exists, but the key is to try to make ϕ unrestricted unless completely specified.

When normality is assumed, the use of the likelihood-ratio test for complex hypotheses and unbalanced data was first proposed by Hartley and Rao (1967), although rigorous justification was not given. Welham and Thompson (1997) showed the equivalence of the likelihood ratio, score, and Wald tests under normality. On the other hand, Richardson and Welsh (1996) considered the likelihood-ratio test without assuming normality, whose approach is similar to our L-test, but their goal was to select the (fixed) covariates. Under the normality assumption, the log-likelihood function for estimating θ is given by

$$l(\psi, y) = \text{constant} - \frac{1}{2}\left\{n \log \lambda + \log(|V|) + \frac{1}{\lambda}(y - X\beta)'V^{-1}(y - X\beta)\right\},$$

where $V = V_\gamma = I + \sum_{i=1}^{s} \gamma_i V_i$ with I being the n-dimensional identity matrix, $V_i = Z_i Z_i'$, $1 \leq i \leq s$, and $|V|$ the determinant of V. The restricted log-likelihood for estimating λ, γ is given by

$$l_R(\lambda, \gamma, y) = \text{constant} - \frac{1}{2}\left\{(n-p)\log \lambda + \log(|K'VK|) + \frac{y'Py}{\lambda}\right\},$$

where K is any $n \times (n-p)$ matrix such that $\text{rank}(K) = n - p$ and $K'X = 0$, and $P = P_\gamma = K(K'VK)^{-1}K' = V^{-1} - V^{-1}X(X'V^{-1}X)^{-1}X'V^{-1}$ (see Appendix B). The restricted log-likelihood is only for estimating the variance components. It is then customary to estimate β by the empirical best linear unbiased estimator:

$$\hat{\beta} = (X'\hat{V}^{-1}X)^{-1}X'\hat{V}^{-1}y,$$

where $\hat{V} = V_{\hat{\gamma}}$, and $\hat{\gamma} = (\hat{\gamma}_i)_{1 \leq i \leq s}$ is the REML estimator of γ. Alternatively, one may define the following "restricted log-likelihood" for ψ.

$$\begin{aligned} l_R(\psi, y) &= \text{constant} \\ &\quad - \frac{1}{2}\left\{(n-p)\log \lambda + \log|K'VK| + \frac{1}{\lambda}(y - X\beta)'V^{-1}(y - X\beta)\right\}. \end{aligned}$$

It is easy to show that the maximizer of $l_R(\psi, y)$ is $\hat{\psi} = (\hat{\beta}'\ \hat{\lambda}\ \hat{\gamma}')'$, where $\hat{\lambda}$ and $\hat{\gamma}$ are the REML estimators, and $\hat{\beta}$ is given above with $\hat{V} = V_{\hat{\gamma}}$. The difference is that, unlike $l(\psi, y)$, $l_R(\psi, y)$ is not a log-likelihood even if normality holds. Nevertheless, it can be shown that both $l(\psi, y)$ and $l_R(\psi, y)$ can be used as the objective function to test (2.17) under a non-Gaussian mixed linear model. The details are given in Section 2.7.1. We now consider an example.

Example 2.2 (Continued). In this case, we have $q = 3$, $\psi_1 = \mu$, $\psi_2 = \lambda = \tau^2$, and $\psi_3 = \gamma = \sigma^2/\tau^2$. Consider the hypothesis H_0: $\lambda = 1$, $\gamma > 1$. Note that under H_0 we have $\mu = \phi_1$, $\lambda = 1$, and $\gamma = 1 + e^{\phi_2}$ for unrestricted ϕ_1, ϕ_2. Thus, (2.21) is satisfied with $d = 2$, $i_1 = 1$, and $i_2 = 3$. The Gaussian log-likelihood is given by (Exercise 2.9)

$$l(\psi, y) = c - \frac{1}{2}\left\{mk\log(\lambda) + m\log(1+k\gamma) + \frac{\text{SSE}}{\lambda} + \frac{\text{SSA}}{\lambda(1+k\gamma)} + \frac{mk(\bar{y}_{\cdot\cdot} - \mu)^2}{\lambda(1+k\gamma)}\right\},$$

where c is a constant, $\text{SSA} = k\sum_{i=1}^m(\bar{y}_{i\cdot} - \bar{y}_{\cdot\cdot})^2$, $\text{SSE} = \sum_{i=1}^m\sum_{j=1}^k(y_{ij} - \bar{y}_{i\cdot})^2$, $\bar{y}_{\cdot\cdot} = (mk)^{-1}\sum_{i=1}^m\sum_{j=1}^k y_{ij}$, and $\bar{y}_{i\cdot} = k^{-1}\sum_{j=1}^k y_{ij}$. Here we have $\psi = (\mu, \lambda, \gamma)'$, $\phi = (\phi_1, \phi_2)'$, and $\psi(\phi) = (\phi_1, 1, 1 + e^{\phi_2})'$, where ϕ_1 and ϕ_2 are unrestricted. The solution to the (Gaussian) ML equation is given by $\hat{\psi}_1 = \hat{\mu} = \bar{y}_{\cdot\cdot}$, $\hat{\psi}_2 = \hat{\lambda} = \text{MSE}$, and $\hat{\psi}_3 = \hat{\gamma} = (1/k)\{(1-1/m)(\text{MSA}/\text{MSE})-1\}$, where $\text{MSA} = \text{SSA}/(m-1)$ and $\text{MSE} = \text{SSE}/m(k-1)$. On the other hand, it is easy to show that the solution to the ML equation under the null hypothesis is given by $\hat{\phi}_1 = \bar{y}_{\cdot\cdot}$, $\hat{\phi}_2 = \log\{(1/k)(1-1/m)\text{MSA} - (1+1/k)\}$, provided that the term inside the logarithm is positive. Because $\text{E(MSA)} = 1 + k\gamma > k + 1$ under H_0 (Exercise 2.9), it is easy to show that, as $m \to \infty$, the logarithm is well defined with probability tending to one.

We now specify the matrices A, C, G, and Σ under the additional assumption that $\text{E}(\alpha_1^3) = \text{E}(\epsilon_{11}^3) = 0$. According to Theorem 2.4, A is given by (2.62), and it can be shown that $X'V^{-1}X/\lambda n = 1/\lambda^2(1+k\gamma)$, $A_1 = \sqrt{k}/2\lambda^2(1+k\gamma)$, and $A_2 = k^2/2\lambda^2(1+k\gamma)^2$. Again, by Theorem 2.4, $G = \text{diag}(\sqrt{mk}, \sqrt{mk}, \sqrt{m})$; C is the 3×2 matrix whose first column is $(1,0,0)'$ and second column is $(0,0,e^{\phi_2})'$. Finally, $\Sigma = A + \Delta$ with Δ given by (2.63), and it can be shown that

$$\frac{\Delta_{00}}{n} = \frac{1}{4\lambda^4(1+k\gamma)^2}[\kappa_0\{1+(k-1)\gamma\}^2 + \kappa_1 k\gamma^2],$$

$$\Delta_1 = \frac{\sqrt{k}}{4\lambda^4(1+k\gamma)^3}[\kappa_0\{1+(k-1)\gamma\} + \kappa_1 k^2\gamma^2],$$

$$\Delta_2 = \frac{k}{4\lambda^4(1+k\gamma)^4}(\kappa_0 + \kappa_1 k^3\gamma^2),$$

where $\kappa_0 = \{\text{E}(\epsilon_{11}^4)/\tau^4\} - 3$ and $\kappa_1 = \{\text{E}(\alpha_1^4)/\sigma^4\} - 3$.

It can be shown that, in this case, the W-test statistic reduces to

$$\hat{\chi}_w^2 = \left(\frac{2k}{k-1} + \hat{\kappa}_0\right)^{-1} mk(\text{MSE} - 1)^2,$$

where $\hat{\kappa}_0$ is the EMM estimator of κ_0 given in Example 2.2 (Continued) below Lemma 2.1. Note that, by the consistency of $\hat{\kappa}_0$ (Exercise 2.7), we have, as $m \to \infty$,

$$\begin{aligned}\frac{2k}{k-1} + \hat{\kappa}_0 &\xrightarrow{\text{P}} \frac{2k}{k-1} + \kappa_0 \\ &\geq \text{E}(\epsilon_{11}^4) - 1 > 0,\end{aligned}$$

under H_0, unless ϵ_{11}^2 is degenerate. Thus, with the exception of this extreme case, the denominator in $\hat{\chi}_w^2$ is positive with probability tending to one under the null hypothesis. By Theorem 2.4, as $m \to \infty$, the asymptotic null distribution of the W-test is χ_1^2 (Exercise 2.10).

As it turns out, the S-test statistic is identical to the W-test statistic in this case, and it has the same asymptotic null distribution (Exercise 2.11).

Finally, the L-test statistic is equal to

$$-2 \log R = m(k-1)\{\text{MSE} - 1 - \log(\text{MSE})\}$$

in this case. Suppose that $m \to \infty$ and k is fixed ($k \geq 2$). Then, it can be shown that $r = 1$ in this case, therefore, by Theorem 2.5, the asymptotic null distribution of $-2 \log R$ is the same as $\lambda_1 \chi_1^2$, where λ_1 is the positive eigenvalue of Q_l given by (2.20) evaluated under H_0. It can be shown that $\lambda_1 = 1 + \{(k-1)/2k\}\kappa_0$, which is estimated by $1 + \{(k-1)/2k\}\hat{\kappa}_0$. Note that if $\kappa_0 = 0$, as will be the case if the errors are normal, the asymptotic null distribution of the L-test is χ_1^2, which is the same as that for the W- and S-tests. Interestingly, the latter result does not require that the random effects are normal (Exercise 2.12).

2.2 Confidence Intervals in Linear Mixed Models

2.2.1 Confidence Intervals in Gaussian Mixed Models

Confidence intervals in linear mixed models include confidence intervals for fixed effects, confidence intervals for variance components, and confidence intervals for functions of variance components. Among the latter, difference and ratio are two simple functions that are frequently used. Other functions such as the heritability, an important quantity in genetics, may be expressed as functions of these two simple functions. For simplicity, the term confidence intervals for variance components is here understood as including functions of variance components. We first consider confidence intervals under Gaussian linear mixed models.

1. Exact confidence intervals for variance components. It is known that in some special cases, mostly balanced cases, exact confidence intervals for variance components can be derived. Here we do not attempt to list all such cases where exact confidence intervals are available. For more details, see Burdick and Graybill (1992). Instead, our approach is to introduce a basic method used to derive exact confidence intervals, so that it may be applied to different cases whenever applicable. The basic idea is to find a *pivotal quantity*, that is, a random variable that depends on both the observations and the variance component, for which an exact confidence interval is to be constructed. Quite often, such a pivotal quantity is in the form of either an "F-statistic" or a "χ^2-statistic". Here the quotes indicate that the quantity is not really a statistic because it involves the variance component. We illustrate the method by examples.

Example 2.2 (Continued). Consider the Hartley–Rao form of variance components $\lambda = \tau^2$ and $\gamma = \sigma^2/\tau^2$. Suppose that one is interested in constructing an exact confidence interval for γ. Consider the following quantity

$$F = \frac{\text{MSA}}{(1+k\gamma)\text{MSE}},$$

where $\text{MSA} = \text{SSA}/(m-1)$ and $\text{MSE} = \text{SSE}/m(k-1)$. It can be shown that, under normality, F has an F-distribution with $m-1$ and $m(k-1)$ degrees of freedom (Exercise 2.13). It follows that, given ρ $(0 < \rho < 1)$, an exact $(1-\rho)\%$ confidence interval for γ is

$$\left[\frac{1}{k}\left(\frac{\text{R}}{F_{\text{U}}} - 1\right), \frac{1}{k}\left(\frac{\text{R}}{F_{\text{L}}} - 1\right)\right],$$

where $\text{R} = \text{MSA}/\text{MSE}$, $F_{\text{L}} = F_{m-1,m(k-1),1-\rho/2}$, and $F_{\text{U}} = F_{m-1,m(k-1),\rho/2}$ (Exercise 2.13).

Example 2.3 (Continued). Suppose that the problem of interest is to construct an exact confidence interval for the variance of any single observation y_{ij}; that is, $\text{var}(y_{ij}) = \sigma^2 + \tau^2$. Let c_{ij}, $1 \leq j \leq k_i$ be constants such that $\sum_{j=1}^{k_i} c_{ij} = 0$ and $\sum_{j=1}^{k_i} c_{ij}^2 = 1 - 1/k_i$. Define $u_i = \bar{y}_{i\cdot} + \sum_{j=1}^{k_i} c_{ij} y_{ij}$, $1 \leq i \leq m$. It can be shown that $u_1, \ldots, u_m$ are independent and normally distributed with mean μ and variance $\sigma^2 + \tau^2$ (Exercise 2.14). Thus, the quantity

$$\chi^2 = \frac{\sum_{i=1}^m (u_i - \bar{u})^2}{\sigma^2 + \tau^2}$$

is distributed as χ^2_{m-1}. It follows that an exact $(1-\rho)\%$ confidence interval for $\sigma^2 + \tau^2$ is

$$\left[\frac{\sum_{i=1}^m (u_i - \bar{u})^2}{\chi^2_{m-1,\rho/2}}, \frac{\sum_{i=1}^m (u_i - \bar{u})^2}{\chi^2_{m-1,1-\rho/2}}\right].$$

The method used in the above example for constructing an exact confidence interval for $\sigma^2 + \tau^2$ is due to Burdick and Sielken (1978). In fact, the authors developed a method that can be used to obtain an exact confidence interval for $a\sigma^2 + b\tau^2$, where a, b are positive constants subject to some additional constraints. One such constraint is that $b \neq 0$. Thus, for example, the method cannot give an exact confidence interval for σ^2 (see Exercise 2.15). This example shows the limitation of the method used to construct exact confidence intervals. In fact, no existing method is known to be able to obtain an exact confidence interval for σ^2 in an analytic form. On the other hand, approximate confidence intervals do exist for σ^2 and other variance components. We discuss such methods next.

2. Approximate confidence intervals for variance components. Satterthwaite (1946) proposed a method, which extended an earlier approach of Smith (1936), for balanced ANOVA models. The goal was to construct a confidence interval for a quantity in the form $\zeta = \sum_{i=1}^h c_i \lambda_i$, where $\lambda_i = \mathrm{E}(S_i^2)$ and S_i^2 is the mean sum of squares corresponding to the ith factor (fixed or random) in the model (e.g., Scheffé 1959). Note that many variance components can be expressed in this form; for example, the variance of y_{ij}, $\sigma^2 + \tau^2$, in Example 2.3 can be expressed as $(1/k)\mathrm{E}(S_1^2) + (1 - 1/k)\mathrm{E}(S_2^2)$, where S_1^2 is the mean sum of squares corresponding to α and S_2^2 that corresponding to ϵ. The idea was to find an appropriate "degrees of freedom," say, d, such that the first two moments of the random variable $d \sum_{i=1}^h c_i S_i^2/\zeta$ match those of a χ_d^2 random variable. This approach is known as Satterthwaite's procedure. Graybill and Wang (1980) proposed a method that improved Satterthwaite's procedure. The authors called their method the modified large sample (MLS) method. The method provides an approximate confidence interval for a nonnegative linear combination of the λ_is, which is exact when all but one of the coefficients in the linear combination are zero. We describe the Graybill–Wang for the special case of balanced one-way random effects model (Example 2.2).

Suppose that one is interested in constructing a confidence interval for $\zeta = c_1\lambda_1 + c_2\lambda_2$, where $c_1 \geq 0$ and $c_2 > 0$. This problem is equivalent to constructing a confidence interval for $\zeta = c\lambda_1 + \lambda_2$, where $c \geq 0$. A uniformly minimum variance unbiased estimator (UMVUE, e.g., Lehmann and Casella 1998) of the quantity is given by $\hat{\zeta} = cS_1^2 + S_2^2$. Furthermore, it can be shown that $\hat{\zeta}$ is asymptotically normal such that $(\hat{\zeta} - \zeta)/\sqrt{\mathrm{var}(\hat{\zeta})}$ has a limiting $N(0,1)$ distribution (Exercise 2.16). Furthermore, the variance of $\hat{\zeta}$ is given by $c^2\{2\lambda_1^2/(m-1)\} + 2\lambda_2^2/m(k-1)$. Again, recall that S_j^2 is an unbiased (and consistent) estimator of λ_j $j = 1, 2$ (Exercise 2.16). This allows one to obtain a large sample confidence interval for ζ as follows.

$$\left[\hat{\zeta} - z_{\rho/2}\sqrt{c^2\left(\frac{2S_1^4}{m-1}\right) + \frac{2S_2^4}{m(k-1)}},\right.$$
$$\left.\hat{\zeta} + z_{\rho/2}\sqrt{c^2\left(\frac{2S_1^4}{m-1}\right) + \frac{2S_2^4}{m(k-1)}}\right], \quad (2.22)$$

where $1 - \rho$ is the confidence coefficient. The confidence interval (2.22) is expected to be accurate when the sample size is large, that is, when $m \to \infty$. However, small sample performance is not guaranteed. Graybill and Wang proposed to modify the constants $z_{\rho/2}$, $2/(m-1)$ and $2/m(k-1)$, so that the confidence interval will be exact when either $\lambda_1 = 0$ or $\lambda_2 = 0$. Their confidence interval is given by

$$\left[\hat{\zeta} - \sqrt{G_1^2 c^2 S_1^4 + G_2^2 S_2^4}, \hat{\zeta} + \sqrt{H_1^2 c^2 S_1^4 + H_2^2 S_2^4}\right],$$

where $G_1 = 1 - (m-1)/\chi^2_{m-1,\rho/2}$, $G_2 = 1 - m(k-1)/\chi^2_{m(k-1),\rho/2}$, $H_1 = (m-1)/\chi^2_{m-1,1-\rho/2} - 1$, and $H_2 = m(k-1)/\chi^2_{m(k-1),\rho/2} - 1$. Using numerical integration, Graybill and Wang compared confidence coefficients of the MLS confidence intervals with those of Satterthwaite and Welch (Welch 1956). They concluded that the confidence coefficients of the MLS are closer to the nominal levels than those of Satterthwaite and Welch. As for the length of the confidence intervals, Graybill and Wang conducted a simulation study. The results showed that average widths of two types of MLS confidence intervals, namely, the shortest unbiased confidence interval and shortest confidence interval, are generally smaller than those of Welch's.

Sometimes, the variance components of interest cannot be expressed as a nonnegative linear combination of the λ_is. For example, in Example 2.2, the variance $\sigma^2 = (\lambda_1 - \lambda_2)/k$, so the coefficients in the linear combination have different signs. It is therefore of interest to obtain confidence intervals for $\zeta = \sum_{i=1}^h c_i \lambda_i$, where the c_is may have different signs. Healy (1961) proposed a procedure that may be used to obtain an exact confidence interval for $c_1\lambda_1 - c_2\lambda_2$, where c_1 and c_2 are nonnegative. However, the procedure requires a randomization device. In other words, the confidence interval is not solely determined by the data. Several authors have proposed (solely data-based) approximate confidence intervals for ζ. For example, Ting et al. (1990) proposed a procedure similar to Graybill and Wang (1980) discussed above. Note that a large sample confidence interval such as (2.22) based on asymptotic normality of $\hat{\zeta}$ does not require that the signs of the c_is be the same. All one has to do is to modify the coefficients of the large sample confidence interval so that it performs better in small sample situations. See Ting et al. (1990) for details. Burdick and Graybill (1992) reviewed several approximate procedures for constructing confidence intervals for ζ. They conclude that there is little difference in terms of performance of the proposed procedures.

Finally, one should bear in mind that, in cases of large samples, a confidence interval as simple as (2.22) can be used without modification. Such a procedure is much easier to derive and calculate. We return to this method in the next section.

3. Simultaneous confidence intervals. Hartley and Rao (1967) derived a simultaneous confidence region for the variance ratios $\gamma_i = \sigma_i^2/\tau^2$, $i = 1, \ldots, s$ (i.e., the Hartley-Rao form of variance components; see Section 1.2.1.1) in a Gaussian mixed ANOVA model. Their derivation is based on maximum likelihood estimation, a method that we visit again in the next section. The Hartley–Rao confidence region is quite general, that is, it applies to a general ANOVA model, balanced or unbalanced. On the other hand, in some special cases different methods may result in confidence intervals that are easier to interpret. For example, Khuri (1981) developed a method of constructing simultaneous confidence intervals for all continuous functions of variance components in the balanced random effects model, a special case of the mixed ANOVA model.

It should be pointed out that, provided one knows how to construct confidence intervals for the individual variance components, by then Bonferroni inequality, a conservative simultaneous confidence interval for the variance components can always be constructed. Suppose that $[L_i, U_i]$ is a $(1-\rho_i)\%$ confidence interval for the variance component θ_i, $i = 1, \dots, q$. Then, by the Bonferroni inequality, the set of intervals $[L_i, U_i]$, $i = 1, \dots, q$ is a (conservative) simultaneous confidence interval for θ_i, $i = 1, \dots, q$ with confidence coefficient greater than or equal to $1 - \sum_{i=1}^{q} \rho_i$. Sometimes, the confidence coefficient may be improved if there is independence among the individual confidence intervals. For example, in the balanced normal random effects model, let n_i be the degrees of freedom associated with S_i^2, the mean sum of squares corresponding to the ith factor (fixed or random). Then, it is known that $n_i S_i^2/\lambda_i$ has a χ^2 distribution with n_i degrees of freedom, where $\lambda_i = \mathrm{E}(S_i^2)$. Furthermore, the random variables $n_i S_i^2/\lambda_i$, $i = 1, \dots, h$ are independent (e.g., Scheffé 1959). It follows that a $(1-\rho)\%$ confidence interval for λ_i is

$$\left[\frac{n_i S_i^2}{\chi^2_{n_i,\rho/2}}, \frac{n_i S_i^2}{\chi^2_{n_i,1-\rho/2}}\right], \tag{2.23}$$

and, furthermore, the set of intervals (2.23) with $i = 1, \dots, h$ is a simultaneous confidence interval for λ_i, $i = 1, \dots, h$ with confidence coefficient $(1-\rho)^h$. Note that $(1-\rho)^h \geq 1 - h\rho$ for any integer $h \geq 1$.

4. Confidence intervals for fixed effects. For the vector of fixed effects β in (1.1), the best linear unbiased estimator, or BLUE, is given by (1.36), provided that the expression does not involve unknown variance components. Furthermore, we have

$$\mathrm{Var}(\hat{\beta}_{\mathrm{BLUE}}) = (X'V^{-1}X)^{-1}. \tag{2.24}$$

In fact, under mild conditions, $\hat{\beta}_{\mathrm{BLUE}}$ is asymptotically normal with mean vector β and asymptotic covariance matrix given by the right side of (2.24). It is known that in some special cases, mostly in the balanced situations, the right side of (1.36) does not depend on the variance components, therefore $\hat{\beta}_{\mathrm{BLUE}}$ can be used as an estimator. However, even in those cases the right side of (2.24) typically depends on the variance components. Of course, in general, both $\hat{\beta}_{\mathrm{BLUE}}$ and its covariance matrix depend on the variance components. Therefore, to construct a confidence interval for a fixed effect, or more generally, any linear function of β, one needs to replaced the unknown variance components by consistent estimators, for example, REML estimators. Except for some special cases (see Example 2.8 below), the resulting confidence interval will be approximate in the sense that its confidence coefficient approaches the nominal level as sample size increases. We consider an example.

Example 2.8. Consider the following model, which is a special case of the so-called *nested error regression model*:

$$y_{ij} = \beta_0 + \beta_1 x_i + \alpha_i + \epsilon_{ij}, \qquad i = 1, \dots, m, \qquad j = 1, \dots, k_i,$$

where β_0, β_1 are unknown regression coefficients, x_is are known covariates, α_is are random effects, and ϵ_{ij}s are errors. Suppose that the random effects and errors are independent and normally distributed such that $\mathrm{E}(\alpha_i) = 0$, $\mathrm{var}(\alpha_i) = \sigma^2$, $\mathrm{E}(\epsilon_{ij}) = 0$, and $\mathrm{var}(\epsilon_{ij}) = \tau^2$.

It can be shown (Exercise 2.17) that, in this case, (1.36) gives the following expressions for the BLUE,

$$\hat{\beta}_{\mathrm{BLUE},0} = \frac{(\sum_{i=1}^m d_i x_i^2)(\sum_{i=1}^m d_i \bar{y}_{i\cdot}) - (\sum_{i=1}^m d_i x_i)(\sum_{i=1}^m d_i x_i \bar{y}_{i\cdot})}{(\sum_{i=1}^m d_i)(\sum_{i=1}^m d_i x_i^2) - (\sum_{i=1}^m d_i x_i)^2}, \quad (2.25)$$

$$\hat{\beta}_{\mathrm{BLUE},1} = \frac{(\sum_{i=1}^m d_i)(\sum_{i=1}^m d_i x_i \bar{y}_{i\cdot}) - (\sum_{i=1}^m d_i x_i)(\sum_{i=1}^m d_i \bar{y}_{i\cdot})}{(\sum_{i=1}^m d_i)(\sum_{i=1}^m d_i x_i^2) - (\sum_{i=1}^m d_i x_i)^2}, \quad (2.26)$$

where $d_i = k_i/(\tau^2 + k_i\sigma^2)$. It follows that, when $k_i = k$, $1 \le i \le m$ (i.e., in the balanced case), we have

$$\hat{\beta}_{\mathrm{BLUE},0} = \frac{(\sum_{i=1}^m x_i^2)(\sum_{i=1}^m \bar{y}_{i\cdot}) - (\sum_{i=1}^m x_i)(\sum_{i=1}^m x_i \bar{y}_{i\cdot})}{m\sum_{i=1}^m x_i^2 - (\sum_{i=1}^m x_i)^2},$$

$$\hat{\beta}_{\mathrm{BLUE},1} = \frac{\sum_{i=1}^m (x_i - \bar{x}_\cdot)(\bar{y}_{i\cdot} - \bar{y}_{\cdot\cdot})}{\sum_{i=1}^m (x_i - \bar{x}_\cdot)^2}.$$

It is seen that in the balanced case, the BLUE does not depend on the variance components but in the unbalanced case it does. Furthermore, $\hat{\beta}_{\mathrm{BLUE}} = (\hat{\beta}_{\mathrm{BLUE},0}, \hat{\beta}_{\mathrm{BLUE},1})'$. It can be shown by (2.24) that

$$\mathrm{Var}(\hat{\beta}_{\mathrm{BLUE}}) = \frac{1}{\tau^2 D}\begin{pmatrix} \sum_{i=1}^m d_i x_i^2 & -\sum_{i=1}^m d_i x_i \\ -\sum_{i=1}^m d_i x_i & \sum_{i=1}^m d_i \end{pmatrix}, \quad (2.27)$$

where $D = (\sum_{i=1}^m d_i)(\sum_{i=1}^m d_i x_i^2) - (\sum_{i=1}^m d_i x_i)^2$ (Exercise 2.17). So even in the balanced case the covariance matrix of BLUE depends on the variance components.

When the variance components involved in BLUE are replaced by their estimators, the resulting estimator is often called empirical BLUE, or EBLUE. It is easy to see that, under normality, EBLUE is the same as the MLE of β, if the variance components are replaced by their MLE. It should be pointed out that EBLUE is more complicated and, in particular, not linear in y. Furthermore, if one replaces the variance components involved on the right side of (2.24) by their estimators, the result would underestimate the true variation of EBLUE. In fact, Kackar and Harville (1981) showed that EBLUE, denoted by $\hat{\beta}$, is still an unbiased estimator of β, that is $\mathrm{E}(\hat{\beta}) = \beta$, provided that the data are normal and estimators of the variance components are even and translation invariant (see Section 2.8 for more detail). In addition, the authors showed that, under normality

$$\mathrm{var}(a'\hat{\beta}) = \mathrm{var}(a'\hat{\beta}_{\mathrm{BLUE}}) + \mathrm{E}\{a'(\hat{\beta} - \hat{\beta}_{\mathrm{BLUE}})\}^2 \tag{2.28}$$

for any given vector a. Because $\mathrm{var}(a'\hat{\beta}_{\mathrm{BLUE}}) = a'\mathrm{Var}(\hat{\beta}_{\mathrm{BLUE}})a$, the first term on the right side of (2.28) can be estimated by the right side of (2.24) with the variance components replaced by their estimators. However, there is a second term on the right side of (2.28) that cannot be estimated this way. Fortunately, for constructing confidence intervals for the fixed effects, this complication does not necessarily cause any problem, at least in the large-sample situation. In fact, for ANOVA models, Jiang (1998b) showed that, when the variance components are estimated by the REML estimators, the asymptotic covariance matrix of $\hat{\beta}$ is still given by the right side of (2.24) (in spite of estimation of the variance components). It is known (e.g., Miller 1977) that, when the variance components are estimated by the MLE, the asymptotic covariance matrix of $\hat{\beta}$ is also given by the right side of (2.24). Thus, in such cases, a (large-sample) confidence interval for $a'\beta$ is given by

$$\Big[a'\hat{\beta} - z_{\rho/2}\{a'(X'\hat{V}^{-1}X)^{-1}a\}^{1/2}, \quad a'\hat{\beta} + z_{\rho/2}\{a'(X'\hat{V}^{-1}X)^{-1}a\}^{1/2}\Big], \tag{2.29}$$

where $\hat{V}$ is V with the variance components replaced by their REML or ML estimators. It is shown in section 2.3 that the complication in EBLUE becomes important in the prediction of a mixed effect, that is, a linear combination of fixed and random effects.

Example 2.8 (Continued). Suppose that one is interested in constructing a confidence interval for $\hat{\beta}_1$. By (2.29) and (2.27), taking $a = (0, 1)'$, a large sample confidence interval is

$$\left[\hat{\beta}_1 - z_{\rho/2}\left(\frac{\sum_{i=1}^m \hat{d}_i}{\hat{\tau}^2\hat{D}}\right)^{1/2}, \ \hat{\beta}_1 + z_{\rho/2}\left(\frac{\sum_{i=1}^m \hat{d}_i}{\hat{\tau}^2\hat{D}}\right)^{1/2}\right],$$

where $\hat{d}_i = k_i/(\hat{\tau}^2 + k_i\hat{\sigma}^2)$, $\hat{\beta}_1$ is given by (2.26) with d_i replaced by $\hat{d}_i$,$1 \leq i \leq m$, and $\hat{D}$ is D with d_i replaced by $\hat{d}_i$, $1 \leq i \leq m$. Here $\hat{\sigma}^2$ and $\hat{\tau}^2$ are understood as the REML (or ML) estimators.

2.2.2 Confidence Intervals in Non-Gaussian Linear Mixed Models

For non-Gaussian linear mixed models, exact confidence intervals for parameters of interest usually do not exist. Therefore, methods of constructing confidence intervals will be based on large sample theory. Suppose that one is interested in obtaining a confidence interval for a linear function of the parameters, which may include fixed effects and variance components. Let ψ be the vector of all fixed parameters involved in a non-Gaussian linear mixed model. Suppose that an estimator of ψ, say $\hat{\psi}$, is available which is consistent and asymptotically normal; that is, (2.7) holds. If a consistent estimator of Σ, the asymptotic covariance matrix of $\hat{\psi}$, is available, say $\hat{\Sigma}$, then, for any linear function $a'\psi$, where a is a known vector, one may be able to show that $a'(\hat{\psi}-\psi)/\sqrt{a'\hat{\Sigma}a}$ is asymptotically standard normal. Therefore, a large-sample $(1-\rho)\%$ confidence interval $(0<\rho<1)$ for $a'\psi$ is

$$\left[a'\hat{\psi} - z_{\rho/2}\sqrt{a'\hat{\Sigma}a},\ a'\hat{\psi} + z_{\rho/2}\sqrt{a'\hat{\Sigma}a}\right].$$

We now consider two special cases of non-Gaussian linear mixed models and discuss how to estimate Σ in those cases.

1. ANOVA models. For ANOVA models, Jiang (1996) derived asymptotic distributions of both REML and ML estimators without the normality assumption. Jiang (1998b) extended these results to include estimators of fixed effects. The main result of the latter is summarized as follows. Consider the Hartley–Rao form of variance components (see Section 1.2.1.1). Let the normalizing constants p_i, $0 \le i \le s$ and matrices $\mathcal{M}$, $\mathcal{J}$ be defined as in Theorem 1.1 of Chapter 1. Define $\mathcal{P} = \mathcal{M}\mathrm{diag}(p_0, p_1, \ldots, p_s)$, $\mathcal{Q} = (X'V^{-1}X)^{1/2}$, $\mathcal{R} = \mathcal{J}^{1/2}\mathcal{T}\mathcal{C}$, where

$$\mathcal{T} = \left(\frac{V_{i,ll}(\gamma)\mathrm{E}(\omega_l^3)}{p_i\lambda^{1_{(i=0)}}}\right)_{0\le i\le s, 1\le l\le n+m},$$

$$\mathcal{C} = \lambda^{1/2}b(\gamma)V^{-1}X\mathcal{Q}^{-1}$$

with $V_{i,ll}$, ω_l and $b(\gamma)$ defined above Theorem 1.1. Then, under suitable conditions, we have

$$\begin{pmatrix} \mathcal{P} & \mathcal{R}\mathcal{Q} \\ \mathcal{R}'\mathcal{P} & \mathcal{Q} \end{pmatrix}\begin{pmatrix} \hat{\theta}-\theta \\ \hat{\beta}-\beta \end{pmatrix} \xrightarrow{\mathcal{D}} N(0, I_{p+s+1}), \tag{2.30}$$

where $\hat{\beta}$ is the EBLUE with REML estimators of variance components (in other words, $\hat{\beta}$ is the REML estimator of β; see Section 1.3.2). Because, under normality, $\mathcal{T} = 0$ hence $\mathcal{R} = 0$, the normalizing matrix on the left side of (2.30) reduces to $\mathrm{diag}(\mathcal{P}, \mathcal{Q})$ in this case. However, for non-Gaussian linear mixed models, the normalizing matrix in (2.30) may involve additional parameters such as the third and fourth moments of the random effects and errors. A method of estimating the higher moments, known as EMM, has been introduced earlier (see Section 2.1.2.1), under the assumption (2.14) (which implies $\mathrm{E}(\omega_l) = 0$, $1 \le l \le n+m$). To see how much difference there may be if one ignores the higher moments, consider the following example.

Example 2.2 (Continued). If normality is not assumed, it can be shown, by (2.30), that the asymptotic variance of $\sqrt{mk}(\hat{\lambda}-\lambda)$ is $\lambda^2/2 + \kappa_0$, that is, $\sqrt{mk}(\hat{\lambda} - \lambda) \to N(0, \lambda^2/2 + \kappa_0)$ in distribution, where κ_0, κ_1 are defined below (2.14). Similarly, the asymptotic variance of $\sqrt{m}(\hat{\gamma} - \gamma)$ is $\gamma^2/2 + \kappa_1/\lambda^2$. Therefore, the difference in asymptotic variance from that under normality is κ_0 for the estimation of λ, and κ_1/λ^2 for the estimation of γ.

If (2.14) is not known to hold, the EMM may not apply. In this case, an alternative method would be that of partially observed information introduced in Section 1.4.2. Note that the latter method applies more generally not only to mixed ANOVA models but also to other types of non-Gaussian linear mixed models for estimating the asymptotic covariance matrix of the REML or ML estimator.

2. Longitudinal models. For longitudinal models, the asymptotic covariance matrix of the vector of parameters of interest, which may include fixed effects and variance components, may be estimated using the jackknife method introduced in Section 1.4.4 [see (1.43)]. Alternatively, the asymptotic covariance matrix may also be estimated by partially observed information. See the remark at the end of Section 1.4.2.

2.3 Prediction

There are two types of prediction problems in the context of linear mixed models. The first is the prediction of a random effect, or, more generally, a mixed effect. Here we focus on a linear mixed effect, which can be expressed as $\eta = a'\alpha + b'\beta$, where a, b are known vectors, and α and β are the vectors of random and fixed effects, respectively, in (1.1). This type of prediction problem has a long history, starting with C. R. Henderson in his early work in the field of animal breeding (e.g., Henderson 1948). The best-known method for this kind of prediction is best linear unbiased prediction, or BLUP. Robinson (1991) gives a wide-ranging account of BLUP with examples and applications. The second type of prediction is that of a future observation. In contrast to the first type, prediction of the second type has received much less attention, although there are plenty of such prediction problems with practical interest (e.g., Jiang and Zhang 2002). In the next two sections we discuss these two types of predictions.

2.3.1 Prediction of Mixed Effect

1. Best prediction when all the parameters are known. When the fixed effects and variance components are known, the best predictor for $\xi = a'\alpha$, in the sense of minimum mean squared error (MSE), is its conditional expectation given the data; that is,

$$\tilde{\xi} = \mathrm{E}(\xi|y) = a'\mathrm{E}(\alpha|y) . \tag{2.31}$$

Assuming normality of the data, we have, by (1.1), that

$$\begin{pmatrix} \alpha \\ y \end{pmatrix} \sim N\left(\begin{pmatrix} 0 \\ X\beta \end{pmatrix}, \begin{pmatrix} G & GZ' \\ ZG & V \end{pmatrix} \right),$$

where $G = \text{Var}(\alpha)$, $R = \text{Var}(\epsilon)$, and $V = \text{Var}(y) = ZGZ' + R$. It follows that

$$\text{E}(\alpha|y) = GZ'V^{-1}(y - X\beta)$$

(see Appendix C). Therefore, by (2.31), the best predictor of ξ is

$$\tilde{\xi} = a'GZ'V^{-1}(y - X\beta).$$

Once the best predictor of $\xi = a'\alpha$ is obtained, the best predictor of $\eta = a'\alpha + b'\beta$ is

$$\hat{\eta}_{\text{B}} = b'\beta + a'GZ'V^{-1}(y - X\beta) . \tag{2.32}$$

Here the subscript B refers to the best predictor.

It can be shown that, without assuming normality, (2.32) gives the best linear predictor of η in the sense that it minimizes the MSE of a predictor that is linear in y. See Searle et al. (1992, §7.3). The following example was given by Mood et al. (1974, pp. 370).

Example 2.9 (IQ tests). Suppose intelligence quotients for students in a particular age group are normally distributed with mean 100 and standard deviation 15. The IQ, say x_1, of a particular student is to be estimated by a test on which he scores 130. It is further given that test scores are normally distributed about the true IQ as a mean with standard deviation 5. What is the best prediction on the student's IQ? (The answer is not 130.)

The solution may be found by applying the method of best prediction. Here we have $y = \mu + \alpha + \epsilon$, where y is the student's test score, which is 130; α is the realization of a random effect corresponding to the student, so that $\mu + \alpha$ is the student's true IQ, which is unobservable. The question is to predict $\mu + \alpha$, a mixed effect. It is known that $\text{IQ} \sim N(100, 15^2)$ and $\text{score}|\text{IQ} \sim N(IQ, 5^2)$. Also, $\mu = 100$ is given. It follows that $Z = 1$, $G = \text{var}(\text{IQ}) = 15^2$, $V = \text{var}(\text{score}) = \text{var}(\text{E}(\text{score}|\text{IQ})) + \text{E}(\text{var}(\text{score}|\text{IQ})) = \text{var}(\text{IQ}) + \text{E}(5^2) = 15^2 + 5^2$. Therefore, by (2.32), the best prediction of the student's IQ is

$$\widetilde{\text{IQ}} = \mu + \frac{15^2}{15^2 + 5^2}(\text{score} - \mu) = 127 .$$

2. Best linear unbiased prediction. If the fixed effects are unknown but the variance components are known, Equation (2.32) is not a predictor. In such a case, it is customary to replace β by $\hat{\beta}$, its maximum likelihood estimator under normality, which is

$$\tilde{\beta} = (X'V^{-1}X)^{-1}X'V^{-1}y . \tag{2.33}$$

Here, for simplicity, we assume that X is of full rank p. (2.33) is also known as the best linear unbiased estimator, or BLUE, whose derivation does not require normality. Henderson (1973) showed that, after β in (2.32) is replaced by the BLUE (2.33), the resulting predictor is the best linear unbiased predictor of η in the sense that (i) it is linear in y, (ii) its expected value is equal to that of η, and (iii) it minimizes the MSE among all linear unbiased predictors, where the MSE of a predictor $\tilde{\eta}$ is defined as $\text{MSE}(\tilde{\eta}) = \text{E}\{(\tilde{\eta}-\eta)(\tilde{\eta}-\eta)'\}$. Again, the result does not require normality. Thus, the BLUP is given by

$$\hat{\eta}_{\text{BLUP}} = b'\tilde{\beta} + a'GZ'V^{-1}(y - X\tilde{\beta}) , \tag{2.34}$$

where $\tilde{\beta}$ is the BLUE given by (2.33). The vector

$$\tilde{\alpha} = GZ'V^{-1}(y - X\tilde{\beta}) \tag{2.35}$$

is also called the BLUP of α.

Henderson's original derivation of BLUP was based on what he called "joint maximum likelihood estimates" of fixed and random effects. Consider a Gaussian mixed model (1.1), where $\alpha \sim N(0, G)$, $\epsilon \sim N(0, R)$, and α and ϵ are independent. Suppose that both G and R are nonsingular. Then, it can be shown that the logarithm of the joint pdf of α and y can be expressed as (Exercise 2.18)

$$c - \frac{1}{2}\left\{(y - X\beta - Z\alpha)'R^{-1}(y - X\beta - Z\alpha) + \alpha'G^{-1}\alpha\right\}, \tag{2.36}$$

where c is a constant. Henderson (1950) proposed to find the "maximum likelihood estimates" of β and α, treating the latter as (fixed) parameters, by differentiating (2.36) with respect to β and α and setting the partial derivatives equal to zero. This leads to Henderson's mixed model equations:

$$\begin{pmatrix} X'R^{-1}X & X'R^{-1}Z \\ Z'R^{-1}X & G^{-1} + Z'R^{-1}Z \end{pmatrix} \begin{pmatrix} \tilde{\beta} \\ \tilde{\alpha} \end{pmatrix} = \begin{pmatrix} X'R^{-1} \\ Z'R^{-1} \end{pmatrix} y , \tag{2.37}$$

the solution to which leads to (2.33) and (2.35) (Exercise 2.18). Later, Henderson (1963) showed that the "maximum likelihood estimates" he derived earlier are indeed the BLUP. A more intuitive approach to show that the BLUP has minimum mean squared error within the class of linear unbiased estimators was given by Harville (1990). Also see Robinson (1991). In particular, this derivation does not require normality assumptions. In other words, the BLUP is well defined for non-Gaussian linear mixed models. The BLUP may also be regarded as the maximum likelihood estimator of the best predictor, because, assuming that the variance components are known, the BLUP may be obtained by replacing β in the expression of the best predictor (2.32) by its maximum likelihood estimator under normality, that is, (2.33). Finally, Jiang (1997b) showed that BLUP is the best predictor based on error contrasts; that is, (2.35) is identical to $\text{E}(\alpha|A'y)$, where A is any $n \times (n-p)$ matrix of full rank such that $A'X = 0$.

Robinson (1991) used the following example to illustrate the calculation of BLUE and BLUP.

Example 2.10. Consider a linear mixed model for the first lactation yields of dairy cows with sire additive genetic merits being treated as random effects and herd effects being treated as fixed effects. The herd effects are represented by β_j, $j = 1, 2, 3$ and sire effects by α_i, $i = 1, 2, 3, 4$, which correspond to sires A, B, C, D. The matrix R is taken to be the identity matrix, while the matrix G is assumed to be 0.1 times the identity matrix. This would be a reasonable assumption, provided that the sires were unrelated and that the variance ratio σ^2/τ^2 had been estimated previously, where $\sigma^2 = \text{var}(\alpha_i)$ and $\tau^2 = \text{var}(\epsilon_{ij})$. Suppose that the data are given below. It can be shown (Exercise 2.20) that

Herd	1	1	2	2	2	3	3	3	3
Sire	A	D	B	D	D	C	C	D	D
Yield	110	100	110	100	100	110	110	100	100

the mixed model equations (2.37) are

$$\begin{pmatrix} 2 & 0 & 0 & 1 & 0 & 0 & 1 \\ 0 & 3 & 0 & 0 & 1 & 0 & 2 \\ 0 & 0 & 4 & 0 & 0 & 2 & 2 \\ 1 & 0 & 0 & 11 & 0 & 0 & 0 \\ 0 & 1 & 0 & 0 & 11 & 0 & 0 \\ 0 & 0 & 2 & 0 & 0 & 12 & 0 \\ 1 & 2 & 2 & 0 & 0 & 0 & 15 \end{pmatrix} \begin{pmatrix} \tilde{\beta}_1 \\ \tilde{\beta}_2 \\ \tilde{\beta}_3 \\ \tilde{\alpha}_1 \\ \tilde{\alpha}_2 \\ \tilde{\alpha}_3 \\ \tilde{\alpha}_4 \end{pmatrix} = \begin{pmatrix} 210 \\ 310 \\ 420 \\ 110 \\ 110 \\ 220 \\ 500 \end{pmatrix},$$

which have the solution

$$\tilde{\beta} = (105.64, 104.28, 105.46)',$$
$$\tilde{\alpha} = (0.40, 0.52, 0.76, -1.67)'.$$

3. *Empirical BLUP.* In practice, the fixed effects and variance components are typically unknown. Therefore, in most cases neither the best predictor nor the BLUP is computable, even though they are known to be best in their respective senses. In such cases, it is customary to replace the vector of variance components, θ, which is involved in the expression of BLUP by a consistent estimator, $\hat{\theta}$. The resulting predictor is often called empirical BLUP, or EBLUP.

Kackar and Harville (1981) showed that, if $\hat{\theta}$ is an even and translation-invariant estimator and the data are normal, the EBLUP remains unbiased. An estimator $\hat{\theta} = \hat{\theta}(y)$ is even if $\hat{\theta}(-y) = \hat{\theta}(y)$, and it is translation invariant if $\hat{\theta}(y - X\beta) = \hat{\theta}(y)$. Some of the well-known estimators of θ, including ANOVA, ML, and REML estimators (see Sections 1.3–1.5), are even and translation invariant. In their arguments, however, Kackar and Harville had assumed the existence of the expected value of EBLUP, which is not obvious because, unlike BLUP, EBLUP is not linear in y. The existence of the expected value of EBLUP was proved by Jiang (1999b, 2000a). See Section 2.7 for details.

Harville (1991) considered the one-way random effects model of Example 1.1, and showed that, in this case, the EBLUP of the mixed effect, $\mu + \alpha_i$, is identical to a parametric empirical Bayes (PEB) estimator. In the meantime, Harville noted some differences between these two approaches, PEB and EBLUP. One of the differences is that much of the work on PEB has been carried out by professional statisticians and has been theoretical in nature. The work has tended to focus on relatively simple models, such as the one-way random effects model, because it is only these models that are tractable from a theoretical standpoint. On the other hand, much of the work on EBLUP has been carried out by practitioners such as researchers in the animal breeding area, and has been applied to relatively complex models.

One problem of practical interest is estimation of the MSE of EBLUP. Such a problem arises, for example, in small area estimation (e.g., Ghosh and Rao 1994). The EBLUP method has been used in small area estimation for estimating small area means, which are in the form of mixed effects. However, the MSE of EBLUP is complicated. A naive estimator of MSE of EBLUP may be obtained by replacing θ by $\hat{\theta}$ in the expression of the MSE of BLUP. However, this is an underestimation. To see this, let $\hat{\eta} = a'\hat{\alpha} + b'\hat{\beta}$ denote the EBLUP of a mixed effect $\eta = a'\alpha + b'\beta$, where $\hat{\alpha}$ and $\hat{\beta}$ are the BLUP of α, given by (2.35), and BLUE of β, given by (2.33), with the variance components θ replaced by $\hat{\theta}$. Kackar and Harville (1981) showed that, under normality assumptions, one has

$$\mathrm{MSE}(\hat{\eta}) = \mathrm{MSE}(\tilde{\eta}) + \mathrm{E}(\hat{\eta} - \tilde{\eta})^2, \tag{2.38}$$

where $\tilde{\eta}$ is the BLUP of η given by (2.34). It is seen that the MSE of BLUP is only the first term on the right side of (2.38). In fact, it can be shown that $\mathrm{MSE}(\tilde{\eta}) = g_1(\theta) + g_2(\theta)$, where

$$\begin{aligned} g_1(\theta) &= a'(G - GZ'V^{-1}ZG)a\ , \\ g_2(\theta) &= (b - X'V^{-1}ZGa)'(X'V^{-1}X)^{-1}(b - X'V^{-1}ZGa) \end{aligned}$$

(e.g., Henderson 1975). It is clear that, using $g_1(\hat{\theta}) + g_2(\hat{\theta})$ as an estimator would underestimate the MSE of $\hat{\eta}$, because it does not take into account the additional variation associated with the estimation of θ, represented by the second term on the right side of (2.38). Such a problem may become particularly important when, for example, large amounts of funds are involved. For example, over $7 billion of funds are allocated annually based on EBLUP estimators of school-age children in poverty at the county and school district levels (National Research Council 2000).

Kackar and Harville (1984) gave an approximation to the MSE of EBLUP under the linear mixed model (1.1), taking account of the variability in $\hat{\theta}$, and proposed an estimator of $\mathrm{MSE}(\hat{\eta})$ based on this approximation. But the approximation is somewhat heuristic, and the accuracy of the approximation and the associated MSE estimator was not studied. Prasad and Rao (1990) studied the accuracy of a second-order approximation to $\mathrm{MSE}(\hat{\eta})$ for two important special cases of longitudinal linear mixed models (see Section 1.2): (i) the Fay–Herriot model (Fay and Herriot 1979), and (ii) the nested error regression model (e.g., Battese et al. 1988). Both models are very popular in the context of small area estimation. Recently, Das et al. (2004) extended the result of Prasad and Rao to general linear mixed models (1.1). For example, for Gaussian mixed ANOVA models with REML estimation of θ, Das *et al.* (2004) showed that $\mathrm{MSE}(\hat{\eta}) = g_1(\theta) + g_2(\theta) + g_3(\theta) + o(d_*^{-2})$, where

$$g_3(\theta) = \mathrm{tr}\left[\{(\partial/\partial\theta')V^{-1}ZGa\}'V\{(\partial/\partial\theta')V^{-1}ZGa\}H^{-1}\right], \tag{2.39}$$

where $H = \mathrm{E}(\partial^2 l_\mathrm{R}/\partial\theta\partial\theta')$ and l_R is the restricted log-likelihood given by (1.17), and $d_* = \min_{1\le i\le s} d_i$ with $d_i = \|Z_i'PZ_i\|_2$ and P given by (1.11). The same result also holds for ML estimation. Based on the approximation, the authors obtained an estimator of $\mathrm{MSE}(\hat{\eta})$ whose bias is corrected to the second order. More specifically, an estimator $\widehat{\mathrm{MSE}}(\hat{\eta})$ was obtained such that $\mathrm{E}\{\widehat{\mathrm{MSE}}(\hat{\eta})\} = \mathrm{MSE}(\hat{\eta}) + o(d_*^{-2})$. See Das et al. (2004) for details.

Alternatively, Jiang et al. (2002) proposed a jackknife method that led to second-order approximation and estimation of the MSE of EBLUP in the case of longitudinal linear mixed models. Denote $\mathrm{MSE}(\tilde{\eta})$ by $b(\theta)$, where $\tilde{\eta}$ is the BLUP given by (2.34). The jackknife estimator of the MSE of $\hat{\eta}$ is given by $\widehat{\mathrm{MSE}}(\hat{\eta}) = \widehat{\mathrm{MSAE}}(\hat{\eta}) + \widehat{\mathrm{MSE}}(\tilde{\eta})$, where

$$\widehat{\mathrm{MSAE}}(\hat{\theta}) = \frac{m-1}{m}\sum_{i=1}^{m}(\hat{\eta}_{-i} - \hat{\eta})^2,$$
$$\widehat{\mathrm{MSE}}(\tilde{\eta}) = b(\hat{\theta}) - \frac{m-1}{m}\sum_{i=1}^{m}\left\{b(\hat{\theta}_{-i}) - b(\hat{\theta})\right\}. \tag{2.40}$$

Here m represents the number of clusters (e.g., number of small areas), $\hat{\theta}_{-i}$ denotes an M-estimator of θ using the data without the ith cluster (e.g., the ith small area), and $\hat{\eta}_{-i}$ the EBLUP of η in which the fixed parameters are estimated using the data without the ith cluster. Jiang et al. (2002) showed that $\mathrm{E}\{\widehat{\mathrm{MSE}}(\hat{\eta})\} = \mathrm{MSE}(\hat{\eta}) + o(m^{-1})$. The result holds, in particular, when $\hat{\theta}$ is either the REML or the ML estimator. Furthermore, the result holds for non-Gaussian (longitudinal) linear mixed models. In fact, the jackknife method also applies to longitudinal generalized linear mixed models, in which EBLUP is replaced by the empirical best predictor (EBP). See Jiang et al. (2002) for details.

Example 2.4 (Continued). Consider, once again, the James–Stein estimator of Example 2.4. Consider the prediction of the random effect $\eta = \alpha_1$. The BLUP is given by $\tilde{\eta} = (1-\omega)y_1$, where $\omega = (1+\psi)^{-1}$. The EBLUP is given by $\hat{\eta} = (1-\hat{\omega})y_1$. Efron and Morris (1973) used the following unbiased estimator, $\hat{\omega} = (m-2)/\sum_{i=1}^m y_i^2$. Note that the MSE of $\tilde{\eta}$ is given by $1-\omega$. The jackknife estimator of the MSE of $\hat{\eta}$ is given by

$$\begin{aligned}\widehat{\text{MSE}} &= 1-\hat{\omega}+\frac{m-1}{m}\sum_{i=1}^m(\hat{\eta}_{-i}-\hat{\eta})^2 \\ &= 1-\hat{\omega}+y_1^2\left(\frac{m-1}{m}\right)\sum_{i=1}^m(\hat{\omega}_{-i}-\hat{\omega})^2.\end{aligned}$$

Note that, because in this case $1-\hat{\omega}$ is an unbiased estimator of $1-\omega$, no bias correction is needed; that is, the second term on the right side of (2.40) is not needed.

Example 2.11 (The baseball example). Efron and Morris (1975) considered a Bayesian model to predict the true 1970 season batting average of each of 18 major league baseball players using the data on batting averages based on the first 45 official at-bats. Their model can be obtained as a simple linear mixed model by adding an unknown μ term to the previous example. The prediction of the true season batting average of player 1 is the same as that of the mixed effect: $\eta = \mu + \alpha_1$. The best predictor of η (see Section 2.3.1.1) is given by $\tilde{\eta} = \mu + (1-\omega)(y_1-\mu)$. The EBLUP is given by $\hat{\eta} = \bar{y} + (1-\hat{\omega})(y_1-\bar{y})$, where $\bar{y}$ is the sample mean. As for $\hat{\omega}$, Morris (1983) suggested a different estimator:

$$\hat{\omega} = \min\left\{\frac{m-3}{m-1}, \frac{m-3}{\sum_{i=1}^m(y_i-\bar{y})^2}\right\}.$$

It can be shown that the bias of $1-\hat{\omega}$ for estimating $1-\omega$, the MSE of $\tilde{\eta}$, is $o(m^{-1})$, thus, again, bias correction is not needed. It follows that the jackknife estimator of the MSE of $\hat{\eta}$ is

$$\widehat{\text{MSE}} = 1-\hat{\omega}+\frac{m-1}{m}\sum_{i=1}^m(\hat{\eta}_{-i}-\hat{\eta})^2,$$

where $\hat{\eta}_{-i} = \bar{y}_{-i} + (1-\hat{\omega}_{-i})(y_1-\bar{y}_{-i})$, $\bar{y}_{-i} = (m-1)^{-1}\sum_{j\neq i} y_j$ and

$$\hat{\omega}_{-i} = \min\left\{\frac{m-4}{m-2}, \frac{m-4}{\sum_{j\neq i}(y_j-\bar{y}_{-i})^2}\right\}.$$

We return to this example later in this chapter.

2.3.2 Prediction of Future Observation

We now consider the problem of predicting a future observation under a non-Gaussian linear mixed model. Because normality is not assumed, the approach is distribution-free; that is, it does not require any specific assumption about the distribution of the random effects and errors. First note that for this type of prediction, it is reasonable to assume that a future observation is independent of the current ones. We offer some examples.

Example 2.12. In longitudinal studies, one may be interested in prediction, based on repeated measurements from the observed individuals, of a future observation from an individual not previously observed. It is of less interest to predict another observation from an observed individual, because longitudinal studies often aim at applications to a larger population (e.g., drugs going to the market after clinical trials).

Example 2.13. In surveys, responses may be collected in two steps: in the first step, a number of families are randomly selected; in the second step, some family members (e.g., all family members) are interviewed for each of the selected families. Again, one may be more interested in predicting what happens to a family not selected, because one already knows enough about selected families (especially when all family members in the selected families are interviewed).

Therefore, we assume that a future observation, y_*, is independent of the current ones. Then, we have $E(y_*|y) = E(y_*) = x_*^t\beta$, so the best predictor is $x_*^t\beta$, if β is known; otherwise, an empirical best predictor (EBP) is obtained by replacing β by an estimator. So the point prediction is fairly straightforward. A question that is often of practical interest but has been so far neglected, for the most part, is that of prediction intervals.

1. Distribution-free prediction intervals. A prediction interval for a single future observation is an interval that will, with a specified coverage probability, contain a future observation from a population. In model-based statistical inference, it is assumed that the future observation has a certain distribution. Sometimes, the distribution is specified up to a finite number of unknown parameters, for example, those of the normal distribution. Then, a prediction interval may be obtained, if the parameters are adequately estimated, and the uncertainty in the parameter estimations is suitably assessed. Clearly, such a procedure is dependent on the underlying distribution in that, if the distributional assumption fails, the prediction interval may be seriously off: it either is wider than necessary, or does not have the claimed coverage probability. An alternative to the parametric method is a distribution-free one, in which one does not assume that the form of the distribution is known.

The problem of prediction intervals is, of course, an old one. One of the earliest works in this field is that of Baker (1935). Patel (1989) provided a review of the literature on prediction intervals when the future observation is independent of the observed sample, including results based on parametric distributions and on distribution-free methods. Hahn and Meeker (1991) reviewed three types of statistical intervals that are used most frequently in practice: the confidence interval, the prediction interval, and the tolerance interval. For a more recent overview, and developments on nonparametric prediction intervals, see Zhou (1997). Although many results on prediction intervals are for the i.i.d. case, the problem is also well studied in some non-i.i.d. cases, such as linear regression (e.g., Sen and Srivastava 1990, §3.8.2). In the context of linear mixed models, Jeske and Harville (1988) considered prediction intervals for mixed effects, assuming that the joint distribution of α and $y - E(y)$ is known up to a vector of unknown parameters. Thus, their approach is not distribution-free.

Note that, even if β is unknown, it is still fairly easy to obtain a prediction interval for y_* if one is willing to make the assumption that the distributions of the random effects and errors are known up to a vector of parameters (e.g., variance components). To see this, consider a simple case: $y_{ij} = x_{ij}'\beta + \alpha_i + \epsilon_{ij}$, where the random effect α_i and error ϵ_{ij} are independent such that $\alpha_i \sim N(0, \sigma^2)$ and $\epsilon_{ij} \sim N(0, \tau^2)$. It follows that the distribution of y_{ij} is $N(x_{ij}'\beta, \sigma^2 + \tau^2)$. Because methods are well developed for estimating fixed parameters such as β, σ^2, and τ^2 (see Section 1.3), a prediction interval with asymptotic coverage probability $1 - \rho$ is easy to obtain. However, it is much more difficult if one does not know the forms of the distributions of the random effects and errors, and this is the case that we consider. In the following, we propose a distribution-free approach to prediction intervals. Our results do not require normality or any specific distributional assumptions about the random effects and errors, and therefore are applicable to non-Gaussian linear mixed models.

First note that to consistently estimate the fixed effects and variance components in a linear mixed model, one does not need to assume that the random effects and errors are normally distributed (see Section 1.4). We categorize (non-Gaussian) linear mixed models into two classes: the standard and the nonstandard ones. A linear mixed model (1.1), (1.2) is standard if each Z_i consists only of 0s and 1s, there is exactly one 1 in each row and at least one 1 in each column. Our approaches are quite different for standard and nonstandard linear mixed models.

2. Standard linear mixed models. For standard linear mixed models, the method is surprisingly simple, and can be described as follows. First, one throws away the middle terms in (1.1) that involve the random effects, that is, (1.2), and pretends that it is a linear regression model with i.i.d. errors: $y = X\beta + \epsilon$. Next, one computes the least squares (LS) estimator $\hat{\beta} = (X'X)^{-1}X'y$ and the residuals $\hat{\epsilon} = y - X\hat{\beta}$. Let $\hat{a}$ and $\hat{b}$ be the $\rho/2$ and $1 - \rho/2$ quantiles of the residuals. Then, a prediction interval for y_* with asymptotic coverage probability $1 - \rho$ is $[\hat{y}_* + \hat{a}, \hat{y}_* + \hat{b}]$, where $\hat{y}_* = x_*'\hat{\beta}$. Note that, although the method sounds almost the same as the residual method in linear regression, its justification is not so obvious because, unlike linear regression, the observations in a (standard) linear mixed model are not independent. The method may be improved if one uses more efficient estimators such as the empirical BLUE (EBLUE; see Section 2.3) instead of the LS estimator. We study this in a simulated example in the sequel.

Let y_* be a future observation that we wish to predict. Suppose that y_* satisfies a standard linear mixed model. Then, y_* can be expressed as

$$y_* = x_*'\beta + \alpha_{*1} + \cdots + \alpha_{*s} + \epsilon_* \ ,$$

where x_* is a known vector of covariates (not necessarily present with the data), α_{*r}s are random effects, and ϵ_* is an error, such that $\alpha_{*i} \sim F_{ir}, \leq i \leq s$, $\epsilon_* \sim F_0$, where the Fs are unknown distributions (not necessarily normal), and $\alpha_{*1}, \ldots, \alpha_{*s}, \epsilon_*$ are independent. According to earlier discussion, we assume that y_* is independent of $y = (y_i)_{1 \leq i \leq n}$. It follows that the best (point) predictor of y_*, when β is known, is $E(y_*|y) = E(y_*) = x_*'\beta$. Because β is unknown, it is replaced by a consistent estimator, $\hat{\beta}$, which may be the OLS estimator or EBLUE (e.g., Jiang and Zhang 2002, Theorem 1; Jiang 1998b). This results in an empirical best predictor:

$$\hat{y}_* = x_*'\hat{\beta} \ . \tag{2.41}$$

Let $\hat{\delta}_i = y_i - x_i'\hat{\beta}$. Define

$$\hat{F}(x) = \frac{\#\{1 \leq i \leq n : \hat{\delta}_i \leq x\}}{n} = \frac{1}{n}\sum_{i=1}^{n} 1_{(\hat{\delta}_i \leq x)} \ . \tag{2.42}$$

Note that, although (2.42) resembles the empirical distribution, it is not one in the classic sense, because the $\hat{\delta}_i$s are not independent (the y_is are dependent, and $\hat{\beta}$ depends on all the data). Let $\hat{a} < \hat{b}$ be any numbers satisfying $\hat{F}(\hat{b}) - \hat{F}(\hat{a}) = 1 - \rho$ $(0 < \rho < 1)$. Then, a prediction interval for y_* with asymptotic coverage probability $1 - \rho$ is given by

$$[\hat{y}_* + \hat{a}, \ \hat{y}_* + \hat{b}] \ . \tag{2.43}$$

See Jiang and Zhang (2002). Note that a typical choice of $\hat{a}$, $\hat{b}$ has $\hat{F}(\hat{a}) = \rho/2$ and $\hat{F}(\hat{b}) = 1 - \rho/2$. Another choice would be to select $\hat{a}$ and $\hat{b}$ to minimize $\hat{b} - \hat{a}$, the length of the prediction interval. Usually, $\hat{a}$, $\hat{b}$ are selected such that the former is negative and the latter positive, so that $\hat{y}_*$ is contained in the interval. Also note that, if one considers linear regression as a special case of the linear mixed model, in which the random effects do not appear, $\hat{\delta}_i$ is the same as $\hat{\epsilon}_i$, the residual, if $\hat{\beta}$ is the least squares estimator. In this case, $\hat{F}$ is the empirical distribution of the residuals, and the prediction interval (2.43) corresponds to that obtained by the bootstrap method (Efron 1979). The difference is that our prediction interval (2.43) is obtained in closed form rather than by a Monte Carlo method. For more discussion on bootstrap prediction intervals, see Shao and Tu (1995, §7.3).

3. Nonstandard linear mixed models. Although most linear mixed models used in practice are standard, nonstandard linear mixed models are also used. First, the method developed for standard models may be applied to some of the nonstandard cases. To illustrate this, consider the following example.

Example 2.14. Suppose that the data are divided into two parts. For the first part, we have $y_{ij} = x_{ij}'\beta + \alpha_i + \epsilon_{ij}$, $i = 1, \ldots, m$, $j = 1, \ldots, n_i$, where $\alpha_1, \ldots, \alpha_m$ are i.i.d. random effects with mean 0 and distribution F_1; ϵ_{ij}s are i.i.d. errors with mean 0 and distribution F_0, and the αs and ϵs are independent. For the second part of the data, we have $y_k = x_k'\beta + \epsilon_k$, $k = N+1, \ldots, N+K$, where $N = \sum_{i=1}^m n_i$, and the ϵ_ks are i.i.d. errors with mean 0 and distribution F_0. Note that the random effects only appear in the first part of the data (and hence there is no need to use a double index for the second part).

For the first part, let the distribution of $\delta_{ij} = y_{ij} - x_{ij}'\beta$ be F $(= F_0 * F_1)$. For the second part, let $\delta_k = y_k - x_k'\beta$. If β were known, the δ_{ij}s (δ_ks) would be sufficient statistics for F (F_0). Therefore it suffices to consider an estimator of F (F_0) based on the δ_{ij}s (δ_ks). Note that the prediction interval for any future observation is determined either by F or by F_0, depending on to which part the observation corresponds. Now, because β is unknown, it is customary to replace it by $\hat{\beta}$. Thus, a prediction interval for y_*, a future observation corresponding to the first part, is $[\hat{y}_* + \hat{a}, \hat{y}_* + \hat{b}]$, where $\hat{y}_* = x_*'\hat{\beta}$, $\hat{a}$, $\hat{b}$ are determined by $\hat{F}(\hat{b}) - \hat{F}(\hat{a}) = 1 - \rho$ with

$$\hat{F}(x) = \frac{1}{N}\#\{(i,j) : 1 \le i \le m, 1 \le j \le n_i, \hat{\delta}_{ij} \le x\}$$

and $\hat{\delta}_{ij} = y_{ij} - x_{ij}'\hat{\beta}$. Similarly, a prediction interval for y_*, a future observation corresponding to the second part, is $[\hat{y}_* + \hat{a}, \hat{y}_* + \hat{b}]$, where $\hat{y}_* = x_*'\hat{\beta}$, $\hat{a}$, $\hat{b}$ are determined similarly with $\hat{F}$ replaced by

$$\hat{F}_0(x) = \frac{1}{K}\#\{k : N+1 \le k \le N+K, \hat{\delta}_k \le x\}$$

and $\hat{\delta}_k = y_k - x_k'\hat{\beta}$. The prediction interval has asymptotic coverage probability $1 - \rho$ (see Jiang and Zhang 2002).

If one looks more carefully, it is seen that the model in Example 2.14 can be divided into two standard submodels, so that the previous method is applied to each submodel. Of course, not every nonstandard linear mixed model can be divided into standard submodels. For such nonstandard models we consider that a different approach may need to be used.

Jiang (1998b) considered estimation of the distributions of the random effects and errors. His approach is the following. Consider the EBLUP of the random effects: $\hat{\alpha}_i = \hat{\sigma}_i^2 Z_i'\hat{V}^{-1}(y - X\hat{\beta})$, $1 \le i \le s$, where $\hat{\beta}$ is the EBLUE (see Section 2.2.1.4). The "EBLUP" for the errors can be defined as $\hat{\epsilon} = y - X\hat{\beta} - \sum_{i=1}^s Z_i\hat{\alpha}_i$. It was shown that, if the REML or ML estimators of the variance components are used, then, under suitable conditions,

$$\hat{F}_i(x) = \frac{1}{m_i}\sum_{u=1}^{m_i} 1_{(\hat{\alpha}_{i,u} \le x)} \xrightarrow{P} F_i(x), \qquad x \in C(F_i),$$

where $\hat{\alpha}_{i,u}$ is the uth component of $\hat{\alpha}_i$, $1 \le i \le s$, and

$$\hat{F}_0(x) = \frac{1}{n}\sum_{u=1}^{n} 1_{(\hat{\epsilon}_u \le x)} \xrightarrow{P} F_0(x), \qquad x \in C(F_0)\,,$$

where $\hat{\epsilon}_u$ is the uth component of $\hat{\epsilon}$. Here $C(F_i)$ represents the set of all continuity points of F_i, $0 \le i \le s$ (see Jiang 1998b).

For simplicity, we assume that all the distributions $F_0, \ldots, F_s$ are continuous. Let y_* be a future observation we would like to predict. As before, we assume that y_* is independent of y and satisfies a mixed linear model, which can be expressed componentwise as

$$y_i = x_i'\beta + z_{i1}'\alpha_1 + \cdots + z_{is}'\alpha_s + \epsilon_i, \qquad i = 1, \ldots, n.$$

This means that y_* can be expressed as

$$y_* = x_*'\beta + \sum_{j=1}^{l} w_j\gamma_j + \epsilon_*\,,$$

where x_* is a known vector of covariates (not necessarily present with the data), w_js are known nonzero constants, γ_js are unobservable random effects, and ϵ_* is an error. In addition, there is a partition of the indices $\{1, \ldots, l\} = \cup_{k=1}^{q} I_k$, such that $\gamma_j \sim F_{r(k)}$ if $j \in I_k$, where $r(1), \ldots, r(q)$ are distinct integers between 1 and s (so $q \le s$); $\epsilon_* \sim F_0$; $\gamma_1, \ldots, \gamma_l, \epsilon_*$ are independent. Define

$$\hat{F}^{(j)}(x) = m_{r(k)}^{-1} \sum_{u=1}^{m_{r(k)}} 1_{(w_j\hat{\alpha}_{r(k),u} \le x)}, \quad \text{if } j \in I_k$$

for $1 \le k \le q$. Let

$$\begin{aligned} \hat{F}(x) &= (\hat{F}^{(1)} * \cdots * \hat{F}^{(l)} * \hat{F}_0)(x) \\ &= \frac{\#\{(u_1, \ldots, u_l, u) : \sum_{k=1}^{q}\sum_{j\in I_k} w_j\hat{\alpha}_{r(k),u_j} + \hat{\epsilon}_u \le x\}}{\left(\prod_{k=1}^{q} m_{r(k)}^{|I_k|}\right) n}\,, \end{aligned} \tag{2.44}$$

where $*$ represents convolution (see Appendix C), and $1 \le u_j \le m_{r(k)}$ if $j \in I_k$, $1 \le k \le q$; $1 \le u \le n$. It can be shown that

$$\sup_x |\hat{F}(x) - F(x)| \xrightarrow{P} 0\,,$$

where $F = F^{(1)} * \cdots * F^{(l)} * F_0$, and $F^{(j)}$ is the distribution of $w_j\gamma_j$, $1 \le j \le l$. Note that F is the distribution of $y_* - x_*'\beta$. Let $\hat{y}_*$ be defined by (2.41) with $\hat{\beta}$ a consistent estimator, and $\hat{a}$, $\hat{b}$ defined by $\hat{F}(\hat{b}) - \hat{F}(\hat{a}) = 1 - \rho$, where $\hat{F}$ is given by (2.44). Then, the prediction interval $[\hat{y}_* + \hat{a}, \hat{y}_* + \hat{b}]$ has asymptotic coverage probability $1 - \rho$ (see Jiang and Zhang 2002).

We conclude this section with a simulated example.

4. A simulated example. Consider the linear mixed model

$$y_{ij} = \beta_0 + \beta_1 x_{ij} + \alpha_i + \epsilon_{ij} \,, \tag{2.45}$$

$i = 1, \ldots, m, j = 1, \ldots, n_i$, where the α_is are i.i.d. random effects with mean 0 and distribution F_1, and ϵ_{ij}s are i.i.d. errors with mean 0 and distribution F_0. The model might be associated with a sample survey, where α_i is a random effect related to the ith family in the sample, and n_i is the sample size for the family (e.g., the family size, if all family members are to be surveyed). The x_{ij}s are covariates associated with the individuals sampled from the family and, in this case, correspond to people's ages. The ages are categorized by the following groups: 0-4, 5-9, ..., 55-59, so that $x_{ij} = k$ if the person's age falls into the kth category (people whose ages are 60 or over are not included in the survey). The true parameters for β_0 and β_1 are 2.0 and 0.2, respectively.

In the following simulations, four combinations of the distributions F_0, F_1 are considered. These are Case I: $F_0 = F_1 = N(0,1)$; Case II: $F_0 = F_1 = t_3$; Case III: F_0 = logistic [the distribution of $\log\{U/(1-U)\}$, where $U \sim \text{Uniform}(0,1)$], F_1 = centralized lognormal [the distribution of $e^X - \sqrt{e}$, where $X \sim N(0,1)$]; Case IV: F_0 = double exponential [the distribution of $X_1 - X_2$, where X_1, X_2 are independent $\sim$ exponential(1)], F_1 = a mixture of $N(-4,1)$ and $N(4,1)$ with equal probability. Note that Cases II–IV are related to the following types of departure from normality: heavy-tail, asymmetry, and bimodal. In each case, the following sample size configuration is considered: $m = 100$, $k_1 = \cdots = k_{m/2} = 2$, and $k_{m/2+1} = \cdots = k_m = 6$. Finally, for each of the above cases, three prediction intervals are considered. The first is the prediction interval based on the OLS estimator of β; the second is that based on the EBLUE of β, where the variance components are estimated by REML (see Section 1.4.1); and the third is the linear regression (LR) prediction interval (e.g., Casella and Berger 2002, pp. 558), which assumes that the observations are independent and normally distributed. The third one is considered here for comparison.

For each of the four cases, 1000 datasets are generated. First, the following are independently generated, (i) x_{ij}, $1 \le i \le m$, $1 \le j \le k_i$, uniformly from the integers $1, \ldots, 12$ (twelve age categories); (ii) α_i, $1 \le i \le m$, from F_1; (iii) ϵ_{ij}, $1 \le i \le m$, $1 \le j \le k_i$, from F_0. Then y_{ij} is obtained by (2.45) with β_0, β_1 being the true parameters. Because of the way that the data are generated, condition on the x_{ij}s, the y_{ij}s satisfy (2.45) with its distributional assumptions. For each dataset generated, and for each of the 12 age categories, three prediction intervals are obtained, where $\rho = .10$ (nominal level 90%): OLS, EBLUE, and LR; then one additional observation is generated, which corresponds to a future observation in that category. The percentages of coverage and average lengths of the intervals over the 1000 data sets are reported.

The results are given in Table 2.1, in which the letters O, E, and L stand for OLS, EBLUE, and LR, respectively. The numbers shown in the table are coverage probabilities based on the simulations, in terms of percentages, and average lengths of the prediction intervals. Note that for OLS and EBLUE the lengths of the prediction intervals do not depend on the covariates, whereas for LR the length of the prediction interval depends on the covariate, but will be almost constant if the sample size is large. This, of course, follows from the definition of the prediction intervals, but there is also an intuitive interpretation. Consider, for example, the normal case. The distribution of a future observation y_* corresponding to a covariate x_* is $N(\beta_0 + \beta_1 x_*, \sigma^2)$, where $\sigma^2 = \text{var}(\alpha_i) + \text{var}(\epsilon_{ij})$ is a constant. So, if the βs were known the length of any prediction interval for y_* would not depend on x_*. If the βs are unknown but replaced by consistent estimators, then if the sample size were large, one would also expect the length of the prediction interval to be almost constant (not dependent on x_*). For such a reason, there is no need to exhibit the lengths of the prediction intervals for different categories, and we only give the averages over all categories.

It is seen that in the normal case there is not much difference among all three methods. This is not surprising. The difference appears in the nonnormal cases. First, the LR prediction intervals are wider than the OLS and EBLUE ones. Second, as a consequence, the coverage probabilities for the LR prediction intervals seem to be higher than 90%. Overall, the OLS and EBLUE perform better than LR in the nonnormal cases. This is not surprising, because the OLS and EBLUE prediction intervals are distribution-free. The EBLUE does not seem to do better than the OLS. This was a bit unexpected. On the other hand, it shows that at least in this special case the OLS, although much simpler than the EBLUE in that one does not need to estimate the variance components, can do just as well as more sophisticated methods such as the EBLUE.

Table 2.1. Coverage probability and average length

Coverage Probability (%)												
	Case I			Case II			Case III			Case IV		
x	O	E	L	O	E	L	O	E	L	O	E	L
1	90	90	90	89	89	92	90	91	93	90	90	94
2	90	90	90	89	89	91	91	91	93	89	90	96
3	88	88	88	91	91	93	90	89	92	88	89	96
4	90	90	89	91	91	93	89	89	91	89	89	97
5	89	89	89	89	89	92	90	90	92	90	90	96
6	89	89	90	89	89	92	91	91	93	90	90	97
7	89	88	89	90	90	92	90	90	93	88	89	96
8	90	90	90	90	90	92	89	89	91	90	90	97
9	90	90	91	89	89	92	89	89	91	89	89	96
10	89	89	90	91	90	93	89	89	93	88	88	95
11	90	90	90	89	89	93	89	89	92	89	89	97
12	89	89	89	89	89	92	91	91	93	89	89	96
Average Length												
	4.6	4.6	4.7	7.0	7.0	7.9	8.1	8.1	9.0	12.1	12.1	14.3

2.4 Model Checking and Selection

The previous sections have been dealing with inference about linear mixed models. For the most part, we have assumed that the basic assumptions about the model, for example, those about the presence of the random effects and their distributions, are correct. In practice, however, these assumptions may also be subject to checking. Methods of model checking are also known as model diagnostics. Sometimes, it is not clear which is the best model to use when there are a number of potential, or candidate, models. Here being best is in the sense that the model is not only correct but also most economical, meaning that it is simplest among all correct models. In this section we deal with the problems of model diagnostics and selection.

2.4.1 Model Diagnostics

Unlike standard regression diagnostics, the literature on diagnostics of linear mixed models involving random effects is not extensive (e.g., Ghosh and Rao 1994, pp. 70–71, Verbeke and Molenberghs 2000, pp. 151–152). Limited methodology is available, mostly regarding assessing the distribution of the random effects and errors. For the most part, the methods may be classified as diagnostic plots and goodness-of-fit tests.

1. Diagnostic plots. Several authors have used the idea of EBLUP or empirical Bayes estimators (EB), discussed in the previous section, for diagnosing distributional assumptions regarding the random effects (e.g., Dempster and Ryan 1985; Calvin and Sedransk 1991). The approach is reasonable because the EBLUP or EB are natural estimators of the random effects. In the following we describe a method proposed by Lange and Ryan (1989) based on a similar idea.

One commonly used assumption regarding the random effects and errors is that they are normally distributed. If such an assumption holds, one has a case of Gaussian mixed models. Otherwise, one is dealing with non-Gaussian linear mixed models. Lange and Ryan considered the longitudinal model (see Section 1.2.1.2), assuming that $G_i = G$, $R_i = \tau^2 I_{k_i}$, $i = 1, \dots, m$, and developed a weighted normal plot for assessing normality of the random effects in a longitudinal model. First, under the model (1.3) and normality, one can derive the best predictors, or Bayes estimators, of the random effects α_i $i = 1, \dots, m$ (see Section 2.3.1.1 and Section 2.5.1.1), assuming that β and θ, the vector of variance components, are known. This is given by

$$\begin{aligned}\tilde{\alpha}_i &= \mathrm{E}(\alpha_i|y_i) \\ &= GZ_i'V_i^{-1}(y_i - X_i\beta),\end{aligned}$$

where $V_i = \mathrm{Var}(y_i) = \tau^2 I_{k_i} + Z_i G Z_i'$. Furthermore, the covariance matrix of $\tilde{\alpha}_i$ is given by

$$\mathrm{Var}(\tilde{\alpha}_i) = GZ_i'V_i^{-1}Z_iG.$$

Lange and Ryan proposed to examine a Q–Q plot of some standardized linear combinations

$$z_i = \frac{c'\tilde{\alpha}_i}{\{c'\mathrm{Var}(\tilde{\alpha}_i)c\}^{1/2}}, \qquad i = 1, \ldots, m, \tag{2.46}$$

where c is a known vector. They argued that, through appropriate choices of c, the plot can be made sensitive to different types of model departures. For example, for a model with two random effects factors, a random intercept and a random slope, one may choose $c_1 = (1, 0)'$ and $c_2 = (0, 1)'$ and produce two Q–Q plots. On the other hand, such plots may not reveal possible nonzero correlations between the (random) slope and intercept. Thus, Lange and Ryan suggested producing a set of plots ranging from one marginal to the other by letting $c = (1 - u, u)'$ for some moderate number of values $0 \leq u \leq 1$.

Dempster and Ryan (1985) suggested that the normal plot should be weighted to reflect the differing sampling variances of $\tilde{\alpha}_i$. Following the same idea, Lange and Ryan proposed a generalized weighted normal plot. They suggested ploting z_i against $\Phi^{-1}\{F^*(z_i)\}$, where F^* is the weighted empirical cdf defined by

$$F^*(x) = \frac{\sum_{i=1}^m w_i 1_{(z_i \leq x)}}{\sum_{i=1}^m w_i},$$

and $w_i = c'\mathrm{Var}(\tilde{\alpha}_i)c = c'GZ_i'V_i^{-1}Z_iGc$.

In practice, however, the fixed effects β and variance components θ are unknown. In such cases, Lange and Ryan suggested using the ML or REML estimators in place of these parameters. They argued that, under suitable conditions, the limiting distribution of $\sqrt{n}\{\hat{F}^*(x) - \Phi(x)\}$ is normal with mean zero and variance equal to the variance of $\sqrt{n}\{F^*(x) - \Phi(x)\}$ minus an adjustment, where $\hat{F}^*(x)$ is $F^*(x)$ with the unknown parameters replaced by their ML (REML) estimators. See Lange and Ryan (1989) for details. This suggests that, in the case of unknown parameters, the Q–Q plot will be $\hat{z}_i$ against $\Phi^{-1}\{\hat{F}^*(\hat{z}_i)\}$, where $\hat{z}_i$ is z_i with the unknown parameters replaced by their ML (REML) estimates. However, the (asymptotic) variance of $\hat{F}^*(x)$ is different from that of $F^*(x)$, as indicated above. Therefore, if one wishes to include, for example, a ± 1 SD bound in the plot, the adjustment for estimation of parameters must be taken into account. See Lange and Ryan (1989). We consider an example.

Example 2.3 (Continued). Consider, again, the one-way random effects model of Example 1.1 with normality assumption. Because α_i is real-valued, $c = 1$ in (2.31). If μ, σ^2, τ^2 are known, the EB estimator of α_i is given by

$$\hat{\alpha}_i = \frac{k_i\sigma^2}{\tau^2 + k_i\sigma^2}(\bar{y}_{i\cdot} - \mu),$$

where $\bar{y}_{i\cdot} = k_i^{-1}\sum_{j=1}^{k_i} y_{ij}$, with

$$w_i = \text{var}(\hat{\alpha}_i) = \frac{k_i\sigma^4}{\tau^2 + k_i\sigma^2}.$$

Therefore, in this case,

$$z_i = \frac{\hat{\alpha}_i}{\text{sd}(\hat{\alpha}_i)} = \frac{\bar{y}_{i\cdot} - \mu}{\sqrt{\sigma^2 + \tau^2/k_i}},$$

$i = 1, \ldots, m$ and

$$F^*(x) = \left(\sum_{i=1}^{m} \frac{k_i\sigma^4}{\tau^2 + k_i\sigma^2}\right)^{-1} \sum_{i=1}^{n} \frac{k_i\sigma^4}{\tau^2 + k_i\sigma^2} 1_{(z_i \le x)}.$$

In practice, μ, σ^2, and τ^2 are unknown and therefore replaced by their REML (ML) estimators when making a Q–Q plot (Exercise 2.20).

2. Goodness-of-fit tests. Recently, several authors have developed tests for checking distributional assumptions involved in linear mixed models. Consider a mixed ANOVA model (1.1), where for $1 \le i \le s$, $\alpha_i = (\alpha_{ij})_{1\le j\le m_i}$, where the α_{ij}s are i.i.d. with mean 0, variance σ_i^2, which is unknown, and continuous distribution $F_i = F_i(\cdot\ |\sigma_i)$; and $\epsilon = (\epsilon_j)_{1\le j\le N}$, where the ϵ_js are i.i.d. with mean 0, variance τ^2, which is unknown, and continuous distribution $G = G(\cdot\ |\tau)$; and $\alpha_1, \ldots, \alpha_s, \epsilon$ are independent. We are interested in testing the following hypothesis,

$$\begin{aligned} H_0 : F_i(\cdot|\sigma_i) &= F_{0i}(\cdot|\sigma_i), \qquad 1 \le i \le s, \\ \text{and}\quad G(\cdot|\tau) &= G_0(\cdot|\tau); \end{aligned} \tag{2.47}$$

that is, the distributions of the random effects and errors, up to a set of unknown variance components $\sigma_1^2, \ldots, \sigma_s^2, \tau^2$, are as assumed.

Such distributional assumptions are vital in many applications of linear mixed models, and this is true even in large sample situations. For example, in many cases the prediction of a mixed effect is of main interest. Consider, for example, a nested error regression model, a special case of linear mixed models, which is useful in small area estimation (e.g., Battese et al. 1988; Prasad and Rao 1990; Ghosh and Rao 1994; Arora, Lahiri, and Mukherjee 1997):

$$y_{ij} = x_{ij}'\beta + \alpha_i + \epsilon_{ij}, \qquad i = 1, \ldots, m, \qquad j = 1, \ldots, k_i, \tag{2.48}$$

where x_{ij} is a known vector of covariates, β is an unknown vector of regression coefficients, α_i is a random effect associated with the ith small area, and ϵ_{ij} is an error. A mixed effect may be in the form $\eta = x'\beta + \alpha_i$, where x is known. If the sample size is large (i.e., m is large), one may consistently estimate β and even obtain an asymptotic confidence interval for it, and this does not rely on distributional assumptions such as normality. However, large sample results may not help, for example, in obtaining a prediction interval for η, because the effective sample size for estimating α_i is k_i, the sample size for the ith small area, which is often very small. Therefore, unless one knows the form of the distribution of α_i (e.g., normal), an accurate prediction interval for η cannot be obtained no matter how large m is (provided that k_i is small). To see another example, consider the estimation of the MSE of the EBLUP. Prasad and Rao (1990) give approximation formulas for MSE of EBLUP in the context of small area estimation, which are correct to the order $o(m^{-1})$. Although their results are asymptotic, assuming that m is large, normality distributional assumption remains critical for the validity of their approximations.

Jiang, Lahiri, and Wu (2001) developed an asymptotic theory of Pearson's χ^2-test with estimated cell frequencies, and applied the method to the case of nested error regression model (2.48) for checking the distributions of α and ϵ. The procedure requires splitting the data into two parts, one used for estimation and the other for testing, and thus raised some concerns about the power of the test. Jiang (2001) developed a method that applies to a general mixed ANOVA model as described above (2.47), which does not require data splitting. The method is described below.

The procedure is similar to Pearson's χ^2-test with estimated cell probabilities (e.g., Moore 1978). Let $E_1, \dots, E_M$ be a partition of R, the real line. Let a_n be a sequence of normalizing constants that is determined later on. Define

$$\hat{\chi}^2 = \frac{1}{a_n} \sum_{j=1}^{M} \{N_j - \mathrm{E}_{\hat{\theta}}(N_j)\}^2, \tag{2.49}$$

where $N_j = \sum_{i=1}^{n} 1_{(y_i \in E_j)} = \#\{1 \le i \le n : y_i \in E_j\}$, and $\hat{\theta}$ is the REML estimator of the vector of parameters involved in the linear mixed model. Despite the similarity of (2.49) to Pearson's χ^2-statistic, there are several major differences. First and most important, the observed count N_k is not a sum of independent random variables. In Pearson's χ^2-test, one deals with i.i.d. observations, so that N_k is a sum of i.i.d. random variables, and hence the asymptotic result follows from the classic central limit theorem (CLT). In a mixed linear model, however, the observations are correlated. Therefore, the classic CLT cannot handle the asymptotics. Second, unlike Pearson's χ^2-statistic, the normalizing constant in (2.49) is the same for all the squares in the sum. The choice of the normalizing constants in Pearson's χ^2-test is such that the asymptotic distribution is χ^2. However, even in the i.i.d. case, the asymptotic distribution of Pearson's χ^2-statistic is not necessarily χ^2, if the parameters are to be estimated (see Moore 1978). In fact, it may never be χ^2 no matter what normalizing constants are used. Thus, for simplicity, we choose a unified normalizing constant a_n. Note that, because of the dependence among the observations, a_n may not increase at the same rate as n, the sample size. Third, in a linear mixed model the number of fixed effects may be allowed to increase with n (e.g., Jiang 1996). As a consequence, the dimension of θ may increase with n. This shows, from another angle, that one can no longer expect an asymptotic distribution such as χ^2_{M-q-1}, where q is the number of (independent) parameters being estimated.

Jiang (2001) showed that, under suitable conditions, the asymptotic distribution of $\hat{\chi}^2$ is a weighted χ^2, that is, the distribution of $\sum_{j=1}^{M} \lambda_j Z_j^2$, where $Z_1, \ldots, Z_M$ are independent $N(0,1)$ random variables, and $\lambda_1 \geq \cdots \geq \lambda_M$ are eigenvalues of some nonnegative definite matrix, which depends on θ. Because the latter is unknown in practice, Jiang (2001) developed a method of estimating the critical value of the asymptotic distribution, and showed that $\mathrm{P}(\hat{\chi}^2 > \hat{c}_\rho) \to \rho$ as $n \to \infty$, where $\rho \in (0,1)$ is the level of the test. The estimated critical value, $\hat{c}_\rho$ is determined as $c_\rho(\hat{\lambda}_1, \ldots, \hat{\lambda}_M)$, where for any given $\lambda_1 \geq \cdots \geq \lambda_M$ and $0 < \rho < 1$, $c_\alpha(\lambda_1, \ldots, \lambda_M)$ is the ρ-critical value of the random variable $\xi = \sum_{j=1}^{M} \lambda_j Z_j^2$, and $\hat{\lambda}_1 \geq \cdots \geq \hat{\lambda}_M$ are the eigenvalues of a matrix $\hat{\Sigma}_n = \Sigma_n(\hat{\theta})$. The definition of $\Sigma_n(\theta)$, which depends on θ, is given in Section 2.7.

It remains to specify the normalizing constant a_n. Jiang (2001) noted that the choice of a_n is not unique. However, in some special cases there are natural choices. For example, in the case of linear regression, which may be regarded as a special case of the linear mixed model [with $s = 0$ in (1.1)], one has $a_n = n$. In the case of the one-way random effects model of Example 1.1, if the k_is are bounded, one has $a_n = m$. The choice is less obvious in the case of multiple random effects factors [i.e., $s > 1$ in (1.1)]. Jiang (2001) proposed the following principle that in many cases either uniquely determines a_n or at least narrows the choices. Note that there are a number of integers that contribute to the total sample size n, for example, m, k in Example 2.2; a, b, c in Example 2.1. Usually, a_n is a function of these integers. It is required that a_n depend on these integers in a way as simple as possible. In particular, no unnecessary constant is allowed in the expression of a_n. This is called a *natural choice* of a_n. A natural choice of a_n can be found by examining the leading term in the expression of the matrix $H_n + \Delta_n$ defined in Section 2.7. The following are some special cases.

Example 2.2 (Continued). In the case of the balanced one-way random effects model, it can be shown that $H_n + \Delta_n = mk^2\{\mathrm{Var}(h_1) + o(1)\}$, where h_1 is some nondegenerate random vector (see Jiang 2001, Section 3). Thus, in this case, a natural choice is $a_n = mk^2$. If, in fact, k is bounded, a natural choice would be $a_n = m$.

Example 2.1 (Continued). Suppose, for simplicity, that $c = 1$; that is, there is a single observation per cell. Similarly, it can be shown that, in this case, a natual choice is $a_n = (ab)^{3/2}$ (see Jiang 2001, Example 4.1).

2.4.2 Model Selection

In a way, model selection and estimation are viewed as two components of a process called model identification. The former determines the form of the model, leaving only some undetermined coefficients or parameters. The latter finds estimators of the unknown parameters. A pioneering work on model selection criteria was Akaike's information criterion (AIC, Akaike 1972). One of the earlier applications of AIC and other procedures such as the Bayesian information criterion (BIC, Schwartz 1978) was determination of the orders of an autoregressive moving-average time series model (e.g., Choi 1992). Similar methods have also been applied to regression model selection (e.g., Rao and Wu 1989; Bickel and Zhang 1992; Shao 1993; and Zheng and Loh 1995). It was shown that most of these model selection procedures are asymptotically equivalent to what is called the generalized information criterion (GIC, e.g., Nishii 1984). Although there is extensive literature on parameter estimation in linear mixed models, so that one component of the model identification has been well studied, the other component, that is, mixed model selection, has received little attention. Only recently have some results emerged in a paper by Jiang and Rao (2003).

Consider a general linear mixed model (1.1), where it is assumed that $\mathrm{E}(\alpha) = 0$, $\mathrm{Var}(\alpha) = G$; $\mathrm{E}(\epsilon) = 0$, $\mathrm{Var}(\epsilon) = R$, where G and R may involve some unknown parameters such as variance components; and α and ϵ are uncorrelated. In the following we first consider the problem of mixed model selection when the random effect factors are not subject to selection.

1. Selection with fixed random factors. Consider the model selection problem when the random part of the model (i.e., $Z\alpha$) is not subject to selection. Let $\zeta = Z\alpha + \epsilon$. Then, the problem is closely related to a regression model selection problem with correlated errors. Consider a general linear model $y = X\beta + \zeta$, where ζ is a vector of correlated errors, and everything else is as above. We assume that there are a number of candidate vectors of covariates, $X_1, \dots, X_l$, from which the columns of X are to be selected. Let $L = \{1, \dots, l\}$. Then, the set of all possible models can be expressed as $\mathcal{B} = \{a : a \subseteq L\}$, and there are 2^l possible models. Let $\mathcal{A}$ be a subset of $\mathcal{B}$ that is known to contain the true model, so the selection will be within $\mathcal{A}$. In an extreme case, $\mathcal{A}$ may be $\mathcal{B}$ itself. For any matrix M, let $\mathcal{L}(M)$ be the linear space spanned by the columns of M; P_M the projection onto $\mathcal{L}(M)$: $P_M = M(M'M)^- M'$; and $P_M^\perp$ the orthogonal projection: $P_M^\perp = I - P_M$ (see Appendix B). For any $a \in \mathcal{B}$, let $X(a)$ be the matrix whose columns are X_j, $j \in a$, if $a \neq \emptyset$, and $X(a) = 0$ if $a = \emptyset$. Consider the following criterion for model selection,

$$\begin{aligned} C_n(a) &= |y - X(a)\hat{\beta}(a)|^2 + \lambda_n |a| \\ &= |P_{X(a)}^\perp y|^2 + \lambda_n |a|, \qquad (2.50) \end{aligned}$$

$a \in \mathcal{A}$, where $|a|$ represents the cardinality of a; $\hat{\beta}(a)$ is the ordinary least squares (OLS) estimator of $\beta(a)$ for the model $y = X(a)\beta(a) + \zeta$; that is,

$\hat{\beta}(a) = \{X(a)'X(a)\}^{-}X(a)'y$, and λ_n is a positive number satisfying certain conditions specified below. Note that $P_{X(a)}$ is understood as 0 if $a = \emptyset$. Denote the true model by a_0. If $a_0 \neq \emptyset$, we denote the corresponding X and β by X and $\beta = (\beta_j)_{1 \leq j \leq p}$ ($p = |a_0|$), and assume that $\beta_j \neq 0$, $1 \leq j \leq p$. This is, of course, reasonable because, otherwise, the model can be further simplified. If $a_0 = \emptyset$, X, β, and p are understood as 0. Let $\nu_n = \max_{1 \leq j \leq q} |X_j|^2$ and $\rho_n = \lambda_{\max}(ZGZ') + \lambda_{\max}(R)$, where $\lambda_{\max}$ means the largest eigenvalue. Let $\hat{a}$ be the minimizer of (2.50) over $a \in \mathcal{A}$, which is our selection of the model. Jiang and Rao (2003) showed that, under suitable conditions, $\hat{a}$ is consistent in the sense that $\mathrm{P}(\hat{a} \neq a_0) \to 0$ as $n \to \infty$, provided that

$$\lambda_n/\nu_n \longrightarrow 0 \quad \text{and} \quad \rho_n/\lambda_n \longrightarrow 0. \tag{2.51}$$

Note 1. If $\rho_n/\nu_n \to 0$, there always exists λ_n that satisfies (2.51). For example, take $\lambda_n = \sqrt{\rho_n \nu_n}$. However, this may not be the best choice of λ_n, as a simulated example in the following shows.

Note 2. Typically, we have $\nu_n \sim n$. To see what the order of ρ_n may turn out to be, consider a special but important case of linear mixed models: the mixed ANOVA model of (1.1) and (1.2). Furthermore, assume that each Z_i ($1 \leq i \leq s$) is a standard design matrix in the sense that it consists only of 0s and 1s, there is exactly one 1 in each row, and at least one 1 in each column. Let n_{ij} be the number of 1s in the jth column of Z_i. Note that n_{ij} is the number of appearance of the jth component of α_i. Also note that $Z_i'Z_i = \mathrm{diag}(n_{ij}, 1 \leq j \leq m_i)$. Thus, we have $\lambda_{\max}(ZGZ') \leq \sum_{i=1}^s \sigma_i^2 \lambda_{\max}(Z_i Z_i') = \sum_{i=1}^s \sigma_i^2 \max_{1 \leq j \leq m_i} n_{ij}$. Also, we have $\lambda_{\max}(R) = \sigma_0^2$. It follows that $\rho_n = O(\max_{1 \leq i \leq s} \max_{1 \leq j \leq m_i} n_{ij})$. Therefore, (2.51) is satisfied provided that $\lambda_n/n \to 0$ and $\max_{1 \leq i \leq s} \max_{1 \leq j \leq m_i} n_{ij}/\lambda_n \to 0$. The following is an example not covered by the above case, because the errors are correlated.

Example 2.15. Consider the following linear mixed model which is a special case of the nested error regression model of (2.48); $y_{ij} = \beta_0 + \beta_1 x_{ij} + \alpha_i + \epsilon_{ij}$, $i = 1, \ldots, m$, $j = 1, \ldots, k$, where β_0, β_1 are unknown coefficients (the fixed effects). It is assumed that the random effects $\alpha_1, \ldots, \alpha_m$ are uncorrelated with mean 0 and variance σ^2. Furthermore, assume that the errors ϵ_{ij}s have the following exchangeable correlation structure: Let $\epsilon_i = (\epsilon_{ij})_{1 \leq j \leq k}$. Then, $\mathrm{Cov}(\epsilon_i, \epsilon_{i'}) = 0$ if $i \neq i'$, and $\mathrm{Var}(\epsilon_i) = \tau^2\{(1-\rho)I + \rho J\}$, where I is the identity matrix and J the matrix of 1s, and $0 < \rho < 1$ is an unknown correlation coefficient. Finally, assume that the random effects are uncorrelated with the errors. Suppose that $m \to \infty$, and

$$0 < \liminf \left[\frac{1}{mk} \sum_{i=1}^{m} \sum_{j=1}^{k} (x_{ij} - \bar{x}_{\cdot\cdot})^2 \right]$$

$$\leq \limsup \left[\frac{1}{mk} \sum_{i=1}^{m} \sum_{j=1}^{k} x_{ij}^2 \right] < \infty,$$

where $\bar{x}_{\cdot\cdot} = (mk)^{-1} \sum_{i=1}^{m} \sum_{j=1}^{k} x_{ij}$. It is easy to see that, in this case, $\rho_n \sim k$ and $\nu_n \sim mk$ (Exercise 2.21).

The above procedure requires selecting $\hat{a}$ from all subsets of $\mathcal{A}$. Note that $\mathcal{A}$ may contain as many as 2^l subsets. When l is relatively large, alternative procedures have been proposed in the (fixed effect) linear model context, which require less computation (e.g., Zheng and Loh 1995). In the following, we consider an approach similar to Rao and Wu (1989). First, note that one can always express $X\beta$ as $X\beta = \sum_{j=1}^{l} \beta_j X_j$ with the understanding that some of the coefficients β_j may be zero. It follows that $a_0 = \{1 \leq j \leq l : \beta_j \neq 0\}$. Let $X_{-j} = (X_u)_{1 \leq u \leq l, u \neq j}$, $1 \leq j \leq l$, $\eta_n = \min_{1 \leq j \leq l} |P_{X_{-j}}^{\perp} X_j|^2$, and δ_n be a sequence of positive numbers satisfying conditions specified below. Let $\hat{a}$ be the subset of $L = \{1, \ldots, l\}$ such that

$$(|P_{X_{-j}}^{\perp} y|^2 - |P_X^{\perp} y|^2)/(|P_{X_{-j}}^{\perp} X_j|^2 \delta_n) > 1 \tag{2.52}$$

for $j \in \hat{a}$. Jiang and Rao (2003) showed that, if $\rho_n/\eta_n \to 0$, where ρ_n is defined earlier, then $\hat{a}$ is consistent, provided that

$$\delta_n \longrightarrow 0 \quad \text{and} \quad \rho_n/(\eta_n \delta_n) \longrightarrow 0.$$

Example 2.15 (Continued). It is easy to show that, in this case, $\eta_n \sim mk$. Recall that $\rho_n \sim k$ in this case. Thus, $\rho_n/\eta_n \to 0$ as $m \to \infty$.

To study the finite sample behavior of the proposed model selection procedures, we consider a simulated example.

Example 2.16 (A simulated example). The model here is similar to Example 2.15 except that it may involve more than one fixed covariate; that is, $\beta_0 + \beta_1 x_{ij}$ is replaced by $x_{ij}'\beta$, where x_{ij} is a vector of covariates and β a vector of unknown regression coefficients. Here we focus on the first model selection procedure, the one defined by (2.50), which we also call GIC (e.g., Nishii 1984). We examine it by simulating the probability of correct selection and also the overfitting ($a1$) and underfitting ($a2$) probabilities,respectively, of various GICs for some given model parameters and sample sizes. Five GICs with different choices of λ are considered: (1) $\lambda = 2$, which corresponds to the C_p method; (2) $\lambda = \log n$. The latter choice satisfies the conditions required for consistency of the model selection. A total of 500 realizations of each simulation were run.

In the simulation the number of fixed factors was $l = 5$ with $\mathcal{A}$ being all subsets of $\{1, \ldots, 5\}$. The first column of X is all ones, corresponding to the intercept, and the other four columns of X are generated randomly from $N(0, 1)$ distributions, then fixed throughout the simulation. Three βs are considered: $(2, 0, 0, 4, 0)'$, $(2, 0, 0, 4, 8)'$, and $(2, 9, 0, 4, 8)'$, which correspond to $a_0 = \{1, 4\}$, $\{1, 4, 5\}$, and $\{1, 2, 4, 5\}$, respectively.

Furthermore, we consider the case where the correlated errors have varying degrees of exchangeable structure as described in Example 2.15, where four values of ρ were considered: $0, 0.2, 0.5, 0.8$. Variance components σ and τ were both taken to be equal to 1. We take the number of clusters (m) to be 50 and 100 and the number of repeats on a cluster to be fixed at $k = 5$. Table 2.2 presents the results.

Table 2.2. Selection probabilities under Example 1.10

Model	ρ	% correct		$a1$		$a2$	
		$\lambda_n = 2$	$\log(n)$	2	$\log(N)$	2	$\log(N)$
$M1(m = 50)$	0	59	94	41	6	0	0
	.2	64	95	36	5	0	0
	.5	59	90	40	9	1	1
	.8	52	93	47	5	1	2
$M1(m = 100)$	0	64	97	36	3	0	0
	.2	57	94	43	6	0	0
	.5	58	96	42	3	0	1
	.8	61	96	39	4	0	0
$M2(m = 50)$	0	76	97	24	3	0	0
	.2	76	97	24	3	0	0
	.5	73	96	27	4	0	0
	.8	68	94	31	4	1	2
$M2(m = 100)$	0	76	99	24	1	0	0
	.2	70	97	30	3	0	0
	.5	70	98	30	2	0	0
	.8	72	98	28	2	0	0
$M3(m = 50)$	0	90	99	10	1	0	0
	.2	87	98	13	2	0	0
	.5	84	98	16	2	0	0
	.8	78	95	21	3	1	2
$M3(m = 100)$	0	87	99	13	1	0	0
	.2	87	99	13	1	0	0
	.5	80	99	20	1	0	0
	.8	78	96	21	3	1	1

2. Selection with random factors. We now consider model selection that involves both fixed and random effects factors. Here we consider the mixed ANOVA model of (1.1), (1.2). If $\sigma_i^2 > 0$, we say that α_i is in the model; otherwise, it is not. Therefore, the selection of random factors is equivalent to simultaneously determining which of the variance components $\sigma_1^2, \dots, \sigma_s^2$ are positive. The true model can be expressed as

$$y = X\beta + \sum_{i \in b_0} Z_i \alpha_i + \epsilon, \tag{2.53}$$

where $X = (X_j)_{j \in a_0}$ and $a_0 \subseteq L$ [defined above (2.50)]; $b_0 \subseteq S = \{1, \ldots, s\}$ such that $\sigma_i^2 > 0$, $i \in b_0$, and $\sigma_i^2 = 0$, $i \in S \setminus b_0$.

There are some differences between selecting the fixed covariates X_j, as we did earlier, and selecting the random effect factors. One difference is that, in selecting the random factors, we are going to determine whether the vector α_i, not a given component of α_i, should be in the model. In other words, the components of α_i are all "in" or all "out". Another difference is that, unlike selecting the fixed covariates, where it is reasonable to assume that the X_i are linearly independent, in a linear mixed model it is possible to have $i \neq i'$ but $\mathcal{L}(Z_i) \subset \mathcal{L}(Z_{i'})$. See Example 2.17 below. Because of these features, the selection of random factors cannot be handled the same way.

To describe the basic idea, first note that we already have a procedure to determine the fixed part of the model, which, in fact, does not require knowing b_0. In any case, we may denote the selected fixed part as $\hat{a}(b_0)$, whether or not it depends on b_0. Now, suppose that a selection for the random part of the model (i.e., a determination of b_0) is $\hat{b}$. We then define $\hat{a} = \hat{a}(\hat{b})$. In other words, once the random part is determined, we may determine the fixed part using the methods developed earlier, treating the random part as known. It can be shown that, if the selection of the random part is consistent in the sense that $\mathrm{P}(\hat{b} \neq b_0) \to 0$, and given b_0, the selection of the fixed part is consistent; that is, $\mathrm{P}(\hat{a}(b_0) \neq a_0) \to 0$, then $\mathrm{P}(\hat{a} = a_0, \hat{b} = b_0) \to 1$; that is, the combined procedure is consistent.

We now describe how to obtain $\hat{b}$. First divide the vectors $\alpha_1, \ldots, \alpha_s$, or, equivalently, the matrices $Z_1, \ldots, Z_s$ into several groups. The first group is called the "largest random factors." Roughly speaking, those are Z_i, $i \in S_1 \subseteq S$ such that $\mathrm{rank}(Z_i)$ is of the same order as n, the sample size. We assume that $\mathcal{L}(X, Z_u, u \in S \setminus \{i\}) \neq \mathcal{L}(X, Z_u, u \in S)$ for any $i \in S_1$, where $\mathcal{L}(M_1, \ldots, M_t)$ represents the linear space spanned by the columns of the matrices $M_1, \ldots, M_t$. Such an assumption is reasonable because Z_i is supposed to be "the largest," and hence should have a contribution to the linear space. The second group consists of Z_i, $i \in S_2 \subseteq S$ such that $\mathcal{L}(X, Z_u, u \in S \setminus S_1 \setminus \{i\}) \neq \mathcal{L}(X, Z_u, u \in S \setminus S_1)$, $i \in S_2$. The ranks of the matrices in this group are of lower order of n. Similarly, the third group consists of Z_i, $i \in S_3 \subseteq S$ such that $\mathcal{L}(X, Z_u, u \in S \setminus S_1 \setminus S_2 \setminus \{i\}) \neq \mathcal{L}(X, Z_u, u \in S \setminus S_1 \setminus S_2)$, and so on. Note that if the first group (i.e., the largest random factors) does not exist, the second group becomes the first, and other groups also move on. As mentioned earlier [see below (2.53)], the selection of random factors cannot be treated the same way as that of fixed factors, because the design matrices $Z_1, \ldots, Z_s$ are usually linearly dependent. Intuitively, a selection procedure will not work if there is linear dependence among the candidate design matrices, because of identifiability problems. To consider a rather extreme example, suppose that Z_1 is a design matrix consisting of 0s and 1s such that there is exactly one 1 in each row, and $Z_2 = 2Z_1$. Then, to have $Z_1\alpha_1$ in the model means that there is a term α_{1i}; whereas to have $Z_2\alpha_2 = 2Z_1\alpha_2$ in the model means that there is a corresponding term $2\alpha_{2i}$. However, it makes no difference in terms of a model, because both α_{1i} and α_{2i} are random effects with mean 0 and certain variances. However, by grouping the random effect factors we have divided the Z_is into several groups such that there is linear independence within each group. This is the motivation behind grouping. To illustrate such a procedure, and also to show that such a division of groups does exist in typical situations, consider the following example.

Example 2.17. Consider the following random effects model,

$$y_{ijkl} = \mu + a_i + b_j + c_k + d_{ij} + f_{ik} + g_{jk} + h_{ijk} + e_{ijkl}, \tag{2.54}$$

$i = 1, \dots, m_1$, $j = 1, \dots, m_2$, $k = 1, \dots, m_3$, $l = 1, \dots, t$, where μ is an unknown mean; a, b, c are random main effects; d, f, g, h are (random) two- and three-way interactions; and e is an error. The model can be written as

$$y = X\mu + Z_1 a + Z_2 b + Z_3 c + Z_4 d + Z_5 f + Z_6 g + Z_7 h + e,$$

where $X = 1_n$ with $n = m_1 m_2 m_3 t$, $Z_1 = I_{m_1} \otimes 1_{m_2} \otimes 1_{m_3} \otimes 1_t, \dots, Z_4 = I_{m_1} \otimes I_{m_2} \otimes 1_{m_3} \otimes 1_t, \dots$, and $Z_7 = I_{m_1} \otimes I_{m_2} \otimes I_{m_3} \otimes 1_t$. It is easy to see that the Z_is are not linearly independent. For example, $\mathcal{L}(Z_i) \subset \mathcal{L}(Z_4)$, $i = 1, 2$, and $\mathcal{L}(Z_i) \subset \mathcal{L}(Z_7)$, $i = 1, \dots, 6$. Also, $\mathcal{L}(X) \subset \mathcal{L}(Z_i)$ for any i. Suppose that $m_j \to \infty$, $j = 1, 2, 3$, and t is bounded. Then, the first group consists of Z_7; the second group Z_4, Z_5, Z_6; and the third group Z_1, Z_2, Z_3. If t also $\to \infty$, the largest random factor does not exist. However, one still has these three groups. It is easy to see that the Z_is within each group are linearly independent.

Suppose that the Z_is are divided into h groups such that $S = S_1 \cup \cdots \cup S_h$. We give a procedure that determines the indices $i \in S_1$ for which $\sigma_i^2 > 0$; then a procedure that determines the indices $i \in S_2$ for which $\sigma_i^2 > 0$; and so on, as follows.

Group one: Write $B = \mathcal{L}(X, Z_1, \dots, Z_s)$, $B_{-i} = \mathcal{L}(X, Z_u, u \in S \setminus \{ij\})$, $i \in S_1$; $r = n - \mathrm{rank}(B)$, $r_i = \mathrm{rank}(B) - \mathrm{rank}(B_{-i})$; $R = |P_B^\perp y|^2$, $R_i = |(P_B - P_{B_{-ij}})y|^2$. Let $\hat{b}_1$ be the set of indices i in S_1 such that

$$(r/R)(R_i/r_i) > 1 + r^{(\rho/2)-1} + r_i^{(\rho/2)-1},$$

where ρ is chosen such that $0 < \rho < 2$. Let $a_{01} = \{i \in L_1 : \sigma_i^2 > 0\}$.

Group two: Let $B_1(b_2) = \mathcal{L}(X, Z_u, u \in (S \setminus S_1 \setminus S_2) \cup b_2)$, $b_2 \subseteq S_2$. Consider

$$C_{1,n}(b_2) = |P_{B_1(b_2)}^\perp y|^2 + \lambda_{1,n}|b_2|, \qquad b_2 \subseteq S_2,$$

where $\lambda_{1,n}$ is a positive number satisfying certain conditions similar to those for λ_n in (2.50) (see Jiang and Rao 2003, Section 3.3 for details). Let $\hat{b}_2$ be the minimizer of $C_{1,n}$ over $b_2 \subseteq S_2$, and $b_{02} = \{i \in S_2 : \sigma_i^2 > 0\}$.

General: The above procedure can be extended to the remaining groups. In general, let $B_t(b_{t+1}) = \mathcal{L}(X, Z_u, u \in (S \setminus S_1 \setminus \cdots \setminus S_{t+1}) \cup b_{t+1})$, $b_{t+1} \subseteq S_{t+1}$, $1 \le t \le h - 1$. Define

$$C_{t,n}(b_{t+1}) = |P^{\perp}_{B_t(b_{t+1})}y|^2 + \lambda_{t,n}|bl_{t+1}|, \qquad b_{t+1} \subseteq S_{t+1},$$

where $\lambda_{t,n}$ is a positive number satisfying certain conditions similar to those for λ_n in (2.50). Let $\hat{b}_{t+1}$ be the minimizer of $C_{t,n}$ over $b_{t+1} \subseteq S_{t+1}$, and $b_{0t+1} = \{i \in S_{t+1} : \sigma_i^2 > 0\}$.

It can be shown that, under suitable conditions, the combined procedure is consistent in the sense that $\mathrm{P}(\hat{b}_1 = b_{01}, \ldots, \hat{b}_h = b_{0h}) \to 1$ as $n \to \infty$. One property of $\hat{b}_t$ is that it does not depend on $\hat{b}_u$, $u < t$. In fact, $\hat{b}_1, \ldots, \hat{b}_h$ can be obtained simultaneously, and $\hat{b} = \cup_{t=1}^h \hat{b}_t$ is a consistent selection for the random part of the model. See Jiang and Rao (2003) for details.

2.5 Bayesian Inference

A linear mixed model can be naturally formulated as a hierarchical model under the Bayesian framework. Such a model usually consists of three levels, or stages of hierarchies. At the first stage, a linear model is set up given the fixed and random effects; at the second stage, the distributions of the fixed and random effects are specified given the variance component parameters; finally, at the last stage, a prior distribution is given for the variance components. Before we further explore these stages, we briefly describe the basic elements of Bayesian inference.

Suppose that y is a vector of observations and θ a vector of parameters that are not observable. Let $f(y|\theta)$ represent the probability density function (pdf) of y given θ, and $\pi(\theta)$ a prior pdf for θ. Then, the posterior pdf of θ is given by

$$\pi(\theta|y) = \frac{f(y|\theta)\pi(\theta)}{\int f(y|\theta)\pi(\theta)d\theta}.$$

Getting the posterior is the goal of Bayesian inference. In particular, some numerical summaries may be obtained from the posterior. For example, a Bayesian point estimator of θ is often obtained as the posterior mean:

$$\begin{aligned} \mathrm{E}(\theta|y) &= \int \theta\pi(\theta|y)d\theta \\ &= \frac{\int \theta f(y|\theta)\pi(\theta)d\theta}{\int f(y|\theta)\pi(\theta)\theta}; \end{aligned}$$

the posterior variance, $\mathrm{var}(\theta|y)$, on the other hand, is often used as a Bayesian measure of uncertainty.

In the first stage of a hierarchical linear model, it is assumed that, given β and α,

$$y = X\beta + Z\alpha + \epsilon,$$

where X and Z are known matrices, and ϵ has distribution F_1. In the second stage, it is assumed that (α, β) has a joint distribution F_2, which depends on some parameters of variance components. Finally, in the last stage, a prior distribution F_3 is assumed for the variance components. Note that a classical linear mixed model essentially involves the first two stages, but not the last one. A hierarchical model that is used most of the time is the so-called normal hierarchy, in which it is assumed that

$$\begin{array}{ll} (1) & \epsilon \sim N(0, R); \\ (2) & \alpha \sim N(0, G),\ \beta \sim N(b, B); \\ (3) & (G, R) \sim \pi, \end{array}$$

where π is a prior distribution. It is often assumed that, in the second stage, α and β are distributed independently, and b and B are known. Thus, a prior for β is, in fact, given in the second stage. The following is an example.

Example 2.18. Consider the one-way random effects model (Example 1.1). A normal hierarchical model assumes that (1) given μ and α_i $(1 \leq i \leq m)$, $y_{ij} = \mu + \alpha_i + \epsilon_{ij}$, $j = 1, \ldots, n_i$, where ϵ_{ij}s are independent and distributed as $N(0, \tau^2)$; (2) μ, $\alpha_1, \ldots, \alpha_m$ are independent such that $\mu \sim N(\mu_0, \sigma_0^2)$, $\alpha_i \sim N(0, \sigma^2)$, where μ_0 and σ_0^2 are known; and (3) σ^2, τ^2 are independent with $\sigma^2 \sim \text{Inverse} - \chi^2(a)$, $\tau^2 \sim \text{Inverse} - \chi^2(b)$, where a, b are known positive constants, and an Inverse$-\chi^2$ distribution with parameter $\nu > 0$ has pdf $\{2^{-\nu/2}/\Gamma(\nu/2)\}x^{-(\nu/2+1)}e^{-1/2x}$, $x > 0$. Alternatively, the priors in (3) may be such that $\sigma^2 \propto 1/\sigma^2$ and $\tau^2 \propto 1/\tau^2$. Note that, in the latter case, the priors are improper.

The inference includes that about the fixed and random effects and that about the variance components. In the following we discuss these two types of inference, starting with the variance components.

2.5.1 Inference about Variance Components

First define the likelihood function under the Bayesian framework. Suppose that, given α, β, and R, $y \sim f(y|\alpha, \beta, R)$. Furthermore, suppose that, given G, α and β are independent such that $\alpha \sim g(\alpha|G)$ and $\beta \sim h(\beta|b, B)$ (b, B known). Then, the full likelihood function, or simply the likelihood, for estimating G and R, is given by

$$L(G, R|y) = \int\int f(y|\alpha, \beta, R) g(\alpha|G) h(\beta|b, B) d\alpha d\beta, \tag{2.55}$$

where the integrals with respect to α and β may be both multivariate. Note that the difference between a likelihood and a posterior is that the prior is not taken into account in obtaining the likelihood. We now consider two special cases under the normal hierarchy.

The first case is when h is a point mass (or degenerate distribution) at β. Then, the limit of (2.55), when $b = \beta$ and $B \to 0$, reduces to

$$L(G, R|y) = \frac{1}{(2\pi)^{n/2}|V|^{1/2}} \exp\left\{-\frac{1}{2}(y - X\beta)'V^{-1}(y - X\beta)\right\}$$

(Exercise 2.22), where $V = ZGZ' + R$. This is simply the (normal) likelihood function given in Section 1.3.1. Under the Bayesian framework, it is also called the conditional likelihood, because a point mass corresponds to being conditional on β.

The second case is when h is a noninformative, or flat, distribution, that is, the prior for β is uniform over $(-\infty, \infty)$. Note that this is an improper prior. Nevertheless, the likelihood (2.55) does exist and has the expression

$$L(G, R|y) = \frac{1}{(2\pi)^{(n-p)/2}|A'VA|^{1/2}} \exp\left\{-\frac{1}{2}z'(A'VA)^{-1}z\right\},$$

where $p = \text{rank}(X)$, $z = A'y$, and A is an $n \times (n - p)$ matrix such that $\text{rank}(A) = n - p$ and $A'X = 0$ (Exercise 2.23). This is exactly the (normal) restricted likelihood function defined in Section 1.3.2. Under the Bayesian framework, it is also called the marginal likelihood, because it has β integrated out with respect to the noninformative prior.

Thus, without taking the prior into account, the likelihood can be used to obtain estimators of G and R, as one does in classical situations. This method is used later to obtain empirical Bayes estimators of the effects.

If the prior is taken into account, then the posterior for G and R can be expressed as

$$\begin{aligned} &\pi(G, R|y) \\ &= \int\int \frac{f(y|\alpha, \beta, R)g(\alpha|G)h(\beta|b, B)\pi(G, R)}{\int\int\int\int f(y|\alpha, \beta, R)g(\alpha|G)h(\beta|b, B)\pi(G, R)d\alpha d\beta dG dR} d\alpha d\beta \\ &= \frac{L(G, R|y)\pi(G, R)}{\int\int L(G, R|y)\pi(G, R)dG dR}, \end{aligned} \tag{2.56}$$

where $\pi(G, R)$ is a prior pdf for G, R. The computation of (2.56) can be fairly complicated even for a simple model (Exercise 2.24). For complex models the computation of (2.56) is typically carried out by Markov chain Monte Carlo (MCMC) methods.

2.5.2 Inference about Fixed and Random Effects

Similar to (2.56), the posterior for β can be expressed as

$$\begin{aligned} &\pi(\beta|y) \\ &= \int\int\int \frac{f(y|\alpha, \beta, R)g(\alpha|G)h(\beta|b, B)\pi(G, R)}{\int\int\int\int f(y|\alpha, \beta, R)g(\alpha|G)h(\beta|b, B)\pi(G, R)d\alpha d\beta dG dR} \\ &\quad d\alpha dG dR, \end{aligned} \tag{2.57}$$

and the posterior for α is

$$\pi(\alpha|y) = \int\int\int \frac{f(y|\alpha,\beta,R)g(\alpha|G)h(\beta|b,B)\pi(G,R)}{\int\int\int\int f(y|\alpha,\beta,R)g(\alpha|G)h(\beta|b,B)\pi(G,R)d\alpha d\beta dG dR} d\beta dG dR. \tag{2.58}$$

If normality is assumed, (2.57) and (2.58) may be obtained in closed forms. In fact, in the case of normal hierarchy, we have

$$\beta|y \sim N(\mathrm{E}(\beta|y), \mathrm{Var}(\beta|y)),$$

where $\mathrm{E}(\beta|y) = (X'V^{-1}X+B^{-1})^{-1}(X'V^{-1}y+B^{-1}b)$, $\mathrm{Var}(\beta|y) = (X'V^{-1}X+B^{-1})^{-1}$; and, similarly,

$$\alpha|y \sim N(\mathrm{E}(\alpha|y), \mathrm{Var}(\alpha|y)),$$

where $\mathrm{E}(\alpha|y) = (Z'LZ + G^{-1})^{-1}Z'L(y - Xb)$, $\mathrm{Var}(\alpha|y) = (Z'LZ + G^{-1})^{-1}$ with $L = R^{-1} - R^{-1}X(B^{-1} + X'R^{-1}X)^{-1}X'R^{-1}$ (Exercise 2.25). It is interesting to note that, when $B^{-1} \to 0$, which corresponds to the case where the prior for β is noninformative, one has $\mathrm{E}(\beta|y) \to (X'V^{-1}X)^{-1}X'V^{-1}y = \tilde{\beta}$, which is the BLUE; similarly, $\mathrm{E}(\alpha|y) \to GZ'V^{-1}(y - X\tilde{\beta})$ (Exercise 2.26), which is the BLUP (see Section 2.3.1.2). Thus, the BLUE and BLUP may be regarded as the posterior means of the fixed and random effects under normal hierarchy and a limiting situation, or noninformative prior for β.

Note that the BLUE and BLUP depend on G and R, which are unknown in practice. Instead of assuming a prior for G and R, one may estimate these covariance matrices, which often depend parametrically on some variance components, by maximizing the marginal likelihood function introduced before (see the early part of Section 2.5.1). This is called the empirical Bayes (EB) method. Harville (1991) showed that in the special case of the one-way random effects model (see Example 1.1), the EB is identical to EBLUP (see Section 2.3.1.3). From the above derivation, it is seen that this result actually holds more generally in a certain sense. Note that when G and R in BLUE and BLUP are replaced by estimators, the results are EBLUE and EBLUP. However, as Harville noted, much of the work on EB has focused on relatively simple models, whereas EBLUP has been carried out by practitioners such as individuals in the animal breeding area and survey sampling to relatively complex models.

2.6 Real-Life Data Examples

2.6.1 Analysis of the Birth Weights of Lambs (Continued)

In this section, we revisit the example of lamb-weight data discussed in section 1.7.1, where estimates of the fixed effects and variance components were obtained.

The BLUPs of the sire effects are obtained by PROC MIXED. The results are shown in Table 2.3. Here Standard Pred Error represents the square root of the estimated mean square prediction error (MSPE) of the EBLUP of s_{ij}, the jth sire effect in line i. The estimated MSPE in PROC MIXED is obtained by substituting the REML estimates of the variance components into the formula for the MSPE assuming known variance components. This is known as the naive method of estimating the MSPE. As discussed earlier (see Section 2.3.1), the naive estimates may underestimate the true MSPE. Methods that improve the accuracy of the MSPE estimation have been proposed. See Section 2.3.1.

Table 2.3. BLUPs of the random effects

Sire	Line	Estimate	Standard Pred Error
11	1	−0.6256	0.6693
12	1	0.3658	0.6693
13	1	0.5050	0.6156
14	1	−0.2452	0.6441
21	2	0.1750	0.6701
22	2	0.1588	0.6296
23	2	−0.0423	0.6717
24	2	−0.2914	0.6457
31	3	−0.2667	0.6184
32	3	−0.2182	0.5850
33	3	0.3212	0.6397
34	3	0.1637	0.6701
41	4	−0.2015	0.6187
42	4	0.0695	0.6454
43	4	0.1319	0.6436
51	5	0.3047	0.6356
52	5	−0.2437	0.6308
53	5	−0.1177	0.6327
54	5	−0.1549	0.6656
55	5	0.3940	0.6684
56	5	−0.6311	0.6318
57	5	0.5762	0.5913
58	5	−0.1274	0.5769

2.6.2 The Baseball Example

In this section, we revisit the Efron–Morris baseball example (Example 2.11) and use it to illustrate methods of diagnostics in linear mixed models. This example is chosen because of its simplicity. The dataset has been analyzed by several authors in the past, including Efron and Morris (1975), Efron (1975), Morris (1983), Datta and Lahiri (2000), Gelman et al. (1995), Rao (2003), and Lahiri and Li (2005), among others. Efron and Morris(1975) used this dataset to demonstrate the performance of their empirical Bayes and limited translation empirical Bayes estimators derived using an exchangeable prior in the presence of an outlying observation. They first obtained the batting average of Roberto Clemente, an extremely good hitter, from the *New York Times* dated April 26, 1970 when he had already batted $n = 45$ times. The batting average of a player is just the proportion of hits among the number at-bats. They selected 17 other major league baseball players who had also batted 45 times from the April 26 and May 2, 1970 issues of the *New York Times*. They considered the problem of predicting the batting averages of all 18 players for the remainder of the 1970 season based on their batting averages for the first 45 at-bats. This is a good example for checking the effect of an outlier on the efficiency of an EB estimation with an exchangeable prior. Gelman et al. (1995) provided additional data for this estimation problem and included important auxiliary data such as the batting average of each player through the end of the 1969 season. Jiang and Lahiri (2005b) reviewed the problem of predicting the batting averages of all 18 players for the entire 1970 season, instead of predicting the batting averages for the remainder of the 1970 season as Efron and Morris (1975) originally considered.

For the player i $(i = 1, \ldots, m)$, let p_i and π_i be the batting average for the first 45 at-bats and the true season batting average of the 1970 season. Note that p_i is the direct maximum likelihood (also unbiased) estimator of π_i under the assumption that conditional on π_i, the number of hits for the first n at-bats, np_i, follows a binomial distribution with number of trials n and success probability π_i, $i = 1, \ldots, m$.

Efron and Morris (1975) considered the following standard arc-sine transformation,

$$y_i = \sqrt{n} \arcsin(2p_i - 1)$$

and then assumed the following model

$$y_i | \theta_i \overset{\text{ind}}{\sim} N(\theta_i, 1), \qquad i = 1, \ldots, m,$$

where $\theta_i = \sqrt{n} \arcsin(2\pi_i - 1)$. There could be a criticism about the validity of the above approximation. However, Efron and Morris (1975) and Gelman et al. (1995) noted that this is not a serious concern given the moderate sample size of 45. The data analysis by Lahiri and Li (2005) supports this conjecture. Efron and Morris (1975) assumed exchangeability of the θ_is and used the two-level Fay–Herriot model, given in Section 2.1, without any covariate and equal sampling variances (i.e., 1).

Gelman et al. (1995) noted the possibility of an extra-binomial variation in the number of hits. The outcomes from successive at-bats could be correlated and the probability of hits may change across at-bats due to injury to the player and other external reasons not given in the dataset. However, there is no way to check these assumptions because of the unavailability of such data.

Assuming Level 1 is reasonable, Lahiri and Li (2005) checked the validity of the above model through graphical tools. To this end, they used the following standardized residual,

$$e_i = \frac{y_i - \bar{y}}{s},$$

where $s^2 = (m-1)^{-1}\sum_{i=1}^{m}(y_i - \bar{y})^2$ is the usual sample variance. Note that marginally $y_i \overset{iid}{\sim} N(\mu, 1 + A)$. Under this marginal model, $\mathrm{E}(e_i) \approx 0$, and $\mathrm{var}(e_i) \approx 1 + A$ for large m. Thus, if the model is reasonable, a plot of the standardized residuals versus the players is expected to fluctuate randomly around 0. Otherwise, one might suspect the adequacy of the two-level model. However, random fluctuation of the residuals may not reveal certain systematic patterns of the data. For example, Lahiri and Li (2005) noted that the residuals, when plotted against players arranged in increasing order of the previous batting averages, did reveal a linear regression pattern, something not apparent when the same residuals were plotted against players arranged in an arbitrary order. This is probably questioning the exchangeability assumption in the Efron–Morris model, a fact we knew earlier because of the intentional inclusion of an extremely good hitter.

Let p_{i0} be the batting average of player i through the end of the 1969 season and $x_i = \sqrt{n}\arcsin(2p_{i0} - 1)$, $i = 1, \ldots, m$. We plot y and θ versus x in Figure 2.1 (a) and (b) respectively. This probably explains the systematic pattern of the residuals mentioned in the previous paragraph. We also note the striking similarity of the two graphs: 1(a) and 1(b). Although Roberto Clemente seems like an outlier with respect to y, θ, or x, player L. Alvarado appears to be an outlier in the sense that his current batting average is much better than his previous batting average. He influences the regression fit quite a bit. For example, the BIC for the two-level model reduced from 55 to 44 when Alvarado was dropped from the model. Further investigation shows that this player is a rookie and batted only 51 times through the end of the 1969 season compared to other players in the dataset, making his previous batting average information not very useful. The BICs for the Fay–Herriot model with and without the auxiliary data are almost the same (54.9 and 55.3, respectively), a fact not expected at the beginning of the data analysis. In spite of more or less similar BIC values and the presence of an outlier in the regression, Figure 2.2 shows that EMReg did a good job in predicting the batting averages of Clemente and Alvarado, two different types of outliers. Further details on this data analysis are given in Lahiri and Li (2005).

2.7 Further Results and Technical Notes

1. Robust versions of classical tests. We first state the following theorems, which also define the matrices A, B, C, and Σ introduced in Section 2.1.2.4.

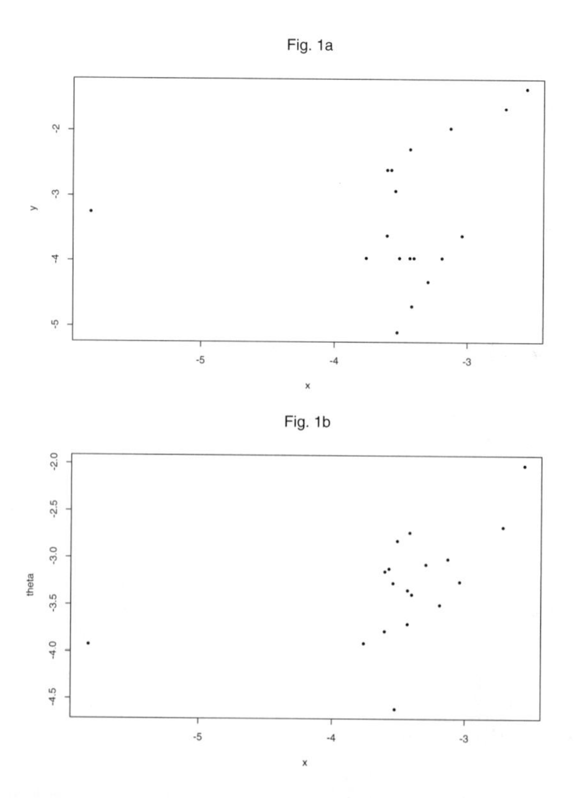

Fig. 2.1. Source: Adapted from Lahiri and Li (2005)

Theorem 2.1. Suppose that the following hold. (i) $l(\cdot, y)$ is twice continuously differentiable for fixed y, and $\psi(\cdot)$ is twice continuously differentiable. (ii) With probability $\to 1$, $\hat{\psi}$, $\hat{\phi}$ satisfy $\partial l/\partial\psi = 0$, $\partial l_0/\partial\phi = 0$, respectively. (iii) There are sequences of nonsingular symmetric matrices $\{G\}$ and $\{H\}$ and matrices A, B, C with A, $B > 0$ such that the following $\to 0$ in probability,

$$\sup_{\mathcal{S}_1} \left\| G^{-1} \left(\left. \frac{\partial^2 l}{\partial\psi_i \partial\psi_j} \right|_{\psi^{(i)}} \right)_{1 \le i,j \le q} G^{-1} + A \right\| ,$$

$$\sup_{\mathcal{S}_2} \left\| H^{-1} \left(\left. \frac{\partial^2 l_0}{\partial\phi_i \partial\phi_j} \right|_{\phi^{(i)}} \right)_{1 \le i,j \le p} H^{-1} + B \right\| ,$$

$$\sup_{\mathcal{S}_3} \left\| G \left(\left. \frac{\partial\psi_i}{\partial\phi_j} \right|_{\phi^{(i)}} \right)_{1 \le i \le q, 1 \le j \le p} H^{-1} - C \right\| ,$$

where $\mathcal{S}_1 = \{|\psi^{(i)} - \psi_0|_v \le |\hat{\psi} - \psi_0|_v \vee |\psi(\hat{\phi}) - \psi(\phi_0)|_v,\ 1 \le i \le q\}$, $\mathcal{S}_2 = \{|\phi^{(i)} - \phi_0|_v \le |\hat{\phi} - \phi_0|_v,\ 1 \le i \le p\}$, $\mathcal{S}_3 = \{|\phi^{(i)} - \phi_0|_v \le |\hat{\phi} - \phi_0|_v,\ 1 \le i \le q\}$

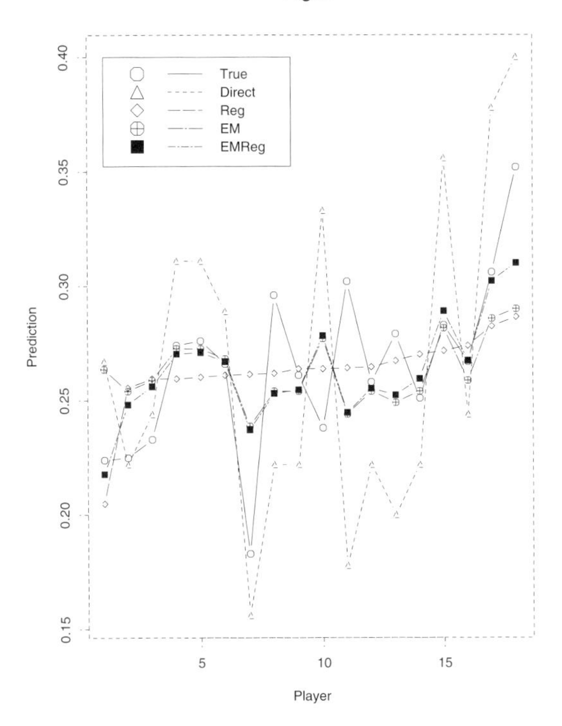

Fig. 2.2. Source: Adapted from Lahiri and Li (2005)

and $|a|_v = (|a_1|, \ldots, |a_k|)'$ for $a = (a_1, \ldots, a_k)'$;
(iv) $D(\partial l/\partial\psi)|_{\psi_0} \to 0$ in probability, where $D = \text{diag}(d_i, 1 \le i \le s)$ with $d_i = \|H^{-1}(\partial^2\psi_i/\partial\phi\partial\phi')|_{\phi_0}H^{-1}\|$, and

$$G^{-1}\left.\frac{\partial l}{\partial\psi}\right|_{\psi_0} \longrightarrow N(0, \Sigma) \quad \text{in distribution.} \tag{2.59}$$

Then, under the null hypothesis, the asymptotic distribution of $\mathcal{W}$ is χ^2_r, where $\mathcal{W}$ is defined in (2.18), and $r = \text{rank}[\Sigma^{1/2}A^{-1/2}(I - P)]$ with $P = A^{1/2}C(C'AC)^{-1}C'A^{1/2}$. In particular, if Σ is nonsingular, then $r = q - p$.

The theorem may be extended to allow the matrices A, B, and so on. to be replaced by sequences of matrices. Such an extension may be useful. For example, suppose G is a diagonal normalizing matrix; then, in many cases, A can be chosen as $-G^{-1}[E(\partial^2 l/\partial\psi\partial\psi')|_{\psi_0}]G^{-1}$, but the latter may not have a limit as $n \to \infty$.

Extension of Theorem 2.1. Suppose that, in Theorem 2.1, A, B, C are replaced by sequences of matrices $\{A\}$, $\{B\}$, and $\{C\}$, such that A, B are symmetric,

$$0 < \liminf[\lambda_{\min}(A) \wedge \lambda_{\min}(B)] \leq \limsup[\lambda_{\max}(A) \vee \lambda_{\max}(B)] < \infty,$$

and $\limsup \|C\| < \infty$. Furthermore, suppose that (2.59) is replaced by

$$\Sigma^{-1/2} G^{-1} \left. \frac{\partial l}{\partial \psi} \right|_{\psi_0} \longrightarrow N(0, I) \quad \text{in distribution,} \tag{2.60}$$

where $\{\Sigma\}$ is a sequence of positive definite matrices such that

$$0 < \liminf \lambda_{\min}(\Sigma) \leq \limsup \lambda_{\max}(\Sigma) < \infty,$$

and I is the p-dimensional identity matrix. Then, the asymptotic distribution of $\mathcal{W}$ is χ^2_{q-p}.

The proofs are given in Jiang (2002). According to the proof, one has $G[\hat{\psi} - \psi(\hat{\phi})] = O_P(1)$, hence

$$\begin{aligned} \hat{\mathcal{W}} &= [\hat{\theta} - \theta(\hat{\phi})]' G [Q_w^- + o_P(1)] G [\hat{\theta} - \theta(\hat{\phi})] \\ &= \mathcal{W} + o_P(1) \, . \end{aligned}$$

Thus, by Theorem 2.1, we conclude the following.

Corollary 2.1. Under the conditions of Theorem 2.1, the asymptotic distribution of $\hat{\mathcal{W}}$ is χ^2_r, where r is the same as in Theorem 2.1. Thus, in particular, if Σ is nonsingular, $r = q - p$. Under the conditions of Extension of Theorem 2.1, the asymptotic distribution of $\hat{\mathcal{W}}$ is χ^2_{q-p}.

We now consider the asymptotic distribution of the S-test defined in (2.19).

Theorem 2.2. Suppose that the conditions of Theorem 2.1 are satisfied with the following changes: (1) in (ii), that $\hat{\psi}$ satisfies $\partial l/\partial \psi = 0$ with probability $\to 1$ is not required; and (2) in (iii), the supremum for the first quantity (involving A) is now over $|\psi^{(i)} - \psi_0|_v \leq |\psi(\hat{\phi}) - \psi(\phi_0)|_v$, $1 \leq i \leq q$. Then, under the null hypothesis, the asymptotic distribution of $\mathcal{S}$ is χ^2_r, where r is the same as in Theorem 2.1. In particular, if Σ is nonsingular, then $r = q - p$.

In exactly the same way, we have the following.

Extension of Theorem 2.2. Suppose that, in Theorem 2.2, A, B, and C are replaced by $\{A\}$, $\{B\}$, and $\{C\}$, and (2.59) by (2.60), where the sequences of matrices $\{A\}$, $\{B\}$, $\{C\}$, and $\{\Sigma\}$ satisfy the conditions of the Extension of Theorem 2.1. Then, the asymptotic distribution of $\mathcal{S}$ is χ^2_{q-p}.

Corollary 2.2. Under the conditions of Theorem 2.2, the asymptotic distribution of $\hat{\mathcal{S}}$ is χ^2_r, where r is the same as in Theorem 2.1. Thus, in particular, if Σ is nonsingular, $r = q - p$. Under the conditions of the Extension of Theorem 2.2, the asymptotic distribution of $\hat{\mathcal{S}}$ is χ^2_{q-p}.

Finally, we consider the asymptotic distribution of the L-test. It is seen that the asymptotic distributions for the W- and S-tests are both χ^2. However, the following theorem states that the asymptotic distribution for the L-test is not χ^2 but a "weighted" χ^2 (e.g., Chernoff and Lehmann 1954). Recall that Q_l is defined near the end of Section 2.1.2.4.

Theorem 2.3. Suppose that the conditions of Theorem 2.1 are satisfied except that the third quantity in (iii) (involving C) $\to 0$ in probability is replaced by $G[(\partial\psi/\partial\phi)|_{\phi_0}]H^{-1} \to C$. Then, under the null hypothesis, the asymptotic distribution of $-2\log R$ is the same as $\lambda_1\xi_1^2 + \cdots + \lambda_r\xi_r^2$, where r is the same as in Theorem 2.1; $\lambda_1, \ldots, \lambda_r$ are the positive eigenvalues of Q_l; and $\xi_1, \ldots, \xi_r$ are independent $N(0,1)$ random variables. In particular, if Σ is nonsingular, then $r = q - p$.

Again, the proofs are given in Jiang (2002). It should be pointed out that if $L(\theta, y)$ is, indeed, the likelihood function, in which case the L-test is the likelihood-ratio test, the asymptotic distribution of $-2\log R$ reduces to χ^2 see Weiss 1975).

Let $\hat{Q}_l$ be a consistent estimator of Q_l. Then, by Weyl's eigenvalue perturbation theorem (see Appendix B), the eigenvalues of $\hat{Q}_l$ are consistent estimators of those of Q_l, and therefore can be used to obtain the asymptotic critical values for the L-test.

We now specify the W-, S-, and L-tests under the non-Gaussian mixed ANOVA model (see Section 1.2.2) with the additional assumption that

$$\mathrm{E}(\epsilon_1^3) = 0, \qquad \mathrm{E}(\alpha_{i1}^3) = 0, \qquad 1 \le i \le s\,. \tag{2.61}$$

As it turns out, this assumption is not essential but simplifies the results considerably. First define

$$\begin{aligned}
A_1 &= (\mathrm{tr}(V^{-1}V_i)/2\lambda\sqrt{nm_i})_{1\le i\le s},\\
A_2 &= (\mathrm{tr}(V^{-1}V_iV^{-1}V_j)/2\sqrt{m_im_j})_{1\le i,j\le s},\\
A &= \begin{pmatrix} X'V^{-1}X/\lambda n & 0 & 0 \\ 0 & 1/2\lambda^2 & A_1' \\ 0 & A_1 & A_2 \end{pmatrix}.
\end{aligned} \tag{2.62}$$

Let $b = (I \ \sqrt{\gamma_1}Z_1 \ \cdots \ \sqrt{\gamma_s}Z_s)$, $B_0 = b'V^{-1}b$, $B_i = b'V^{-1}V_iV^{-1}b$, $1 \le i \le s$. Furthermore, we define

$$\begin{aligned}
D_{0,ij} &= \sum_{l=1}^{n} B_{i,ll}B_{j,ll},\\
D_{1,ij} &= \sum_{l=n+1}^{n+m_1} B_{i,ll}B_{j,ll},\\
&\ \ \vdots\\
D_{s,ij} &= \sum_{l=n+m_1+\cdots+m_{s-1}+1}^{n+m_1+\cdots+m_s} B_{i,ll}B_{j,ll},
\end{aligned}$$

where $B_{i,kl}$ is the (k,l) element of B_i, $0 \le i,j \le s$. The kurtoses of the errors and random effects are defined by $\kappa_0 = (E\epsilon_1^4/\sigma_0^4) - 3$, and $\kappa_i = (E\alpha_{i1}^4/\sigma_i^4) - 3$, $1 \le i \le s$. Let $\Delta_1 = (\Delta_{0i}/\sqrt{nm_i})_{1 \le i \le s}$, $\Delta_2 = (\Delta_{ij}/\sqrt{m_i m_j})_{1 \le i,j \le s}$, and

$$\Delta = \begin{pmatrix} 0 & 0 & 0 \\ 0 & \Delta_{00}/n & \Delta_1' \\ 0 & \Delta_1 & \Delta_2 \end{pmatrix}, \tag{2.63}$$

where $\Delta_{ij} = \{4\lambda^{1_{(i=0)}+1_{(j=0)}}\}^{-1} \sum_{t=0}^{s} \kappa_t D_{t,ij}$, $0 \le i,j \le s$. Let

$$W = b'V^{-1}X(X'V^{-1}X)^{-1/2},$$

and W_l' be the lth row of W, $1 \le l \le n+m$, where $m = m_1 + \cdots + m_s$.

Theorem 2.4. Suppose that the following hold.
(i) $\theta(\cdot)$ is three-times continuously differentiable and satisfies (2.21), and $\partial\theta_{i_k}/\partial\phi_k \neq 0$, $1 \le k \le d$.
(ii) $E\epsilon_1^4 < \infty$, $\mathrm{var}(\epsilon_1^2) > 0$, $E\alpha_{i1}^4 < \infty$, $\mathrm{var}(\alpha_{i1}^2) > 0$, $1 \le i \le s$, and (2.61) holds.
(iii) $n \to \infty$, $m_i \to \infty$, $1 \le i \le s$, $0 < \liminf \lambda_{\min}(A) \le \limsup \lambda_{\max}(A) < \infty$, and $\max_{1 \le l \le n+m} |W_l| \to 0$;
Then, for $l(\theta, y)$ there exist $\hat{\theta}$ and $\hat{\phi}$ such that the conditions of the Extensions of Theorems 2.1 and 2.2 are satisfied with

$$\begin{aligned} G &= \mathrm{diag}(\sqrt{n}, \ldots, \sqrt{n}, \sqrt{m_1}, \ldots, \sqrt{m_s}) \\ &= \mathrm{diag}(g_i, 1 \le i \le q), \end{aligned}$$

$H = \mathrm{diag}(g_{i_k}, 1 \le k \le a)$, A is given by (2.62), $C = \partial\theta/\partial\phi$, $B = C'AC$, and $\Sigma = A + \Delta$, where Δ is given by (2.63). Therefore, the asymptotic null distribution of both $\hat{\chi}_w^2$ and $\hat{\chi}_s^2$ is χ_{q-d}^2. The same conclusion holds for $l_R(\theta, y)$ as well.

Note that the ith row of $\partial\theta/\partial\phi$ is $\partial\theta_i/\partial\phi'$, which is $(0, \ldots, 0)$ if $i \notin \{i_1, \ldots, i_a\}$, and $(0, \ldots, 0, \partial\theta_{i_k}/\partial\phi_k, 0, \ldots, 0)$ (kth component nonzero) if $i = i_k$, $1 \le k \le a$ under (2.21).

Theorem 2.5. Suppose that the conditions of Theorem 2.4 are satisfied except that, in (iii), the condition about A is strengthened to that $A \to A_0$, where $A_0 > 0$, and $\Sigma \to \Sigma_0$. Then, the conditions of Theorem 2.3 are satisfied with $A = A_0$, $\Sigma = \Sigma_0$, and everything else given by Theorem 2.4. Therefore, the asymptotic null distribution of $-2\log R$ is the same as $\sum_{j=1}^{r} \lambda_j \xi_j^2$, where $r = \mathrm{rank}\{\Sigma^{1/2}A^{-1/2}(I-P)\}$, evaluated under H_0 with $P = A^{1/2}C(C'AC)^{-1}C'A^{1/2}$; λ_js are the positive eigenvalues of Q_l given by (2.20), again evaluated under H_0; and ξ_js are independent $N(0,1)$ random variables. In particular, if Σ is nonsingular under H_0, then $r = q - d$. The same conclusion holds for $l_R(\theta, y)$ as well.

The proof of Theorems 2.1–2.5 can be found in Jiang (2005c).

It is seen from (2.63) that Δ, and hence Σ, depends on the kurtoses κ_i, $0 \leq i \leq s$, in addition to the variance components σ_i^2, $0 \leq i \leq s$. One already has consistent estimators of σ_i^2, $0 \leq i \leq s$ (e.g., the ML or REML estimators). As for κ_i, $0 \leq i \leq s$, they can be estimated by the empirical method of moments (EMM) of Jiang (2003b).

The extension of Theorem 1 and Theorem 2 without assuming (2.61) is fairly straightforward, although the results will not be as simple. Note that Theorems 2.1–2.3 (and their extensions) do not require (2.61). However, there is a complication in estimating the additional parameters involved in Σ. This is because, without (2.61), the matrix Δ also involves the third moments of the random effects and errors (on the off-diagonal). In such a case, the EMM of Jiang (2003b) is not directly applicable. Alternatively, Σ can be consistently estimated by the POQUIM method (see Sections 1.4.2 and 1.8.5), which does not require (2.61).

2. Existence of moments of ML and REML estimators. Jiang (2000a) established the existence of moments of ML and REML estimators under non-Gaussian linear mixed models (see Section 1.4.1) as an application of a matrix inequality. Let $A_1, \ldots, A_s$ be nonnegative definite matrices. Then, there are positive constants depending on the matrices such that for all positive numbers $x_1, \ldots, x_s$,

$$A_i \leq \frac{c_i}{x_i^2}\left(I + \sum_{j=1}^{s} x_j A_j\right)^2, \qquad 1 \leq i \leq s.$$

Now consider a non-Gaussian mixed ANOVA model (see Section 1.2.2.1), where $y = (y_i)_{1 \leq i \leq n}$. The ML and REML estimators are defined in Sections 1.3.1 and 1.3.2, respectively, and EBLUE and EBLUP in Sections 2.2.1.4 and 2.3.1.3, respectively.

Theorem 2.6. The kth moments ($k > 0$) of the ML or REML estimators of $\sigma_1^2, \ldots, \sigma_s^2, \tau^2$ are finite, provided that the $2k$th moments of y_i, $1 \leq i \leq n$ are finite.

3. Existence of moments of EBLUE and EBLUP. In the same paper, Jiang (2000a) established the existence of moments of EBLUE and EBLUP as another application of the same matrix inequality. Again, no normality assumption is made. Note that here the only requirement for the variance components estimators is that they are nonnegative. In the following theorem, the abbreviations EBLUEs and EBLUPs stand for the components of EBLUE and EBLUP, respectively.

Theorem 2.7. The kth moments ($k > 0$) of EBLUEs and EBLUPs are finite, provided that the kth moments of y_i, $1 \leq i \leq n$ are finite, and the variance components estimators are nonnegative.

Because it is always assumed that the second moments of the data are finite, we have the following conclusion.

Corollary 2.3. The means and MSEs of EBLUE and EBLUP exist as long as the variance components estimators are nonnegative.

Note 1. Kackar and Harville (1984) showed that the EBLUE and EBLUP remain unbiased if the variance components are estimated by nonnegative, even, and translation-invariant estimators (see Section 2.3.1.3). In deriving their results, Kackar and Harville avoided the existence of the means of EBLUE and EBLUP. Jiang (1999b) considered a special case of linear mixed models corresponding to $s = 1$ in (1.2) and proved the existence of the means. The above corollary has solved the problem for the general case.

Note 2. The ML and REML estimators are nonnegative by their definitions (see Section 1.4.1). However, for example, the ANOVA estimators may take negative values (see Section 1.5.1).

4. The definition of $\Sigma_n(\theta)$ in Section 2.4.1.2. First consider the case $s = 0$, that is, the case of linear regression. In this case, we have $y_i = x_i'\beta + \epsilon_i$, $i = 1, \dots, n$, where x_i' is the ith row of X, which has full rank p, and ϵ_is are i.i.d. errors with mean 0, variance τ^2, and an unknown distribution $G(\cdot|\tau)$. Thus, in this case, $\theta = (\beta', \tau^2)'$. The matrix $\Sigma_n(\theta)$ is defined as $n^{-1}\sum_{i=1}^n \mathrm{Var}(h_i)$, where

$$h_i = (1_{(y_i \in E_k)} - p_{ik}(\theta))_{1 \le k \le M} - \left(\sum_{i=1}^n \frac{\partial p_i(\theta)}{\partial \beta'}\right)(X'X)^{-1}x_i\epsilon_i$$
$$- \frac{1 - x_i'(X'X)^{-1}x_i}{n-p}\left(\sum_{i=1}^n \frac{\partial p_i(\theta)}{\partial \tau^2}\right)\epsilon_i^2$$

with $p_i(\theta) = (p_{ik}(\theta))_{1 \le k \le M}$ and $p_{ik}(\theta) = \mathrm{P}_\theta(y_i \in E_k)$. Jiang (2001) gives a more explicit expression of $\Sigma_n(\theta)$. On the other hand, it may be more convenient to compute $\hat{\Sigma}_n = \Sigma_n(\hat{\theta})$ by a Monte Carlo method, where $\hat{\theta} = (\hat{\beta}', \hat{\tau}^2)'$ with $\hat{\beta}$ being the least squares estimator and $\hat{\tau}^2 = |y - X\hat{\beta}|^2/(n-p)$.

We now consider another special case, the case $s = 1$, such that $y_{ij} = x_{ij}'\beta + \alpha_i + \epsilon_{ij}$, $i = 1, \dots, m$, $j = 1, \dots, k_i$, where the α_is are i.i.d. with mean 0, variance σ^2, and an unknown distribution $F(\cdot|\sigma)$, ϵ_{ij}s are i.i.d. with mean 0, variance τ^2, and an unknown distribution $G(\cdot|\tau)$, and α, ϵ are independent. In other words, we consider the nested error regression model (2.48). Write the model in the standard form $y = X\beta + Z\alpha + \epsilon$. Let $\theta = (\beta', \tau^2, \gamma)'$, where $\gamma = \sigma^2/\tau^2$. Define

$$\Sigma_n(\theta) = a_n^{-1}\left\{\sum_{i=1}^m \mathrm{Var}(h_i) + 2\Phi'(\mathcal{I} - \mathcal{R})\Phi\right\},$$

where $\mathcal{I}$ is defined in Section 1.8.3, and h_i, Φ, $\mathcal{R}$ are defined as follows. Recall the notation introduced in Section 1.8.3. Redefine $p_1 = [\mathrm{tr}\{(Z'V(\gamma)Z)^2\}]^{1/2}$.

Recall $p_0 = \sqrt{n-p}$. Let $\rho = \mathrm{tr}\{Z'V(\gamma)Z\}/p_0 p_1$. Let $P_{ij}(\theta)$ be the $M \times (p+2)$ matrix whose (k, r) element is

$$\frac{\partial}{\partial \theta_r} \int \{G(c_k - x'_{ij}\beta - u|\tau) - G(c_{k-1} - x'_{ij}\beta - u|\tau)\} dF(u|\sigma)$$

(θ_r is the rth component of θ). Let $P_{ij}[r](\theta)$ be the rth column of $P_{ij}(\theta)$, and $P_{ij}[1,p](\theta)$, the matrix, consist of the first p columns of $P_{ij}(\theta)$. Define

$$\begin{aligned} \Phi &= \frac{1}{1-\rho^2} \begin{pmatrix} \tau^4 & -\tau^2\rho \\ -\tau^2\rho & 1 \end{pmatrix} \begin{pmatrix} p_0^{-1} \sum_{i,j} P_{ij}[p+1](\theta)' \\ p_1^{-1} \sum_{ij} P_{ij}[p+2](\theta)' \end{pmatrix} \\ &= \begin{pmatrix} \Phi'_0 \\ \Phi'_1 \end{pmatrix}, \\ \Psi &= \tau b(\gamma) V_\gamma^{-1} X (X' V_\gamma^{-1} X)^{-1} \sum_{i,j} P_{ij}[1,p](\theta)' \\ &= (\Phi'_l)_{1 \le l \le m+n}, \end{aligned}$$

where $V_\gamma = V/\tau^2$. Let $S_i = \{l : \sum_{i'<i} k_{i'} + 1 \le l \le \sum_{i' \le i} k_{i'}\} \cup \{n+i\}$. Write $\omega(i) = (\omega_l)_{l \in S_i}$, $V_j(i,i') = (V_j(\gamma)_{l,l'})_{l \in S_i, l' \in S_{i'}}$, $j = 0, 1$, $\Psi(i) = (\Phi'_l)_{l \in S_i}$. Let

$$\begin{aligned} h_i = &\left(\sum_{j=1}^{k_i} \{1_{(y_{ij} \in E_k)} - p_{ijk}(\theta)\} \right)_{1 \le k \le M} - \Psi(i)'\omega(i) \\ &- \sum_{j=0}^{1} \frac{\omega(i)' V_j(i,i) \omega(i)}{\tau^{2(1-j)} p_j} \Phi_j, \end{aligned}$$

where $p_{ijk}(\theta) = \mathrm{P}_\theta(y_{ij} \in E_k)$. Finally, let $\mathcal{R} = (r_{j,j'})_{j,j'=0,1}$, where

$$r_{j,j'} = \frac{\sum_{i=1}^m \mathrm{tr}\{V_j(i,i) V_{j'}(i,i)\}}{\tau^{2(2-j-j')} p_j p_{j'}}.$$

Finally, in the case of multiple random effect factors, that is, $s \ge 2$, $\Sigma_n(\theta)$ is defined in a similar way; that is, $\Sigma_n(\theta) = a_n^{-1}\{\sum_{l=1}^L \mathrm{Var}(h_l) + 2\Phi'(\mathcal{I} - \mathcal{R})\Phi\}$. We omit the definitions of h, Φ, and $\mathcal{R}$ here and refer the details to Jiang (2001, Section 4) ($\mathcal{I}$ is the same as before).

2.8 Exercises

2.1. Derive explicit expressions of the test statistic (2.3) (in terms of the y_{ijk}s) for the two cases considered in Example 2.1 where the exact F-test applies: (i) testing $\sigma_1^2 = 0$ under the model without interaction; and (ii) testing $\sigma_3^2 = 0$ under the model with interaction.

2.2. Consider the following random effects model,

$$y_{ijkl} = \mu + f_i + g_j + u_{ij} + v_{jk} + w_{ijk} + e_{ijkl}$$

(see, e.g., Searle 1971, for notation), $i = 1, \ldots, a$, $j = 1, \ldots, b$, $k = 1, \ldots, c$, $l = 1, \ldots, d$, where μ is an unknown mean, e_{ijkl} is an error, and all the others are random effects. Assume that the random effects and errors are independent such that $f_i \sim N(0, \sigma_1^2)$, $g_j \sim N(0, \sigma_2^2)$, $u_{ij} \sim N(0, \sigma_3^2)$, $v_{jk} \sim N(0, \sigma_4^2)$, $w_{ijk} \sim N(0, \sigma_5^2)$, and $e_{ijkl} \sim N(0, \tau^2)$. Do exact or optimal tests exist for testing H_0: $\sigma_2^2 = 0$? Please explain. (Hint: Consider Result 2 of Mathew and Sinha (1988) described in Section 2.1.1.2).

2.3. Derive an expression for $-2 \log \mathcal{R}$, where $\mathcal{R}$ is the likelihood ratio (2.6), under the one-way random effects model of Example 2.3 for testing H_0: $\sigma^2 = 0$. What is the asymptotic distribution of the likelihood-ratio test, that is, the asymptotic distribution of $-2 \log \mathcal{R}$? Study empirically the (asymptotic) size of the likelihood-ratio test and compare it with the nominal levels. For the empirical study, let the true parameters be $\mu = 0.5$ and $\tau^2 = 1.0$; and consider sample sizes $m = 50, 100, 200$ and $k_i = 5$ for all i in all cases.

2.4. Suppose that $X_1, \ldots, X_n$ are i.i.d. observations from a population with mean μ and variance σ^2, and the problem of interest is to estimate μ. A well-known estimator is the sample mean, $\hat{\mu} = \bar{X}$. However, because $\text{var}(\bar{X}) = \sigma^2/n$, in order to evaluate the precision of $\hat{\mu}$, one needs knowledge about σ^2. Show that an EMM estimator of σ^2 is given by $\hat{\sigma}^2 = n^{-1} \sum_{i=1}^n (X_i - \bar{X})^2$, which is the same as the ML estimator when the data are normal.

2.5. Consider a linear regression model

$$y_i = x_i'\beta + \epsilon_i, \qquad i = 1, \ldots, n ,$$

where $x_i = (x_{i1}, \ldots, x_{ip})'$ is a vector of known covariates; β is a vector of unknown regression coefficients that are of main interest; and $\epsilon_1, \ldots, \epsilon_n$ are i.i.d. errors with mean 0 and variance σ^2. The model can be expressed as $y = X\beta + \epsilon$, where the ith row of X is x_i'. Assume that $\text{rank}(X) = p$. Then, the least squares (LS) estimator of β is given by

$$\hat{\beta} = (X'X)^{-1}X'y .$$

Although β is of main interest, because $\text{Var}(\hat{\beta}) = \sigma^2(X'X)^{-1}$, to find the standard errors of the estimators one needs knowledge about σ^2. Show that an EMM estimator of σ^2 is $\hat{\sigma}^2 = n^{-1} \sum_{i=1}^n (y_i - x_i'\hat{\beta})^2$, which, again, is the ML estimator when normality is assumed.

2.6. Show that the estimating function $M(\beta, \sigma^2, \kappa, y)$ defined above (2.16) is unbiased in the sense that $\text{E}\{M(\beta, \sigma^2, \kappa, y)\} = 0$ when β, σ^2, κ correspond to the true parameters.

2.7. Show that the EMM estimators derived in closed form in Example 2.2 (continued) below Lemma 2.1 are consistent, provided that $m \to \infty$ and $k \geq 2$. You may assume that $\hat{\sigma}_0^2$ and $\hat{\sigma}_1^2$ are the REML estimators and that they are consistent.

2.8. Show that, in the balanced one-way random effects model with the Hartley–Rao form of variance components, the POQUIM estimator of the asymptotic variance of the REML estimator of γ, that is, the diagonal element of the POQUIM estimator of the asymptotic covariance matrix of the REML estimator corresponding to $\hat{\gamma}$, is given by $\hat{\Sigma}_{\mathrm{R},11}$ in Example 2.2 (Continued) in Section 2.1.2.2.

2.9. This and the next three exercises concern Example 2.2 (Continued) in Section 2.1.2.4. Verify the expression for the Gaussian log-likelihood, $l(\psi, y)$, given there. Show that $\mathrm{E(MSA)} = 1+k\gamma$, therefore, under the null hypothesis, the probability approaches one as $m \to \infty$, so that the estimator $\hat{\phi}_2$ is well defined.

2.10. Continuing with the previous exercise, verify that the W-test statistic for H_0: $\lambda = 1$ and $\gamma > 1$ is given by

$$\hat{\chi}_w^2 = \left(\frac{2k}{k-1} + \hat{\kappa}_0\right)^{-1} mk(\mathrm{MSE}-1)^2,$$

where $\hat{\kappa}_0$ may be chosen as the EMM estimator of κ_0 given in Example 2.2 (Continued) below Lemma 2.1. Also show that $2k/(k-1)+\kappa_0 > 0$ unless ϵ_{11}^2 is a constant with probability one.

2.11. Continuing with the previous exercise, show that the S-test statistic is identical to the *W*-test statistic in this case.

2.12. Continuing with the previous exercise, show that the L-test statistic is equal to

$$-2\log R = m(k-1)\{\mathrm{MSE} - 1 - \log(\mathrm{MSE})\}$$

in this case. Furthermore, show that the asymptotic null distribution of the test statistic is $\lambda_1\chi_1^2$, where $\lambda_1 = 1 + \{(k-1)/2k\}\kappa_0$, which is estimated by $1 + \{(k-1)/2k\}\hat{\kappa}_0$. Note that the asymptotic null distribution is χ_1^2 if the errors are normal but regardless of the normality of the random effects. (Hint: Use Theorem 2.5.)

2.13. Consider the balanced one-way random effects model of Example 2.2. Consider the Hartley–Rao form of variance components $\lambda = \tau^2$ and $\gamma = \sigma^2/\tau^2$. Suppose that one is interested in constructing an exact confidence interval for γ. Consider the following quantity

$$F = \frac{\mathrm{MSA}}{(1+k\gamma)\mathrm{MSE}},$$

where $\mathrm{MSA} = \mathrm{SSA}/(m-1)$ and $\mathrm{MSE} = \mathrm{SSE}/m(k-1)$. Show that, under normality, F has an F-distribution with $m-1$ and $m(k-1)$ degrees of freedom. Furthermore, show that, given ρ $(0 < \rho < 1)$, an exact $(1-\rho)\%$ confidence interval for γ is

$$\left[\frac{1}{k}\left(\frac{\mathrm{R}}{F_\mathrm{U}} - 1\right), \frac{1}{k}\left(\frac{\mathrm{R}}{F_\mathrm{L}} - 1\right)\right],$$

where $\mathrm{R} = \mathrm{MSA}/\mathrm{MSE}$, $F_\mathrm{L} = F_{m-1,m(k-1),1-\rho/2}$ and $F_\mathrm{U} = F_{m-1,m(k-1),\rho/2}$.

2.14. Consider the one-way random effects model of Example 2.3. Let c_{ij}, $1 \leq j \leq k_i$ be constants such that $\sum_{j=1}^{k_i} c_{ij} = 0$ and $\sum_{j=1}^{k_i} c_{ij}^2 = 1 - 1/k_i$. Define $u_i = \bar{y}_{i\cdot} + \sum_{j=1}^{k_i} c_{ij} y_{ij}$, $1 \leq i \leq m$. Prove the following.

a. The random variables $u_1, \ldots, u_m$ are independent and normally distributed with mean μ and variance $\sigma^2 + \tau^2$.

b. The quantity $\chi^2 = \sum_{i=1}^{m} (u_i - \bar{u})^2 / (\sigma^2 + \tau^2)$ is distributed as χ_{m-1}^2.

2.15. In Exercise 2.14, find an exact confidence interval for τ^2, the variance of the error ϵ_{ij}.

*2.16**. In the balanced one-way random effects model of Example 2.2, it is known that a UMVU estimator of $\zeta = c\lambda_1 + \lambda_2$ is $\hat{\zeta} = cS_1^2 + S_2^2$, where S_1^2 and S_2^2 are MSA and MSE, respectively, defined in Example 1.1 (continued) in Section 1.5.1.1.

a. Show that S_j^2 is a consistent estimator of λ_j, $j = 1, 2$.

b. Show that $(\hat{\zeta} - \zeta)/\sqrt{\text{var}(\hat{\zeta})}$ converges in distribution to the standard normal distribution.

2.17. Show that, in Example 2.8, the BLUE is given by (2.25) and (2.26) and its covariance matrix is given by (2.27). How do these formulae compare with the corresponding expressions under a linear regression model, that is, those for the least squares estimators? and when do the former reduce to the latter?

2.18. Show that, in Section 2.3.1.2, the logarithm of the joint pdf of α and y can be expressed as (2.36). Furthermore, derive Henderson's mixed model equations (2.37).

2.19. For the following linear mixed models determine the order of d_* above (2.39).

a. One-way random effects model (Example 1.1)

b. Two-way random effects model (Example 1.2)

c. Example 2.8, which is a special case of the nested error regression model

2.20. In Example 2.3 (continued) in Section 2.4.1.1, let the true parameters be $\mu = -0.5$, $\sigma^2 = 2.0$, and $\tau^2 = 1.0$. Also, let $m = 100$ and $k_i = 5$, $1 \leq i \leq m$. In the following, the errors are always generated from a normal distribution.

a. Generate the random effects from a normal distribution. Make a Q–Q plot to assess normality of the random effects, using REML estimators of the parameters.

b. Generate the random effects from a double-exponential distribution (with the same variance). Make a Q–Q plot to assess normality of the random effects, using REML estimators of the parameters.

c. Generate the random effects from a centralized-exponential distribution (with the same variance). Here a centralized-exponential distribution is the distribution of $\xi - \text{E}(\xi)$, where ξ has an exponential distribution. Make a Q–Q plot to assess normality of the random effects, using REML estimators of the parameters.

d. Compare the plots in a, b, and c. What do you conclude?

2.21. Show that, in Example 2.15, $\rho_n \sim k$ and $\nu_n \sim mk$ as $m \to \infty$ (k may or may not go to ∞). Also show that, in Example 2.15 (continued) below (2.37), $\eta_n \sim mk$.

2.22. Show that, in Section 2.5.1, under normal hierarchy and when $b = \beta$ and $B \to 0$, the likelihood (2.55) reduces to the normal likelihood of Section 1.3.1 when the prior for β is a point mass at β.

2.23. Show that, in Section 2.5.1, under normal hierarchy the likelihood (2.55) reduces to the normal restricted likelihood of Section 1.3.21 when the prior for β is noninformative.

2.24. Consider Example 2.18. Let the priors be such that $\sigma^2 \propto 1/\sigma^2$, $\tau^2 \propto 1/\tau^2$, and σ^2, τ^2 independent. Derive the likelihood (2.55) and posterior (2.56). Is the posterior proper (even though the priors are improper)?

2.25. Show that, under normal hierarchy, the posterior of β is multivariate normal with $\mathrm{E}(\beta|y) = (X'V^{-1}X + B^{-1})^{-1}(X'V^{-1}y + B^{-1}b)$ and $\mathrm{Var}(\beta|y) = (X'V^{-1}X + B^{-1})^{-1}$. Similarly, the posterior of α is multivariate normal with $\mathrm{E}(\alpha|y) = (Z'LZ + G^{-1})^{-1}Z'L(y - Xb)$ and $\mathrm{Var}(\alpha|y) = (Z'LZ + G^{-1})^{-1}$, where $L = R^{-1} - R^{-1}X(B^{-1} + X'R^{-1}X)^{-1}X'R^{-1}$.

2.26. Show that, under normal hierarchy and when $B^{-1} \to 0$, which corresponds to the case where the prior for β is noninformative, one has $\mathrm{E}(\beta|y) \to (X'V^{-1}X)^{-1}X'V^{-1}y = \tilde{\beta}$, which is the BLUE; similarly, $\mathrm{E}(\alpha|y) \to GZ'V^{-1}(y - X\tilde{\beta})$.

3

Generalized Linear Mixed Models: Part I

3.1 Introduction

For the most part, linear mixed models have been used in situations where the observations are continuous. However, there are cases in practice where the observations are discrete, or categorical. For example, the number of heart attacks of a potential patient during the past year takes the values 0, 1, 2, ..., and therefore is a discrete random variable. McCullagh and Nelder (1989) proposed an extension of linear models, called generalized linear models, or GLM. They noted that the key elements of a classical linear model, that is, a linear regression model, are (i) the observations are independent, (ii) the mean of the observation is a linear function of some covariates, and (iii) the variance of the observation is a constant. The extension to GLM consists of modification of (ii) and (iii) above; by (ii)$'$ the mean of the observation is associated with a linear function of some covariates through a link function; and (iii)$'$ the variance of the observation is a function of the mean. Note that (iii)$'$ is a result of (ii)$'$. See McCullagh and Nelder (1989) for details. Unlike linear models, GLMs include a variety of models that includes normal, binomial, Poisson, and multinomial as special cases. Therefore, these models are applicable to cases where the observations may not be continuous. The following is an example.

Example 3.1. On January 27, 1986, hours before the launch of the space shuttle *Challenger*, a three-hour teleconference was under way. The discussions had focused on a single topic, that is, whether the scheduled launch next morning should be called off, because of the unusually cold temperature forecast for launch time, 31 degrees fahrenheit, or if it should go ahead as scheduled. After numerous conversations, and examining data from the previous launches, a decision was made that gave the green light for the launch.

The Rogers Commission, which was formed after the Challenger disaster, concluded that the accident was caused by the failure of an unknown number of O-rings, which resulted in a combustion gas leak through a joint in one of the booster rockets. There was a total of six primary O-rings and six secondary O-rings that were supposed to seal the field joints of the rockets. Failures of O-rings had been reported in the previous launches and, during the three-hour teleconference, a plot was presented that showed a possible association between the number of O-ring failures and the temperature at launch time. However, an important piece of information was missing. The missing data should have indicated the many previous launches, in which there were no O-ring failures, at all (see figures on page 946 of Dalal et al. 1989).

Dalal et al. (1989) proposed a logistic regression model, which is a special case of GLM, to analyze the risk associated with O-ring failures in the space shuttle. Their studies focused on the primary O-ring failures, because data of previous launches had suggested that they were the majority of O-ring failures. In fact, there was only one incident of secondary damage among the 23 previous launches. In other words, a substantial amount of primary O-ring failures would have doomed the space shuttle regardless of the secondaries. It was assumed that, given the temperature t and pressure s, the number of thermally distressed primary O-rings X is a binomial random variable with $n = 6$ and $p = p(t, s)$. Here n denotes the total number of independent trials and p the probability of success in a single trial. Furthermore, it was assumed that the probability $p(t, s)$ is associated with the temperature and pressure through a logistic link function:

$$\mathrm{logit}\{p(t, s)\} = \alpha + \beta t + \gamma s,$$

where $\mathrm{logit}(p) = \log\{p/(1-p)\}$. Using this model, Dalal et al. calculated the estimated probability of at least one complete joint failure at the temperature of 31°F and pressure of 200 psi as 13%, which is 600% higher than that at the temperature of 60°F and same pressure.

One element that GLMs have in common with linear models is that the observations are assumed to be independent. In many cases, however, the observations, or responses, are correlated, as well as discrete or categorical. For example, if $y_{i1}, \dots, y_{i10}$ indicate whether the ith individual (person) visited a doctor during each of the past ten years, that is, $y_{ij} = 1$ if the ith individual visited a doctor within the jth year in the past, and $y_{ij} = 0$ otherwise, then the responses from the same individual are likely to be correlated. On the other hand, the responses are binary and hence not continuous. As mentioned earlier, the linear mixed models discussed in the previous chapters do not apply to such cases. It is clear now that what one needs is an extension of the linear mixed model to cases where the responses are both correlated and, at the same time, discrete or categorical. We consider such an extension next.

3.2 Generalized Linear Mixed Models

To motivate the extension, let us first consider an alternative expression of the Gaussian linear mixed model introduced in Chapter 1. Suppose that, given a vector of random effects α, the observations $y_1, \ldots, y_n$ are (conditionally) independent such that $y_i \sim N(x_i'\beta + z_i'\alpha, \tau^2)$, where x_i and z_i are known vectors, β is an unknown vector of regression coefficients, and τ^2 is an unknown variance. Furthermore, suppose that α is multivariate normal with mean 0 and covariance matrix G, which depends on a vector θ of unknown variance components. Let X and Z be the matrices whose ith rows are x_i' and z_i', respectively. It is easy to see (Exercise 3.1) that this leads to the linear mixed model (1.1) with normality and $R = \tau^2 I$.

The two key elements in the above that define a Gaussian linear mixed model are (i) conditional independence (given the random effects) and a conditional distribution and (ii) the distribution of the random effects. These two elements may be used to define a generalized linear mixed model, or GLMM. Suppose that, given a vector of random effects α, the responses $y_1, \ldots, y_n$ are (conditionally) independent such that the conditional distribution of y_i given α is a member of the exponential family with pdf

$$f_i(y_i|\alpha) = \exp\left\{ \frac{y_i\xi_i - b(\xi_i)}{a_i(\phi)} + c_i(y_i, \phi) \right\}, \tag{3.1}$$

where $b(\cdot)$, $a_i(\cdot)$, $c_i(\cdot, \cdot)$ are known functions, and ϕ is a dispersion parameter which may or may not be known. The quantity ξ_i is associated with the conditional mean $\mu_i = \mathrm{E}(y_i|\alpha)$, which, in turn, is associated with a linear predictor

$$\eta_i = x_i'\beta + z_i'\alpha, \tag{3.2}$$

where x_i and z_i are known vectors and β a vector of unknown parameters (the fixed effects), through a known link function $g(\cdot)$ such that

$$g(\mu_i) = \eta_i. \tag{3.3}$$

Furthermore, it is assumed that $\alpha \sim N(0, G)$, where the covariance matrix G may depend on a vector θ of unknown variance components.

Note that, according to the properties of the exponential family, one has $b'(\xi_i) = \mu_i$. In particular, under the so-called canonical link, one has

$$\xi_i = \eta_i;$$

that is, $g = h^{-1}$, where $h(\cdot) = b'(\cdot)$. Here h^{-1} represents the inverse function (not reciprocal) of h. For example, a table of canonical links is given in McCullagh and Nelder (1989, pp. 32). We now consider some special cases.

Example 3.2 (Normal linear mixed model). As mentioned, the normal linear mixed model (1.1), in which $R = \tau^2 I$, is a special case of the GLMM, in which the (conditional) exponential family is normal with mean μ_i and variance τ^2, and the link function is $g(\mu) = \mu$. Note that, in this case, the dispersion parameter $\phi = \tau^2$, which is unknown.

Example 3.3 (Mixed logistic model). Suppose that, given the random effects α, binary responses $y_1, \dots, y_n$ are conditionally independent *Bernoulli*. Furthermore, with $p_i = \mathrm{P}(y_i = 1|\alpha)$, one has

$$\mathrm{logit}(p_i) = x_i'\beta + z_i'\alpha,$$

where x_i and z_i are as in the definition of GLMM. This is a special case of the GLMM, in which the (conditional) exponential family is Bernoulli, and the link function is $g(\mu) = \mathrm{logit}(\mu)$. Note that in this case the dispersion parameter $\phi = 1$.

Example 3.4 (Poisson log-linear mixed model). The Poisson distribution is often used to model responses that are counts. Supposed that, given the random effects α, the counts $y_1, \dots, y_n$ are conditionally independent such that $y_i|\alpha \sim \mathrm{Poisson}(\lambda_i)$, where

$$\log(\lambda_i) = x_i'\beta + z_i'\alpha,$$

and x_i, z_i are as in the definition of GLMM. Again, this is a special case of GLMM, in which the (conditional) exponential family is Poisson and the link function is $g(\mu) = \log(\mu)$. The dispersion parameter ϕ in this case is again equal to 1.

Note that in all three examples above the link function is canonical. However, noncanonical links are, indeed, used in practice. For example, in Example 3.3, another link function that is often used is Probit; that is, $g(\mu) = \Phi^{-1}(\mu)$, where Φ is the cdf of standard normal distribution.

Unlike GLM, the responses under a GLMM are (marginally) correlated. For such reason, GLMM is often used to model correlated discrete or categorical responses. In the following we give some examples of applications.

3.3 Real-Life Data Examples

3.3.1 The Salamander Mating Experiments

A well-known example that was also one of the first published in the context of GLMM was in McCullagh and Nelder's book, *Generalized Linear Models* (1989, §14.5). The example involved data from mating experiments regarding two populations of salamanders, Rough Butt and Whiteside. These populations, which are geographically isolated from each other, are found in the southern Appalachian mountains of the eastern United States. The question whether the geographic isolation had created barriers to the animals' interbreeding was thus of great interest to biologists studying speciation.

Three experiments were conducted during 1986, one in the summer and two in the autumn. In each experiment there were 10 males and 10 females from each population. They were paired according to the design given by

Table 14.3 in McCullagh and Nelder (1989). The same 40 salamanders were used for the summer and first autumn experiments. A new set of 40 animals was used in the second autumn experiment. For each pair, it was recorded whether a mating occurred, 1, or not, 0.

The responses are binary and clearly correlated, so that neither linear mixed models nor GLM would apply. McCullagh and Nelder (1989) proposed the following mixed logistic model with crossed random effects. For each experiment, let u_i and v_j be the random effects corresponding to the ith female and jth male involved in the experiment. Then, on the logistic scale, the probability of successful mating is modeled in term of fixed effects$+u_i + v_j$. It was further assumed that the random effects are independent and normally distributed with means 0 and variances σ^2 for the females and τ^2 for the males. Under these assumptions, a GLMM may be formulated as follows. Note that there are 40 different animals of each sex. Suppose that, given the random effects $u_1, \dots, u_{40}$ for the females, and $v_1, \dots, v_{40}$ for the males, the binary responses y_{ijk} are conditionally independent such that $\text{logit}\{\text{P}(y_{ijk} = 1|u, v)\} = x'_{ij}\beta + u_i + v_j$. Here y_{ijk} represents the kth binary response corresponding to the same pair of ith female and jth male, x_{ij} is a vector of fixed covariates, and β is an unknown vector of regression coefficients. More specifically, x_{ij} consists of an intercept; an indicator of Whiteside female WS_f, an indicator of Whiteside male WS_m, and the product $\text{WS}_f \cdot \text{WS}_m$, representing an interaction.

It should be pointed out that there is a simplification of the potential correlations among the responses. More specifically, the binary responses y_{ijk} may not be conditionally independent given the random effects. To see this, note that the same group of animals was involved in two of the three experiments (summer and first autumn experiments). It is unclear whether serial correlations exist between the two experiments. Note that conditional independence is an essential part of the definition of GLMM given in the previous section. Alternatively, one could pool the responses from the two experiments involving the same group of animals, as suggested by McCullagh and Nelder (1989, §4.1), so let $y_{ij\cdot} = y_{ij1} + y_{ij2}$, where y_{ij1} and y_{ij2} represent the responses from the summer and first autumn experiments, respectively. This may avoid the issue of conditional independence, however, a new problem emerges. The problem is that, given the female and male (random) effects, the conditional distribution of $y_{ij\cdot}$ is not an exponential family. Note that $y_{ij\cdot}$ is not necessarily binomial given the random effects, because of the potential serial correlation. Although the (conditional) exponential family assumption is another important part of the GLMM defined in Section 3.2, it may be replaced by weaker assumptions. See Section 4.2.4 for details. Such an extension of GLMM is similar to the extension of Gaussian linear mixed models to non-Gaussian ones, which we have extensively discussed in the previous chapters.

This example is further discussed several times in the sequel.

3.3.2 A Log-Linear Mixed Model for Seizure Counts

As mentioned earlier, Poisson distribution is often used to model responses that are counts. However, in many cases there is overdispersion (or underdispersion), so that the variance of the response does not follow that of a Poisson distribution. Thall and Vail (1990) provided an example of such cases. In Table 2 of their article, the authors presented data from a clinical trial involving 59 epileptics. These patients were randomized to a new drug (treatment) or a placebo (control). The number of epileptic seizures was recorded for each patient during an eight-week period, namely, one seizure count during the two-week period before each of four clinic visits. Baseline seizures and the patient's age were available and treated as covariates. An interesting feature of this dataset is that there is apparent overdispersion, heteroscedasticity, and within-patient correlation, as demonstrated by Table 3 of Thall and Vail (1990). Another feature of the data is that they are longitudinal; that is, the responses were collected over time.

Breslow and Clayton (1993) reanalyzed the data by proposing a Poisson log-linear mixed model. They assumed that the seizure count y_{ij} for the ith patient on the jth visit ($i = 1, \ldots, 59$, $j = 1, \ldots, 4$) was associated with an individual-specific bivariate random effect $(\alpha_{1i}, \alpha_{2i})$, and that, given the random effect, y_{ij} was conditionally Poisson distributed with mean μ_{ij}. Furthermore, the conditional mean μ_{ij} satisfies

$$\log(\mu_{ij}) = x_{ij}'\beta + \alpha_{1i} + \alpha_{2i}(\text{Visit}_k/10) + \epsilon_{ij},$$

where x_{ij} is a vector of covariates including indicators of the treatment, visit, the logarithm of 1/4 times the number of baseline seizures (Base), the logarithm of age (Age) and some interactions; Visit_k is the visit code that equals -3, -1, 1, and 3, respectively, for $k = 1, \ldots, 4$, and ϵ_{ij} is a random error that represents overdispersion in addition to that introduced by the individual-specific random effects. It was assumed that the random effects α_{1i} and α_{2i} are bivariate normal with zero means, unknown variances, and correlation, and that the error ϵ_{ij} is also normal with zero mean and unknown variance.

Breslow and Clayton (1993) fitted the model, which is a special case of GLMM introduced in the previous section, using a method of approximate inference. This method is discussed later.

3.3.3 Small Area Estimation of Mammography Rates

One area of application of GLMM is small area estimation. In surveys, direct estimates for small geographic areas or subpopulations are likely to yield inaccurate results, because of the small sample sizes for such areas or subpopulations. Therefore, it is necessary to "borrow strength" from related areas or other sources to find more accurate estimates for a given small area or, simultaneously, for several small areas. One method of borrowing strength is via statistical modeling. For continuous responses, such an idea has led to a linear mixed model approach, in which there is a random effect corresponding to each small area. See Ghosh and Rao (1994) and Rao (2003) for methods of small area estimation and, in particular, the use of linear mixed models in this field. Recently, several authors have used GLMM for small area estimation in cases of binary responses. See, for example, Malec el al. (1997), Ghosh et al. (1998), Jiang and Lahiri (2001). The following is a similar example given by Jiang et al. (2001).

The Behavioral Risk Factor Surveillance System (BRFSS) is a Center for Disease Control and Prevention coordinated, state-based random-digit-dialing telephone survey. One data set of particular interest involved the use of mammography among women aged 40 or older, from 1993 to 1995, and for areas from three federal regional offices: Boston (including Maine, Vermont, Massachusetts, Connecticut, Rhode Island, and New Hampshire), New York (including New York and New Jersey) and Philadelphia (including Pennsylvania, Delaware, Washington DC, Maryland, Virginia, and West Virginia). Overall, there are 118 health service areas (HSAs) in the region. Initial analysis of the data suggested that mammography rates gradually increase from age groups 40–44 to 50–54, and then decrease. To catch this curvature phenomena, Jiang et al. (2001) proposed a mixed logistic model for the proportion of women having had mammography. Under this model, there is a random effect corresponding to each HSA and, given a HSA, the proportion of women having had mammography, p, satisfies

$$\text{logit}(p) = \beta_0 + \beta_1 * \text{age} + \beta_2 * \text{age}^2 + \beta_3 * \text{Race} + \beta_4 * \text{Edu} + \text{HSA effect},$$

where Age is grouped as 40–44, 45–49, ..., 75–79, and 80 and over; Race as white and others; and Edu as the percentage of people in the HSA aged 25 or older with at least a high school education.

In Jiang et al. (2001), the authors did not assume that the random effects are normally distributed. In fact, a Q–Q plot had suggested otherwise. Nevertheless, the method that the authors used for inference about the model which, again, is a special case of LMM, did not require such an assumption. The method is discussed in the sequel.

3.4 Likelihood Function under GLMM

The preceding sections demonstrated the usefulness of GLMM in statistical applications. The rest of this chapter, and also the next chapter, are devoted to inference about these models. Unlike linear mixed models, the likelihood function under a GLMM typically does not have a closed-form expression (with, of course, the exception of the normal case). In fact, such a likelihood may involve high-dimensional integrals that cannot be evaluated analytically. To understand the computational difficulties, we consider a simple example.

Example 3.5. Suppose that, given the random effects $u_1, \dots, u_{m_1}$ and $v_1, \dots, v_{m_2}$, binary responses y_{ij}, $i = 1, \dots, m_1$, $j = 1, \dots, m_2$ are conditionally independent such that, with $p_{ij} = \mathrm{P}(y_{ij} = 1|u, v)$,

$$\mathrm{logit}(p_{ij}) = \mu + u_i + v_j,$$

where μ is an unknown parameter, $u = (u_i)_{1 \leq i \leq m_1}$, and $v = (v_j)_{1 \leq j \leq m_2}$. Furthermore, the random effects $u_1, \dots, u_{m_1}$ and $v_1, \dots, v_{m_2}$ are independent such that $u_i \sim N(0, \sigma_1^2)$, $v_j \sim N(0, \sigma_2^2)$, where the variances σ_1^2 and σ_2^2 are unknown. Thus, the unknown parameters involved in this model are $\psi = (\mu, \sigma_1^2, \sigma_2^2)'$. It can be shown (Exercise 3.3) that the likelihood function under this model for estimating ψ can be expressed as

$$\begin{aligned}
& c - \frac{m_1}{2}\log(\sigma_1^2) - \frac{m_2}{2}\log(\sigma_2^2) + \mu y_{\cdot\cdot} \\
& + \log \int \cdots \int \left[\prod_{i=1}^{m_1}\prod_{j=1}^{m_2}\{1 + \exp(\mu + u_i + v_j)\}^{-1}\right] \\
& \times \exp\left(\sum_{i=1}^{m_1} u_i y_{i\cdot} + \sum_{j=1}^{m_2} v_j y_{\cdot j} - \frac{1}{2\sigma_1^2}\sum_{i=1}^{m_1} u_i^2 - \frac{1}{2\sigma_2^2}\sum_{j=1}^{m_2} v_j^2\right) \\
& du_1 \cdots du_{m_1} dv_1 \cdots dv_{m_2},
\end{aligned} \tag{3.4}$$

where c is a constant, $y_{\cdot\cdot} = \sum_{i=1}^{m_1}\sum_{j=1}^{m_2} y_{ij}$, $y_{i\cdot} = \sum_{j=1}^{m_2} y_{ij}$, and $y_{\cdot j} = \sum_{i=1}^{m_1} y_{ij}$. The multidimensional integral involved in (3.4) has no closed-form expression, and it cannot be further simplified. Furthermore, such an integral is difficult to evaluate even numerically. For example, if $m = n = 40$, the dimension of the integral will be 80. To make it even worse, the integrand involves a product of 1600 terms with each term less than one. This makes it almost impossible to evaluate the integral using a naive Monte Carlo method. To see this, suppose that $u_1, \dots, u_{40}$ and $v_1, \dots, v_{40}$ are simulated random effects (from the normal distributions). Then, the product in the integrand (with $m = n = 40$) is numerically zero. Therefore, numerically, the law of large numbers, which is the basis of the (naive) Monte Carlo method, will not yield anything but zero without a huge Monte Carlo sample size.

The example shows that, although maximum likelihood and restricted maximum likelihood methods have become standard procedures in linear mixed models, likelihood-based inference in GLMM is still computationally challenging. For such a reason, there have been several approaches to inference about GLMM, trying either to solve, or to avoid, the computational difficulties. The approach aiming at developing computational methods for the maximum likelihood estimation was represented by Monte Carlo EM algorithm (McCulloch 1997; Booth and Hobert 1999) and a method of estimation by parts (Song et al. 2003). The nonlikelihood-based computationally attractive methods may be classified as approximate inference (Breslow and Clayton

1993; Lin and Breslow 1996; Lee and Nelder 1996; among others) and generalized estimating equation (GEE; e.g., Diggle et al. 1996, §7.5; Jiang 1998a; Jiang and Zhang 2001). The Bayesian method based on the Gibbs sampler is similar to the Monte-Carlo EM method (e.g., Zeger and Karim 1991). In this chapter, we discuss the approximate inference and prediction of random effects. Monte Carlo EM, Bayesian, and GEE methods as well as other related topics are left to the next chapter.

3.5 Approximate Inference

3.5.1 Laplace Approximation

When the exact likelihood function is difficult to compute, approximation becomes one of the natural alternatives. A well-known method of approximate integrals is named after Laplace. Suppose that one wishes to approximate an integral of the form

$$\int \exp\{-q(x)\}dx, \tag{3.5}$$

where $q(\cdot)$ is a "well-behaved" function in the sense that it achieves its minimum value at $x = \tilde{x}$ with $q'(\tilde{x}) = 0$ and $q''(\tilde{x}) > 0$. Then, we have, by Taylor expansion,

$$q(x) = q(\tilde{x}) + \frac{1}{2}q''(\tilde{x})(x - \tilde{x})^2 + \cdots,$$

which yields an approximation to (3.5) (Exercise 3.4),

$$\int \exp\{-q(x)\}dx \approx \sqrt{\frac{2\pi}{q''(\tilde{x})}} \exp\{-q(\tilde{x})\}. \tag{3.6}$$

There is a multivariate extension of (3.6), which is more useful in our case. Let $q(\alpha)$ be a well-behaved function that attains its minimum value at $\alpha = \tilde{\alpha}$ with $q'(\tilde{\alpha}) = 0$ and $q''(\tilde{\alpha}) > 0$, where q' and q'' denote the gradient (i.e., the vector of first derivatives) and Hessian (i.e., the matrix of second derivatives) of q, respectively, and the notation $A > 0$ means that the matrix A is positive definite. Then, we have

$$\int \exp\{-q(\alpha)\}d\alpha \approx c|q''(\tilde{\alpha})|^{-1/2} \exp\{-q(\tilde{\alpha})\}, \tag{3.7}$$

where c is a constant depending only on the dimension of the integral (Exercise 3.4), and $|A|$ denotes the determinant of matrix A.

3.5.2 Penalized Quasi-Likelihood Estimation

With Laplace approximation, one may proceed as in maximum likelihood, treating the approximated likelihood function as the true likelihood function. The method may be illustrated under a more general framework as an approximate qausi-likelihood estimation approach. Suppose that the conditional mean of the response y_i $(1 \leq i \leq n)$, given the random effects $\alpha = (\alpha_1, \ldots, \alpha_m)'$, satisfies

$$\mathrm{E}(y_i|\alpha) = h(x_i'\beta + z_i'\alpha), \tag{3.8}$$

where β is a vector of unknown parameters (the fixed effects), x_i, z_i are known vectors, and $h(\cdot)$ is the inverse function of a known link function $g(\cdot)$. Furthermore, write $\mu_i = \mathrm{E}(y_i|\alpha)$ and $\eta_i = g(\mu_i) = x_i'\beta + z_i'\alpha$. It is assumed that the conditional variance satisfies

$$\mathrm{var}(y_i|\alpha) = a_i(\phi)v(\mu_i), \tag{3.9}$$

where ϕ is an additional dispersion parameter, $a_i(\cdot)$ is a known function that is often equal to ϕ/w_i with w_i being a known weight, and $v(\cdot)$ is a known variance function. Note that the assumptions made so far are weaker than that in the definition of GLMM (see section 3.2), in which it is assumed that the conditional distribution of y_i given α is a member of the exponential family; that is, (3.1) holds.

1. Derivation of PQL. Under the additional assumption that $y_1, \ldots, y_n$ are conditionally independent given α, and that α has a multivariate normal distribution with mean 0 and covariance matrix G, that is, $\alpha \sim N(0, G)$, where G is specified up to a vector θ of dispersion parameters, a quasi-likelihood function based on $y = (y_1, \ldots, y_n)'$ may be expressed as

$$L_{\mathrm{Q}} \propto |G|^{-1/2} \int \exp\left\{ -\frac{1}{2}\sum_{i=1}^{n} d_i - \frac{1}{2}\alpha' G^{-1}\alpha \right\} d\alpha, \tag{3.10}$$

where the subscript Q indicates quasi-likelihood, and

$$d_i = -2\int_{y_i}^{\mu_i} \frac{y_i - u}{a_i(\phi)v(u)} du$$

is known as the (quasi-) deviance. The term is drawn from generalized linear models, because under the assumption that the conditional distribution of y_i given α is a member of the exponential family with conditional pdf (3.1), in which $a_i(\phi) = \phi/w_i$, d_i is equal to the scaled difference $2\phi\{l(y_i; y_i) - l(y_i; \mu_i)\}$, where $l(y; \mu)$ denotes the conditional likelihood of the observation y given its mean μ (e.g., McCullagh and Nelder 1989, §2.3).

Now, using the Laplace approximation (3.7), the logarithm of L_{Q}, denoted by l_{Q}, may be expressed as

$$l_{\rm Q} \approx c - \frac{1}{2}\log|G| - \frac{1}{2}\log|q''(\tilde{\alpha})| - q(\tilde{\alpha}), \tag{3.11}$$

where c does not depend on the parameters,

$$q(\alpha) = \frac{1}{2}\left(\sum_{i=1}^{n} d_i + \alpha' G^{-1}\alpha\right),$$

and $\tilde{\alpha}$ minimizes $q(\alpha)$. Typically, $\tilde{\alpha}$ is the solution to the equation $q'(\alpha) = 0$, that is,

$$G^{-1}\alpha - \sum_{i=1}^{n} \frac{y_i - \mu_i}{a_i(\phi)v(\mu_i)g'(\mu_i)} z_i = 0, \tag{3.12}$$

where $\mu_i = x_i'\beta + z_i'\alpha$. It can be shown that

$$q''(\alpha) = G^{-1} + \sum_{i=1}^{n} \frac{z_i z_i'}{a_i(\phi)v(\mu_i)\{g'(\mu_i)\}^2} + r, \tag{3.13}$$

where the remainder term r has expectation 0 (Exercise 3.5). If we denote the term in the denominator of (3.13) by w_i^{-1}, and ignore the term r, assuming that it is in probability of lower order than the leading terms, then we have a further approximation

$$q''(\alpha) \approx Z'WZ + G^{-1}, \tag{3.14}$$

where Z is the matrix whose ith row is z_i', and $W = \text{diag}(w_1, \dots, w_n)$. Note that the quantity w_i is known as the GLM iterated weights (e.g., McCullagh and Nelder 1989, §2.5). By combining the approximations (3.11) and (3.13), one obtains

$$l_{\rm Q} \approx c - \frac{1}{2}\left(\log|I + Z'WZG| + \sum_{i=1}^{n} \tilde{d}_i + \tilde{\alpha}' G^{-1}\tilde{\alpha}\right), \tag{3.15}$$

where $\tilde{d}_i$ is d_i with α replaced by $\tilde{\alpha}$.

A further approximation may be obtained by assuming that the GLM iterated weights vary slowly as a function of the mean (Breslow and Clayton 1993, pp. 11). Then, because the first term inside the $(\cdots)$ in (3.15) depends on β only through W (the estimation of θ is considered later), one may ignore this term and thus approximate $l_{\rm Q}$ by

$$l_{\rm PQ} \approx c - \frac{1}{2}\left(\sum_{i=1}^{n} \tilde{d}_i + \tilde{\alpha}' G^{-1}\tilde{\alpha}\right). \tag{3.16}$$

Equation (3.16) is related to the penalized quasi-log-likelihood, or PQL (Green 1987), as the notation has indicated, by the following observation. Recall that $\tilde{\alpha}$ minimizes $q(\alpha)$ defined below (3.11). This means that, given β, $\tilde{\alpha}$ is the maximizer of l_{PQ}. Because this maximizer depends on β, we may write $\tilde{\alpha} = \tilde{\alpha}(\beta)$. For fixed θ, let $\hat{\beta}$ be the maximizer of l_{PQ} as a function of β. Then, it is easy to see that $\hat{\beta}$, $\hat{\alpha}$ jointly maximize Green's PQL (Green 1987),

$$l_{\mathrm{PQ}}(\beta, \alpha) = -\frac{1}{2}\left(\sum_{i=1}^{n} d_i + \alpha' G^{-1}\alpha\right) \tag{3.17}$$

as a function of β and α, where $\hat{\alpha} = \tilde{\alpha}(\hat{\beta})$. Note that $l_{\mathrm{PQ}}(\beta, \alpha)$ is the negative of $q(\alpha)$ defined below (3.11).

2. Computational procedures. The standard method of maximizing (3.17) involves solving a system of nonlinear equations, namely, $\partial l_{\mathrm{PQ}}/\partial\beta = 0$ and $\partial l_{\mathrm{PQ}}/\partial\alpha = 0$, or, equivalently,

$$\sum_{i=1}^{n} \frac{(y_i - \mu_i)x_i}{a_i(\phi)v(\mu_i)g'(\mu_i)} = 0, \tag{3.18}$$

$$\sum_{i=1}^{n} \frac{(y_i - \mu_i)z_i}{a_i(\phi)v(\mu_i)g'(\mu_i)} - G^{-1}\alpha = 0. \tag{3.19}$$

In practice, there are often a large number of random effects involved in a GLMM. For example, in the salamander mating experiments discussed in Section 3.3.1, the number of random effects associated with the female and male animals is 80. In the BRFSS survey considered in Section 3.3.3, the number of random effects corresponding to the small areas is 118. This means that the solution of (3.18) and (3.19) is in a high-dimensional space. In other words, one has to simultaneously solve a large number of nonlinear equations. It is well known that standard procedures of solving nonlinear systems such as Newton-Raphson may be inefficient and extremely slow when the dimension of the solution is high. In fact, even in the linear case, directly solving a large equation system, such as the BLUP equations discussed below, may involve inverting a large matrix, which may still be computationally burdensome. Due to such concerns, Jiang (2000b) developed a nonlinear Gauss–Seidel algorithm for solving (3.18) and (3.19). The author showed that the algorithm converges globally in virtually all typical situations of GLMM.

Alternatively, Breslow and Clayton (1993) proposed an iterative procedure for solving (3.18) and (3.19) by modifying the Fisher scoring algorithm developed by Green (1987). An attractive feature of the Breslow–Clayton procedure is that it exploits a close correspondence with the well-known mixed model equations (Henderson et al. 1959), which leads to the BLUP in linear mixed models (see Section 2.3.1.2). First define a working vector $\tilde{y} = (\tilde{y}_i)_{1\leq i\leq n}$, where $\tilde{y}_i = \eta_i + g'(\mu_i)(y_i - \mu_i)$, and where η_i and μ_i are evaluated at the current estimators of β and α. Then, the solution to (3.18) and (3.19) via Fisher scoring may be expressed as the iterative solution to the system

$$\begin{pmatrix} X'WX & X'WZ \\ Z'WX & G^{-1} + Z'WZ \end{pmatrix} \begin{pmatrix} \beta \\ \alpha \end{pmatrix} = \begin{pmatrix} X'W \\ Z'W \end{pmatrix} \tilde{y}. \tag{3.20}$$

It is seen that (3.20) is just (2.37) with R^{-1} replaced by W and y by $\tilde{y}$. Note that, because W depends on β and α, it has to be updated at each iteration. Equivalently, the solution to (3.20) may be expressed in the following way

$$\beta = (X'V^{-1}X)^{-1}X'V^{-1}\tilde{y}, \tag{3.21}$$
$$\alpha = GZ'V^{-1}(\tilde{y} - X\beta), \tag{3.22}$$

where $V = W^{-1} + ZGZ'$, assuming that inverse matrices exist. These suggest that one may first use (3.21) to update β, and then (3.22) to update α, and so on. Although (3.21) and (3.22) look simple, some potential computational difficulties may still exist. To see this, note that V has dimension n, which is the total number of observations. Thus, for a large dataset, the inversion of V may be computationally burdensome, unless V has a certain special structure, such as blockdiagonal.

3. Variance components. Typically, a GLMM involves a vector θ of dispersion parameters, or variance components. In practice, these variance components are unknown, and therefore have to be estimated, before any inference can be made. Note that in the above derivations θ has been held fixed. For example, the right sides of (3.21) and (3.22) depend on θ. Therefore, these are not estimators unless θ is known or estimated. Breslow and Clayton (1993) proposed that one substitutes the maximizer of (3.17), say, $\tilde{\beta}(\theta)$ and $\tilde{\alpha}(\theta)$, into (3.15), and thus obtains a profile quasi-log-likelihood function. Furthermore, the authors suggested further approximations that led to a similar form of REML in linear mixed models (see Section 1.3.2). See Breslow and Clayton (1993, pp. 11–12) for details.

4. Consistency of PQL estimators. It is clear that there are a number of approximations involved in deriving the PQL. Therefore, it may not be surprising to know that the approximations have brought bias to the resulting estimators. In fact, they did. The question is whether such a bias is, in some sense, ignorable. It is now known that the PQL estimators are inconsistent (e.g., Jiang 1998). In other words, the bias due to the approximations will not vanish, no matter how large the sample size. Recognizing the problem of bias, Lin and Breslow (1996) proposed a bias correction to PQL based on second-order Laplace approximation. The second-order approximation improves the first-order one. As a result, the bias in PQL is significantly reduced, as was demonstrated by Lin and Breslow. However, like the first-order method, the second-order approximation does not eliminate the bias asymptotically. In other words, the bias-corrected PQL estimator is still inconsistent. In fact, no matter to what order the Laplace approximation is carried, the bias-corrected PQL estimator will never be consistent. Of course, as the Laplace approximation is carried to even higher order, the bias may be reduced to such a level that is acceptable from a practical point of view. On the other hand, one advantage of PQL is that it is computationally easy to operate. As the Laplace approximation is carried to higher order, the computational difficulty increases. Note that, if the computation required for an approximate method is comparable to that for the exact (maximum likelihood) method, which is discussed later, the approximate method may not be worth pursuing, at least from a computational point of view.

On the other hand, there is a situation where PQL is expected to work well, that is, the situation where the variance components are small. This is because the Laplace approximation becomes accurate when the variance components are close to zero. To see this, note that the Laplace approximation is, for the most part, based on an expansion at the mode of the distribution of the random effects. If the variance component is close to zero, the distribution of the random effects, which is assumed normal, is concentrated near its mode (i.e., zero). In such a case, approximation to the integral by the expansion is accurate. In particular, the Laplace approximation gives the exact value of the integral, if the variance component is equal to zero (Exercise 3.6). One application of this simple fact is used in testing hypotheses of zero variance components, which we discuss next.

3.5.3 Tests of Zero Variance Components

There is considerable interest, in practice, in testing for overdispersion, heteroscedasticity, and correlation among responses. In some cases, the problem is equivalent to testing for zero variance components. Lin (1997) considered two classes of GLMMs. The first is the so-called longitudinal GLMM, in which the conditional mean vector μ_i of the responses in the ith cluster given the random effects satisfies

$$g(\mu_i) = X_i\beta + Z_i\alpha_i, \tag{3.23}$$

where $g(\cdot)$ is the link function, X_i, Z_i are known covariate matrices, β is a vector of fixed effects, and α_i is a q-dimensional vector of random effects whose distribution depends on an s-dimensional vector θ of dispersion parameters. Here for any vector $a = (a_1, \dots, a_k)'$, $g(a) = (g(a_1), \dots, g(a_k))'$. The second class is the so-called ANOVA GLMM, in which the conditional mean vector, $\mu = (\mu_i)_{1 \leq i \leq n}$, of the responses given the random effects satisfies the equation

$$g(\mu) = X\beta + Z_1\alpha_1 + \cdots + Z_s\alpha_s, \tag{3.24}$$

where X is a matrix of known covariates, $Z_1, \dots, Z_s$ are known design matrices, β is a vector of fixed effects, $\alpha_1, \dots, \alpha_s$ are independent vectors of random effects such that the components of α_j are i.i.d. with distribution F_j whose mean is 0 and variance is θ_j. The hypothesis to be tested is H_0: $\theta = 0$, where $\theta = (\theta_1, \dots, \theta_s)'$. Note that, under this null hypothesis, there are no random effects involved in the GLMM, therefore the model becomes a GLM. The first step is to obtain an approximate expansion of the quasi-log-likelihood function. Lin (1997) proposed using the second-order Laplace approximation (Breslow and Lin 1995; Lin and Breslow 1996). Let $l(\beta, \theta)$ denote the approximate quasi-log-likelihood. A global score statistic is constructed as follows,

$$\chi_{\mathrm{G}}^2 = U_\theta(\hat{\beta})' \tilde{I}(\hat{\beta})^{-1} U_\theta(\hat{\beta}),$$

where $\hat{\beta}$ is the MLE under the null hypothesis, that is, the MLE under the GLM, assuming that the responses are independent; $U_\theta(\beta)$ is the gradient vector with respect to θ (i.e., $\partial l / \partial \theta$) and $\tilde{I}$ is the information matrix of θ evaluated under H_0, which takes the form

$$\tilde{I} = I_{\theta\theta} - I'_{\beta\theta} I_{\beta\beta}^{-1} I_{\beta\theta}$$

with

$$\begin{aligned} I_{\theta\theta} &= \mathrm{E}\left(\frac{\partial l}{\partial \theta} \frac{\partial l}{\partial \theta'} \right), \\ I_{\beta\theta} &= \mathrm{E}\left(\frac{\partial l}{\partial \beta} \frac{\partial l}{\partial \theta'} \right), \\ I_{\beta\beta} &= \mathrm{E}\left(\frac{\partial l}{\partial \beta} \frac{\partial l}{\partial \theta'} \right). \end{aligned}$$

Note that, given the estimator $\hat{\beta}$ and under the null hypothesis, the information matrix can be estimated, using the properties of the exponential family (McCullagh and Nelder 1989, pp. 350). In fact, Lin (1997) showed that the information matrix may be estimated when the exponential-family assumption is replaced by some weaker assumptions on the cumulants of the responses. Furthermore, Lin (1997) established the following results. Under some regularity conditions, the global score test based on χ_{G}^2 follows a χ_s^2 distribution asymptotically under H_0; it is a locally asymptotically most powerful test if $s = 1$, and is a locally asymptotically most stringent test if $s > 1$ (Bhat and Nagnur 1965).

In Lin (1997), the author also studied the problem of testing for the individual variance component, namely, H_{0j}: $\theta_j = 0$, under the ANOVA GLMM (3.24). However, the result is less satisfactory. Note that, unlike the test for $\theta = 0$, under H_{0j} the rest of the random effects, α_k, $k \neq j$, do not vanish, so the model does not reduce to a GLM with independent observations. In such a case, Lin's approach was to estimate the rest of the variance components by PQL with a bias correction (Breslow and Lin 1995; Lin and Breslow 1996). However, the PQL is known to result in inconsistent estimators (e.g., Jiang 1998a, Booth and Hobert 1999) and so does its bias-corrected version, which is based on the second-order Laplace approximation. For example, the testing method for H_{0j} developed by Lin seems to be too conservative in terms of the size in the case of binary responses, although the test seems to work reasonably well when the responses are binomial means with a moderate denominator (i. e., group size; Lin 1997, pp. 321).

3.5.4 Maximum Hierarchical Likelihood

Lee and Nelder (1996) proposed a method, which they called maximum hierarchical likelihood. The method may be regarded as an extension of PQL (Breslow and Clayton 1993) in that it allows the random effects to have certain nonnormal distributions. Let y be the response and u an (unobserved) random component. Lee and Nelder defined a hierarchical GLM, or HGLM, as follows.

(a) The conditional (log-)likelihood for y given u has the GLM form

$$l(\xi, \phi; y|u) = \frac{y\xi - b(\xi)}{a(\phi)} + c(y, \phi), \tag{3.25}$$

where ξ denotes the canonical parameter, ϕ is the dispersion parameter, and $a(\cdot)$ and $c(\cdot, \cdot)$ are known functions. Write μ for the conditional mean of y given u, and $\eta = g(\mu)$, where $g(\cdot)$ is the link function for the conditional GLM. It is assumed that the linear predictor η takes the form $\eta = \zeta + v$, where $\zeta = x'\beta$, and $v = v(u)$ for some strictly monotonic function of u.

(b) The distribution of u is assumed appropriately.

Example 3.6 (Poisson-gamma HGLM). Suppose that the distribution of y given u is Poisson with mean $\mu = \mathrm{E}(y|u) = \exp(\zeta)u$. Then, with the log-link function, one has $\eta = \zeta + v$ with $v = \log(u)$. The distribution of u is assumed to be *gamma* with shape parameter ψ and mean 1.

Some may wonder how that distribution of u in Example 3.6 is picked. This is what Lee and Nelder called *conjugate distribution.* They prefer to assume such a distribution instead of normality for the random effects, and this is the difference between HGLM and GLMM discussed earlier. To define a conjugate HGLM, consider, for simplicity, the case where the responses may be expressed as y_{ij}, $i = 1, \dots, t$, $j = 1, \dots, n_i$, and u_i is a random effect associated with the ith cluster. Consider the canonical link function such that $\xi_{ij} = \xi(\mu_{ij}) = \xi(g^{-1}(\zeta_{ij})) + v_i$ with $v_i = \xi(u_i)$. The hierarchical likelihood, or h-likelihood, is defined as the logarithm of the joint density function of y and u; that is,

$$h = l(\xi, \phi; y|v) + l(\psi; v), \tag{3.26}$$

where $l(\xi, \phi; y|v) = \sum_{ij} l_{ij}$ with l_{ij} given by (3.25) after replacing y and ξ by y_{ij} and ξ_{ij}, respectively, and ψ is an additional parameter. As for the second term on the right side of (3.26), under the conjugate distribution, it is assumed that the kernel of $l(\psi; v)$ has the following form,

$$\sum_i \{a_1(\psi)v_i - a_2(\psi)b(v_i)\}, \tag{3.27}$$

where $a_1(\cdot)$ and $a_2(\cdot)$ are some functions. Note that the function $b(\cdot)$ is the same as that in (3.24). Lee and Nelder noted that, although expression (3.27) takes the form of the Bayesian conjugate prior (Cox and Hinkley 1974, pp. 370), it is only for v; no priors were specified for β, ϕ, or ψ. By maximizing the h-likelihood, one obtains the maximum h-likelihood estimators (MHLEs) of the fixed and random effects, which are solutions to the following equations,

$$\frac{\partial h}{\partial \beta} = 0, \tag{3.28}$$

$$\frac{\partial h}{\partial v} = 0. \tag{3.29}$$

It is clear that, when normality, instead of conjugate distribution, is assumed for the random effects, HGLM is the same as the GLMM of Breslow and Clayton (1993), among others, that were discussed earlier. Furthermore, MHLE is the same as the method of joint estimation of fixed and random effects, which was first proposed by Henderson (1950) in the case of linear mixed models. In the latter case, the method is known to result in the BLUE and BLUP. See discussions in Section 2.3. In the case of GLMM, the method of joint estimation of fixed and random effects is equivalent to the PQL of Breslow and Clayton (1993); that is, MHLE is equivalent to PQL in the case of GLMM. One advantage of the conjugate distribution is that the MHLE for the random effects has a simple form on the u-scale. To see this, note that under the assumed model, and using properties of the exponential family, one has $(\partial/\partial\xi)b(\xi(\mu)) = \mu$, so that $(\partial/\partial v)b(v) = u$. Thus, by differentiating the h-likelihood with respect to v_i and letting the derivative equal zero, one obtains the following expression,

$$u_i = \frac{y_{i\cdot} - \mu_{i\cdot} + \phi a_1(\psi)}{\phi a_2(\psi)}, \tag{3.30}$$

where $y_{i\cdot} = \sum_j y_{ij}$ and $\mu_{i\cdot} = \sum_j \mu_{ij}$. Note that (3.30) is not a closed-form expression for u_i, because $\mu_{i\cdot}$ also involves u_i. Still, the expression is useful in solving equations (3.28) and (3.29) iteratively. There is more discussion on this in the next section.

Lee and Nelder showed that, in some cases, the conjugate MHLE for β is the same as the (marginal) MLE for β. One such case is, of course, the normal–normal case, or Gaussian mixed models (see Section 1.3). Another example is the Poisson-gamma HGLM of Example 3.6 (Exercise 3.7). In general, Lee and Nelder showed that the MHLEs of the fixed effects are asymptotically equivalent to the (marginal) MLE of the fixed effects. However, the asymptotics is in the sense that $n_i \to \infty$, $1 \le i \le t$ at the same rate whereas t, the number of clusters, remains constant. Such a condition is often not satisfied in a mixed model situation. For example, in small area estimation (e.g., Ghosh and Rao 1994), n_i corresponds to the sample size for the ith small area, which may be quite small, whereas the number of small areas t can be quite large. In fact, as Lee and Nelder pointed out, the MHLE equations are the first-order approximation to the ML equations. Such an approximation becomes accurate when the cluster sizes n_i become large. As for the MHLE of random effects, Lee and Nelder showed that they are asymptotically best unbiased predictors, under the same asymptotic assumption. The latter sometimes may not be realistic, however, it would seem reasonable if the objective were to consistently estimate the random effects. See further discussion in the next section.

3.6 Prediction of Random Effects

In many cases the problem of main interest is to estimate, or predict, the random effects or, more generally, a mixed effect, in a GLMM. Here a mixed effect is defined as a (possibly nonlinear) function of the fixed and random effects. In the special case of linear mixed models, the prediction problem has been extensively studied. See Robinson (1991). The prediction of mixed effects was also an outstanding problem in small area estimation with binary responses. See, for example, Jiang and Lahiri (2001). In the latter cases, the prediction was in the context of GLMM. For the most part, there have been two main approaches in predicting the random effects. The first approach, which by far represents the majority of the literature in this area, is based on joint estimation of fixed and random effects. See, for example, Breslow and Clayton (1993), Lee and Nelder (1996), and Jiang et al. (2001a). The method is an extension of the BLUP method in linear mixed models that was first proposed by Henderson (1950). The second approach is a recently developed empirical best prediction method by Jiang and Lahiri (2001, 2005a) in the context of small area estimation. In the following sections we describe these two methods and related results.

3.6.1 Joint Estimation of Fixed and Random Effects

1. *Maximum a posterior.* Jiang et al. (2001a) took another look at Henderson's method of predicting the random effects in the linear mixed models (Henderson 1950). Let y be a vector of responses, and θ a vector of dispersion parameters. Write $L_J(\alpha, \beta) = f(y, \alpha|\beta, \theta)$, the joint density function of y and α given β and θ, where α is a vector of random effects, and β a vector of fixed effects. Because

$$\max_{\alpha,\beta} L_J(\alpha, \beta) = \max_{\beta} \max_{\alpha} L_J(\alpha, \beta),$$

the maximization can be in steps. In the first step one finds $\tilde{\alpha} = \tilde{\alpha}(\beta)$ that maximizes $L_J(\alpha, \beta)$ for fixed β. In the second step, one finds $\hat{\beta}$ that maximizes $L_J(\tilde{\alpha}, \beta)$, and lets $\hat{\alpha} = \tilde{\alpha}(\hat{\beta})$. Now consider the first step. Observe that

$$f(y, \alpha|\beta, \theta) = f(y|\beta, \theta) f(\alpha|y, \beta, \theta). \tag{3.31}$$

The first factor on the right side of (3.31) corresponds to the likelihood function for estimating β and θ, and the second factor to the posterior density of

α given y (if one would like, assuming that a noninformative prior has been assigned to α). Henderson's (1950) idea was to find α and β that jointly maximize $f(y, \alpha|\beta, \theta)$. Because the first factor does not depend on α, maximizing $f(y, \alpha|\beta, \theta)$ is equivalent to maximizing the posterior, and then the profile joint density $f(\tilde{\alpha}|\beta, \theta)$. Note that, although in linear mixed models the maximizers $\hat{\alpha}$ and $\hat{\beta}$ correspond to the BLUP and BLUE, they are no longer such predictor and estimator in nonlinear cases, such as GLMM. Still, the method is intuitive in the sense that $\hat{\alpha}$ maximizes the posterior. For such a reason, Jiang et al. called $\hat{\alpha}$ and $\hat{\beta}$ maximum posterior estimators, or MPE.

2. Computation of MPE. The MPE are typically obtained by solving a system of equations similar to (3.28) and (3.29); that is,

$$\frac{\partial l_{\mathrm{J}}}{\partial \beta} = 0, \tag{3.32}$$

$$\frac{\partial l_{\mathrm{J}}}{\partial \alpha} = 0, \tag{3.33}$$

where $l_{\mathrm{J}} = \log(L_{\mathrm{J}})$. In practice, there are often a large number of random effects involved in a GLMM. For example, in the salamander mating problem that was discussed earlier (McCullagh and Nelder 1989, §14.5), the number of random effects associated with the female and male salamanders is 80. In an NHIS problem considered by Malec et al. (1997), the number of random effects corresponding to the small areas is about 600. This means that the first step of MPE (i.e., the maximization of $L_{\mathrm{J}}(\alpha, \beta)$ for fixed β) is over a high-dimensional space. In other words, one has to simultaneously solve a large number of nonlinear equations (3.32) and (3.33). It is well known that standard methods of solving nonlinear systems such as Newton–Raphson (N–R) may be inefficient and extremely slow when the dimension of the solution is high. In fact, even in the linear case directly solving the BLUP equations may involve inversion of a large matrix, which can be computationally burdensome. There are other disadvantages of N–R. First, convergence of the N–R is sensitive to the initial values. When the dimension of the solution is high, it can be very difficult to find the initial values that will result in convergence. Second, N–R requires computation of partial derivatives, the analytic derivation of which can be tedious, and errors are often made in the process as well as in programming.

Jiang (2000b) proposed a nonlinear Gauss–Seidel algorithm (NLGSA) for computing the MPE, which is an extension of the Gauss–Seidel algorithm in numerical analysis for solving large linear systems. We use an example to illustrate the algorithm, and leave further details to Section 3.7.

Example 3.5 (Continued). Consider, once again, Example 3.5. To compute the MPE, one needs to solve the following system of nonlinear equations (Exercise 3.8),

$$\frac{u_i}{\sigma_1^2} + \sum_{j=1}^{m_2} \frac{\exp(\mu + u_i + v_j)}{1 + \exp(\mu + u_i + v_j)} = y_{i\cdot}, \qquad 1 \le i \le m_1, \tag{3.34}$$

$$\frac{v_j}{\sigma_2^2} + \sum_{i=1}^{m_1} \frac{\exp(\mu + u_i + v_j)}{1 + \exp(\mu + u_i + v_j)} = y_{\cdot j}, \qquad 1 \le j \le m_2, \tag{3.35}$$

where $y_{i\cdot} = \sum_{j=1}^{m_2} y_{ij}$ and $y_{\cdot j} = \sum_{i=1}^{m_1} y_{ij}$. Note that given the v_js, each equation in (3.34) is univariate, which can be easily solved (e.g., by bisection or one-dimensional N–R). A similar fact is observed in (3.35). This motivates the following algorithm. Starting with initial values $v_j^{(0)}$, $1 \le j \le m_2$, solve (3.34) with $v_j^{(0)}$ in place of v_j, $1 \le j \le m_2$ to get $u_i^{(1)}$, $1 \le i \le m_1$; then (3.35) with $u_i^{(1)}$ in place of u_i, $1 \le i \le m_1$ to get $v_j^{(1)}$, $1 \le j \le m_2$; and so on.

It is clear that the algorithm does not require the calculation of derivatives. Each step of the algorithm is easy to operate and, in fact, has a unique solution. Finally, it can be shown that the convergence of the algorithm is not affected by initial values. In other words, one has global convergence. See Section 3.7.1 for details.

3. Penalized generalized WLS. Jiang (1999) extended the weighted least squares (WLS) method in linear (mixed) models (see Section 1.4.3) to GLMMs for estimating the fixed and random effects. He noted that the (fixed effects) linear model is a special case of GLM (McCullagh and Nelder 1989) only when normality is assumed. On the other hand, the definition of linear models does not have to be associated with normality. A similar paradox exists between the linear mixed model and GLMM, because the former does not have to be Gaussian. See Section 1.2.2. He then extended the definition of GLMM so that it includes a linear mixed model as a special case regardless of normality. In the extended definition, it is assumed that, given a vector α of random effects, which satisfy

$$\mathrm{E}(\alpha) = 0, \tag{3.36}$$

the responses $y_1, \ldots, y_n$ are conditional independent with conditional expectation

$$\mathrm{E}(y_i|\alpha) = b_i'(\eta_i), \tag{3.37}$$

where $b_i(\cdot)$ is a known differentiable function. Furthermore, assume that (3.2) holds, where β, x_i, and z_i are the same as before. Note that no assumption of exponential family (3.1) for the conditional distribution is made here; only the form of the conditional mean is assumed.

Now consider inference about the extended GLMM. In linear models, which correspond to (3.37) and (3.2) with $b_i(x) = x^2/2$ and no random effect, the parameters β may be estimated by WLS, namely, by minimizing

$$\sum_{i=1}^{n} w_i (y_i - \eta_i)^2,$$

where w_is are weights, or, equivalently, by maximizing

$$\sum_{i=1}^{n} w_i \left(y_i \eta_i - \frac{\eta_i^2}{2} \right).$$

A straight generalization of the WLS to GLMM would suggest the maximizer of the following as the estimators of β and α:

$$\sum_{i=1}^{n} w_i \{ y_i \eta_i - b_i(\eta_i) \}. \tag{3.38}$$

However, conditionally, the individual fixed and random effects in a GLMM may not be identifiable. For example, in Example 3.5, we have

$$\text{logit}(p_{ij}) = (\mu + c + d) + (u_i - c) + (v_j - d)$$

for any c and d. Of course, such a problem occurs in linear models as well, in which case there are two remedies: reparameterization and constraints. Here we focus on the second. A set of linear constraints on α may be expressed as $P\alpha = 0$ for some matrix P. By Lagrange's method of multipliers, maximizing (3.38) subject to $P\alpha = 0$ is equivalent to maximizing

$$\sum_{i=1}^{n} w_i \{ y_i \eta_i - b_i(\eta_i) \} - \lambda |P\alpha|^2 \tag{3.39}$$

without constraints, where λ is the multiplier. On the other hand, for fixed λ the last term in (3.39) may be regarded as a penalizer. The only thing that needs to be specified is the matrix P. For any matrix M and vector space V, let $\mathcal{B}(V) = \{B : B$ is a matrix whose columns constitute a base for $V\}$; $\mathcal{N}(M) =$ the null-space of $M = \{v : Mv = 0\}$; $P_M = M(M'M)^- M'$, where $-$ means generalized inverse (see Appendix B); and $P_{M^\perp} = I - P_M$. Let $A \in \mathcal{B}\{\mathcal{N}(P_{X^\perp} Z)\}$, where the ith row of X and Z are x_i' and z_i', respectively. The penalized generalized WLS (PGWLS) estimator of $\gamma = (\beta', \alpha')'$ is defined as the maximizer of

$$l_{\text{P}}(\gamma) = \sum_{i=1}^{n} w_i \{ y_i \eta_i - b_i(\eta_i) \} - \frac{\lambda}{2} |P_A \alpha|^2, \tag{3.40}$$

where λ is a positive constant. The choice of the penalizer is explained in Section 3.7.2.

It might appear that the method is not using the information about the distribution of the random effects. However, as Jiang (1999) pointed out, the only information about the distribution of α; that is, (3.36), is indeed used. This is because the true random effects satisfy, on average, the constraint $P_A \alpha = 0$. Furthermore, in PGWLS, the random effects are somewhat treated as fixed. A question then arises as to whether the individual random effects can be estimated consistently, because in practice there is often not sufficient information about the individual random effects. This issue is addressed in

Section 3.7.2, but, roughly speaking, the answer is the following. The random effects can be consistently estimated in some overall sense, if the total number of random effects increases at a slower rate than the sample size; that is, $m/n \to 0$. Another feature of PGWLS is that, unlike MPE that was discussed earlier, here the estimation of the fixed and random effects is separated from that of the variance components. In fact, the latter are not even defined under the extended GLMM. Furthermore, it is shown that the consistency of the PGWLS estimators is not affected by ϕ, the additional dispersion parameter, at which the estimators are computed.

Note. The latest fact is similar to the facts that in linear models, consistency of the WLS estimator is not affected by the choice of the weights; in GLM, consistency of the generalized estimating equation (GEE) estimator is not affected by the choice of the working covariance matrix (Liang and Zeger 1986). Furthermore, Jiang et al.(2001a) showed that, in certain large sample situations, consistency of the MPE (see Section 3.6.1.1) is not affected by the variance components at which the MPE are computed. Note that equations (3.32) and (3.33) depend on θ, the vector of variance components.

The PGWLS estimators are typically obtained by solving the equations

$$\frac{\partial l_{\mathrm{P}}}{\partial \gamma} = 0. \tag{3.41}$$

The NLGSA proposed earlier to compute the MPE can be used here to obtain a solution to (3.41).

3. Maximum conditional likelihood. Quite often in situations where GLMMs are used, the information is not sufficient for some or, perhaps, all of the individual random effects. In the case where there is insufficient information about all the random effects, consistent estimation of any individual random effect is, of course, impossible. On the other hand, in some cases there may be sufficient information for some of the random effects. For example, consider the following.

Example 3.7. Suppose that, given the random effects a_i, b_{ij}, $1 \leq i \leq m_1$, $1 \leq j \leq m_2$, binary responses y_{ijk} are (conditionally) independent with $\mathrm{logit}(p_{ijk}) = \mu + a_i + b_{ij}$, where $p_{ijk} = \mathrm{P}(y_{ijk} = 1|a, b)$, $k = 1, \dots, r$. If $m_1, m_2 \to \infty$ but r remains fixed, there is sufficient information about the a_is but not the b_{ij}s.

In situations like Example 3.7, it is desirable to develop a method that can consistently estimate the random effects for which the data have provided sufficient information, as well as the fixed parameters. Jiang (1999) proposed such a method, which he called the maximum conditional likelihood. To illustrate the method, consider a special case, in which

$$\eta = X\beta + Z\alpha + U\zeta,$$

where $\eta = (\eta_i)_{1\le i\le n}$, and the random effects $\alpha = (\alpha_k)_{1\le k\le K}$ and $\zeta = (\zeta_j)_{1\le j\le J}$ are independent. Here ζ represents a subset of the random effects for which there is insufficient information. Furthermore, suppose that U is a standard design matrix (see Section 2.4.2.1, Note 2), and that ζ_j, $1 \le j \le J$ are i.i.d. with density function $\psi(\cdot/\tau)$, where $\psi(\cdot)$ is a known density and $\tau > 0$ an unknown scale parameter. We also assume that

$$f(y_i|\alpha,\zeta) = f(y_i|\eta_i), \qquad 1 \le i \le n. \tag{3.42}$$

Here $f(\xi_2|\xi_1)$ denotes the conditional density of ξ_2 given ξ_1. Let u_i' be the ith row of U, and $e_{J,j}$ the J-dimensional vector whose jth component is 1 and other components are 0. Let $S_j = \{1 \le j \le n : u_i = e_{J,j}\}$, and $y_{(j)} = (y_i)_{i\in S_j}$, $1 \le j \le J$. Then, it is easy to show that

$$f(y|\alpha) = \prod_{j=1}^{J} f\left(y^{(j)}\middle|\alpha\right), \tag{3.43}$$

where

$$f\left(y^{(j)}\middle|\alpha\right) = \mathrm{E}\left\{\prod_{i\in S_j} f(y_i|x_i'\beta + z_i'\alpha + \tau\xi)\right\}, \tag{3.44}$$

and the expectation in (3.44) is taken with respect to ξ whose density function is $\psi(\cdot)$. We now consider estimation of $\tilde{\beta}$ and $\tilde{\alpha}$, which are reparameterizations of β and α such that $X\beta + Z\alpha = \tilde{X}\tilde{\beta} + \tilde{Z}\tilde{\alpha}$ for some known matrices $\tilde{X}$ and $\tilde{Z}$. Because the estimation is based on (3.43), which is the likelihood conditional on a subset of random effects, the method is referred to as the maximum conditional likelihood, or MCL.

We assume that there are no random effects nested within ζ. In notation, this means that z_i is the same for all $i \in S_j$, say, $z_i = z_{*j} = (z_{*jk})_{1\le k\le K}$, $1 \le j \le J$. Jiang (1999, Lemma 2.4) showed that there is a map $\beta \to \tilde{\beta}$, $\gamma \to \tilde{\alpha}$ with the following properties: (i) $X\beta + Z\alpha = \tilde{X}\tilde{\beta} + \tilde{Z}\tilde{\alpha}$, where $(\tilde{X}\ \tilde{Z})$ is a known matrix of full rank; and (ii) $\tilde{z}_i = \tilde{z}_{*j}$, $i \in S_j$ for some known vector $\tilde{z}_{*j}$, where $\tilde{z}_i'$ is the ith row of $\tilde{Z}$. With this result, we have

$$\eta = W\tilde{\gamma} + U\zeta,$$

where $W = (\tilde{X}\ \tilde{Z})$, $\tilde{\gamma} = (\tilde{\beta}', \tilde{\alpha}')'$. Let $\varphi = (\tilde{\alpha}', \tilde{\beta}', \tau)'$. Note that, unlike γ, the vector φ is identifiable. By (3.42), we have

$$f(y|\varphi) = \prod_{j=1}^{J} f\left(y^{(j)}\middle|\varphi\right).$$

Furthermore, it can be shown that $f\left(y^{(j)}\middle|\varphi\right) = g_j(\tilde{z}_{*j}\tilde{\alpha}, \tilde{\beta}, \tau)$, where

$$g_j(s) = \mathrm{E}\left\{\prod_{i\in S_j} f(y_i|s_1 + \tilde{x}_i' s_{(2)} + s_{r+2}\xi)\right\}$$

for $s = (s_1, \dots, s_{r+2})'$. Here $s_{(2)} = (s_2, \dots, s_{r+1})'$, $r = \dim(\tilde{\beta})$, and $\tilde{x}_i'$ is the ith row of $\tilde{X}$. Let $h_j(s) = \log\{g_j(s)\}$, $l_{\mathrm{C}}(\varphi) = \log f(y|\varphi)$, and $l_{\mathrm{C},j}(\varphi) = h_j(\tilde{z}_{*j}\tilde{\alpha}, \tilde{\beta}, \tau)$. Then, the conditional log-likelihood can be expressed as

$$l_{\mathrm{C}}(\varphi) = \sum_{j=1}^{J} l_{\mathrm{C},j}(\varphi).$$

The MCL estimator of φ, $\hat{\varphi}$, is defined as the maximizer of $l_{\mathrm{C}}(\varphi)$. Typically, $\hat{\varphi}$ is obtained by solving the equations

$$\frac{\partial l_{\mathrm{C}}}{\partial \varphi} = 0. \tag{3.45}$$

Once again, the NLGSA proposed earlier can be used for obtaining the solution.

Jiang (1999) studied asymptotic properties of the MCL estimators. It was shown that, under suitable conditions, with probability tending to one there is a solution to (3.45) which is consistent.

3.6.2 Empirical Best Prediction

In this section, we restrict our attention to a special class of GLMM, the so-called longitudinal GLMM. The characteristic of this class of models is that the responses may be divided into independent clusters, or groups. There are two major areas of applications of these models. The first is the analysis of longitudinal data; the second is small area estimation. In most cases of longitudinal data, the problem of main interest is the estimation of mean responses, which are usually associated with the fixed effects in the GLMM. Such estimation problems are further discussed in Section 4.2.1. On the other hand, many problems in small area estimation are closely related to the prediction of mixed effects (e.g., Ghosh and Rao 1994; Rao 2003).

Example 3.8. Jiang and Lahiri (2001) considered the following mixed logistic model for small area estimation. Suppose that, conditional on α_i, binary responses y_{ij}, $j = 1, \dots, n_i$ are independent with $\mathrm{logit}\{\mathrm{P}(y_{ij} = 1|\alpha_i)\} = x_{ij}'\beta + \alpha_i$, where x_{ij} is a vector of known covariates and β a vector of unknown regression coefficients. Furthermore, $\alpha_1, \dots, \alpha_m$ are independent and distributed as $N(0, \sigma^2)$, where σ^2 is an unknown variance. It is easy to show that the model is a special case of GLMM (Exercise 3.10). Here α_i is a random effect associated with the ith small area.

A mixed effect of interest is the conditional probability $\mathrm{P}(y_{ij} = 1|\alpha_i)$, which may represent, for example, the proportion of women (aged 40 or older) having had mammography in the ith health service area (Jiang et al. 2001a). Note that the mixed effect can be expressed as $h(x_{ij}'\beta + \alpha_i)$, where $h(x) = e^x/(1+e^x)$, so, in particular, the mixed effect is a nonlinear function of the fixed and random effects.

In the following we introduce two methods for predicting a mixed effect in the context of small area estimation.

1. Empirical best prediction under GLMM. We first introduce a GLMM that is suitable for small area estimation. Suppose that, conditional on a vector of random effects, $\alpha_i = (\alpha_{ij})_{1\le j\le r}$, responses $y_{i1},\ldots,y_{in_i}$ are independent with density

$$f(y_{ij}|\alpha_i) = \exp\left[\left(\frac{a_{ij}}{\phi}\right)\{y_{ij}\xi_{ij} - b(\xi_{ij})\} + c\left(y_{ij}, \frac{\phi}{a_{ij}}\right)\right],$$

where $b(\cdot)$ and $c(\cdot,\cdot)$ are functions associated with the exponential family (McCullagh and Nelder 1989, §2), ϕ is a dispersion parameter, a_{ij} is a weight such that $a_{ij} = 1$ for ungrouped data; $a_{ij} = l_{ij}$ for grouped data when the average is considered as response and l_{ij} is the group size; and $a_{ij} = l_{ij}^{-1}$ when the sum of individual responses is considered. Furthermore, ξ_{ij} is associated with a linear function

$$\eta_{ij} = x_{ij}'\beta + z_{ij}'\alpha_i$$

through a link function $g(\cdot)$; that is, $g(\xi_{ij}) = \eta_{ij}$, or $\xi_{ij} = h(\eta_{ij})$, where $h = g^{-1}$. Here x_{ij} and z_{ij} are known vectors, and β is a vector of unknown regression coefficients. In the case of a canonical link, we have $\xi_{ij} = \eta_{ij}$. Finally, suppose that $v_1,\ldots,v_m$ are independent with density $f_\theta(\cdot)$, where θ is a vector of variance components. Let $\psi = (\beta', \theta')'$, and $\vartheta = (\psi', \phi)$. Note that in some cases such as binomial and Poisson the dispersion parameter ϕ is known, so ψ represents the vector of all unknown parameters.

Consider the problem of predicting a mixed effect of the following form,

$$\zeta = \zeta(\beta, \alpha_S),$$

where S is a subset of $\{1,\ldots,m\}$, and $\alpha_S = (\alpha_i)_{i\in S}$. Let $y_S = (y_i)_{i\in S}$, where $y_i = (y_{ij})_{1\le j\le n_i}$ and $y_{S-} = (y_i)_{i\notin S}$. Under the above model, the best predictor (BP) of ζ, in the sense of minimum MSE, is given by

$$\begin{aligned}
\tilde{\zeta} &= \mathrm{E}(\zeta|y) \\
&= \mathrm{E}(\zeta(\beta,\alpha_S)|y_S) \\
&= \frac{\int \zeta(\beta,\alpha_S) f(y_S|\alpha_S) f_\theta(\alpha_S) d\alpha_S}{\int f(y_S|\alpha_S) f_\theta(\alpha_S) d\alpha_S} \\
&= \frac{\int \zeta(\beta,\alpha_S)\exp\{\phi^{-1}\sum_{i\in S} s_i(\beta,\alpha_i)\}\prod_{i\in S} f_\theta(\alpha_i)\prod_{i\in S} d\alpha_i}{\prod_{i\in S}\int \exp\{\phi^{-1}s_i(\beta,v)\} f_\theta(v)dv},
\end{aligned}$$

where $s_i(\beta,v) = \sum_{j=1}^{n_i} a_{ij}[y_{ij}h(x_{ij}'\beta + z_{ij}'v) - b\{h(x_{ij}\beta + z_{ij}v)\}]$. The dimension of integrals involved in the denominator on the right side of (3.46) is $r =$

$\dim(\alpha_i)$, and that of the numerator is at most sr, where $s = |S|$, the cardinality of S. When r and s are relatively small, such integrals may be evaluated by Monte-Carlo methods, provided that ψ (ϑ) is known. For example, suppose that $\alpha_i \sim N\{0, V(\theta)\}$, where $V(\theta)$ is a covariance matrix depending on θ, and that $S = \{i\}$. Then, we have

$$\begin{aligned}\tilde{\zeta} &= \frac{\int \xi(\beta, v) \exp\{\phi^{-1} s_i(\beta, v)\} f_\theta(v) dv}{\int \exp\{\phi^{-1} s_i(\beta, v)\} f_\theta(v) dv} \\ &\approx \frac{\sum_{l=1}^L \zeta(\beta, v_l) \exp\{\phi^{-1} s_i(\beta, v_l)\}}{\sum_{l=1}^L \exp\{\phi^{-1} s_i(\beta, v_l)\}},\end{aligned}$$

where $f_\theta(v)$ is the density of $N\{0, V(\theta)\}$, and $v_1, \dots, v_L$ are generated independently from $N\{0, V(\theta)\}$.

Note that the BP depends on both y_S and ψ, that is, $\tilde{\zeta} = u(y_S, \psi)$. Inasmuch as ψ is usually unknown, it is customary to replace ψ by a consistent estimator, say, $\hat{\psi}$. The result is called the empirical best predictor (EBP), given by

$$\hat{\zeta} = u(y_S, \hat{\psi}). \tag{3.46}$$

In practice, it is desirable not only to compute the EBP but also to assess its variation. A measure of the variation is the MSE, defined by $\mathrm{MSE}(\hat{\zeta}) = \mathrm{E}(\hat{\zeta} - \zeta)^2$. Unfortunately, the latter may be difficult to evaluate. In some cases, an expression of the MSE of $\tilde{\zeta}$, not that of $\hat{\zeta}$, may be obtained, say, $\mathrm{MSE}(\tilde{\zeta}) = b(\psi)$. Then, a naive estimator of the MSE of $\hat{\zeta}$ is obtained as $b(\hat{\psi})$. However, this is an underestimator of the true MSE. To see this, note the following decomposition of the MSE,

$$\begin{aligned}\mathrm{MSE}(\hat{\zeta}) &= \mathrm{MSE}(\tilde{\zeta}) + \mathrm{E}(\hat{\zeta} - \tilde{\zeta})^2 \\ &= b(\psi) + \mathrm{E}(\hat{\zeta} - \tilde{\zeta})^2. \end{aligned} \tag{3.47}$$

It is clear that the naive estimator simply ignores the second term on the right side of (3.47), and therefore underestimates the true MSE.

Jiang (2003a) developed a method based on Taylor series expansion that gives an estimate whose bias is corrected to the second order. The method may be regarded as an extension of the Prasad–Rao method for estimating the MSE of EBLUP in linear mixed models (see Section 2.3.1.3). Consider, for simplicity, the case that ϕ is known (e.g., binomial, Poisson), so that $b(\psi) = b(\theta)$ in (3.47). Then, the estimator may be expressed as

$$\widehat{\mathrm{MSE}}(\hat{\zeta}) = b(\hat{\theta}) + m^{-1}\{e(\hat{\theta}) - B(\hat{\theta})\}, \tag{3.48}$$

where the functions $e(\cdot)$ and $B(\cdot)$ are given in Section 3.9.3. Here by second-order correctness we mean that the estimator has the property that

$$\mathrm{E}\{\widehat{\mathrm{MSE}}(\hat{\zeta}) - \mathrm{MSE}(\hat{\zeta})\} = o(m^{-1}). \tag{3.49}$$

Note that, if $\widehat{\mathrm{MSE}}(\hat{\zeta})$ in (3.49) is replaced by the naive estimator $b(\hat{\theta})$, which is the first term on the right side of (3.48), the right side of (3.49) will have to be replaced by $O(m^{-1})$. In other words, the naive estimator is correct to the first order, not the second one.

Example 3.8 (Continued). As a special case, we consider, again, the mixed logistic model introduced earlier. Suppose that the problem of interest is to predict α_i, the small-area specific random effect. By (3.46), the EBP is $\hat{\alpha}_i = u_i(y_{i\cdot}, \hat{\theta})$, where $y_{i\cdot} = \sum_{j=1}^{n_i} y_{ij}$, $\theta = (\beta', \sigma)'$,

$$u_i(y_{i\cdot}, \theta) = \sigma \frac{\mathrm{E}[\xi \exp\{s_i(y_{i\cdot}, \sigma\xi, \beta)\}]}{\mathrm{E}[\exp\{s_i(y_{i\cdot}, \sigma\xi, \beta)\}]}$$

with $s_i(k, v, \beta) = kv - \sum_{j=1}^{n_i} \log\{1 + \exp(x_{ij}'\beta + v)\}$ and $\xi \sim N(0, 1)$.

To see the behavior of u_i, note that $u_i(y_{i\cdot}, \theta)/\sigma \to 0$ as $\sigma \to 0$. Now consider a special case in which $x_{ij} = x_i$, that is, the covariates are at the small area (e.g., county) level. Then, it can be shown (Jiang and Lahiri 2001) that, as $\sigma \to \infty$,

$$u_i(y_{i\cdot}, \theta) \longrightarrow \sum_{k=1}^{y_{i\cdot}-1} \left(\frac{1}{k}\right) - \sum_{k=1}^{n_i - y_{i\cdot}-1} \left(\frac{1}{k}\right) - x_i'\beta.$$

To see what the expression means, note that when n is large, $\sum_{k=1}^{n-1}(1/k) \sim \log(n) + C$, where C is Euler's constant. Therefore, as $\sigma \to \infty$, we have (Exercise 3.12)

$$u_i(y_{i\cdot}, \theta) \approx \mathrm{logit}(\bar{y}_{i\cdot}) - x_i'\beta.$$

Finally, it can be shown that, as $m \to \infty$ and $n_i \to \infty$, we have

$$\hat{\alpha}_i - \alpha_i = O_{\mathrm{P}}(m^{-1/2}) + O_{\mathrm{P}}(n_i^{-1/2}).$$

Now consider the estimation of the MSE of $\hat{\alpha}_i$. It can be shown (Exercise 3.13) that, in this case, the terms $b(\theta)$ and $e(\theta)$ in (3.48) have the following expressions,

$$b(\theta) = \sigma^2 - \sum_{k=0}^{n_i} u_i^2(k, \theta) p_i(k, \theta),$$

$$e(\theta) = \sum_{k=0}^{n_i} \left(\frac{\partial u_i}{\partial \theta'}\right) V(\theta) \left(\frac{\partial u_i}{\partial \theta}\right) p_i(k, \theta),$$

where

$$\begin{aligned} p_i(k, \theta) &= \mathrm{P}(y_{i\cdot} = k) \\ &= \sum_{z \in S(n_i, k)} \exp\left(\sum_{j=1}^{n_i} z_j x_{ij}'\beta\right) \mathrm{E}[\exp\{s_i(z_{\cdot}, \sigma\xi, \beta)\}] \end{aligned}$$

with $S(l,k) = \{z = (z_1, \dots, z_l) \in \{0,1\}^l : z_\cdot = z_1 + \cdots + z_l = k\}$.

Next, we consider the prediction of the mixed effect $p_i = \mathrm{P}(y_{ij} = 1|\alpha_i)$. For simplicity, suppose that the covariates are at the small area level; that is, $x_{ij} = x_i$. Then, we have

$$p_i = \frac{\exp(x_i'\beta + \alpha_i)}{1 + \exp(x_i'\beta + \alpha_i)}.$$

The EBP of p_i is given by

$$\begin{aligned} \hat{p}_i &= u_i(y_{i\cdot}, \hat{\theta}) \\ &= \exp(x_i'\hat{\beta}) \\ &\quad \times \frac{\mathrm{E}\exp[(y_{i\cdot} + 1)\hat{\sigma}\xi - (n_i + 1)\log\{1 + \exp(x_i'\hat{\beta} + \hat{\sigma}\xi)\}]}{\mathrm{E}\exp[y_{i\cdot}\hat{\sigma}\xi - n_i\log\{1 + \exp(x_i'\hat{\beta} + \hat{\sigma}\xi)\}]}, \end{aligned} \tag{3.50}$$

where the expectations are taken with respect to $\xi \sim N(0,1)$. Note that the EBP is not p_i with β and α_i replaced, respectively, by $\hat{\beta}$ and $\hat{\alpha}_i$.

On the other hand, a naive predictor of p_i is $\bar{y}_{i\cdot} = y_{i\cdot}/n_i$. Although the EBP given by (3.50) is not difficult to compute (e.g., by the Monte Carlo method), it does not have a closed form. So, one question is: just how much better is the EBP than the naive predictor? To answer this question, we consider the relative savings loss (RSL) introduced by Efron and Morris (1973). In the current case, the RSL is given by

$$\begin{aligned} \mathrm{RSL} &= \frac{\mathrm{MSE}(\hat{p}_i) - \mathrm{MSE}(\tilde{p}_i)}{\mathrm{MSE}(\bar{y}_{i\cdot}) - \mathrm{MSE}(\tilde{p}_i)} \\ &= \frac{\mathrm{E}(\hat{p}_i - \tilde{p}_i)^2}{\mathrm{E}(\bar{y}_{i\cdot} - \tilde{p}_i)^2}, \end{aligned} \tag{3.51}$$

where $\tilde{p}_i$ is the BP of p_i. It can be shown (Exercise 3.14) that the numerator on the right side of (3.51) is $O(m^{-1})$, and

$$\begin{aligned} \text{the denominator} &= \sum_{k=0}^{n_i} \left\{ \frac{k}{n_i} - u_i(k,\theta) \right\}^2 p_i(k,\theta) \\ &\geq \{u_i(0,\theta)\}^2 p_i(0,\theta). \end{aligned} \tag{3.52}$$

If n_i is bounded, the right side of (3.52) has a positive lower bound. Therefore, RSL $\to 0$ as $m \to \infty$. In fact, the convergence rate is $O(m^{-1})$. So, the complication of EBP is worthwhile.

2. EBP with design consistency. An important feature of the EBP method is that it is a model-based method. If the assumed model fails, the predictor is no longer the EBP. In fact, the EBP may perform poorly when the assumed model fails (see a simulated example below). In the following, we propose a new model assisted EBP approach. The development of the method was motivated by the estimation of the mean of a finite population domain within a large population covered by a complex survey. Domain estimation is an important problem encountered by many government agencies. For example, the Bureau of Labor Statistics produces monthly estimates of unemployment rates not only for the entire United States but also for different small and large domains (e.g., the 50 states and the District of Columbia). A direct expansion estimator due to Brewer (1963) and Hajek (1971) has been frequently used in domain estimation. Such an estimator is typically design consistent under many sampling designs in common use; that is, the estimator approaches in probability induced by the sampling design to the true domain finite population mean when the domain sample size is large. The new method we propose here also has the property of design consistency. In other words, the new method protects the large domain estimators from possible model failure. The method essentially amounts to obtaining an EBP assuming a (linear or generalized linear) mixed model on the commonly used design-consistent estimator of domain means. However, no explicit model is assumed for the unobserved units of the finite population. Under the assumed model, the predictor corresponds to the EBP. On the other hand, even under model failure, the predictor is approximately equal to the commonly used design-consistent estimator as long as the domain sample size is large.

We begin with a brief description of the problem of estimating a finite population mean. More details about the problem are given in Chapter 5. Consider a finite population divided into m domains. Let N_i be the population size of the ith domain. Let y_{ij} denote the value of a variable of interest for the jth unit in the ith domain. We are interested in the estimation of the i domain finite population mean given by

$$\bar{Y}_i = \frac{1}{N_i}\sum_{j=1}^{N_i} y_{ij}$$

based on a sample, say, y_{ij}, $i = 1, \ldots, m$, $j = 1, \ldots, n_i$. Let $\tilde{w}_{ij}$ denote the corresponding sampling weights, which are defined as the inverse of the first-order inclusion probability under the sampling design employed. When the sampling weights vary within a domain of interest, an estimator popular among survey practitioners is given by

$$\hat{y}_{iw} = \sum_{j=1}^{n_i} w_{ij} y_{ij},$$

where $w_{ij} = \tilde{w}_{ij}/\sum_{j=1}^{n_i}\tilde{w}_{ij}$. The estimator was proposed by Brewer (1963) and Hajek (1971). Under many commonly used designs, one has

$$\bar{y}_{iw} \longrightarrow \bar{Y}_i \quad \text{in } \mathrm{P_d},$$

as $n_i \to \infty$, where $\mathrm{P_d}$ is the probability measure induced by the sampling design.

The problem with $\bar{y}_{iw}$ is that it is not very efficient for small n_i. One way to improve the efficiency is to "borrow strength" from other similar domains. First, we need an explicit model for the sampled units, for example, a linear mixed model or a GLMM. However, the explicit model is not needed for the unobserved units of the finite population: only the existence of a random effect v_i is assumed, which is associated with the ith domain, such that

$$\mathrm{E}_{\mathrm{m}}(y_{ij}|v_i, \bar{y}_{iw}) = \mathrm{E}_{\mathrm{m}}(y_{ij}|v_i), \tag{3.53}$$

where E_{m} means expectation with respect to the assumed model. Assumption (3.53) holds, for example, for the linear mixed model considered by Prasad and Rao (1999) and for the mixed logistic model considered by Jiang and Lahiri (2001). We define the mean squared prediction error (MSPE), which is similar to the MSE, of an arbitrary predictor of $\bar{Y}_i$, say, $\hat{\bar{Y}}_i$, as $\mathrm{MSPE}(\hat{\bar{Y}}_i) = \mathrm{E}(\hat{\bar{Y}}_i - \bar{Y}_i)^2$. Jiang and Lahiri (2005a) showed that, under the assumed mixed model and the sampling design,

$$\hat{\bar{Y}}_i^{\mathrm{BP}} = \frac{1}{N_i}\sum_{j=1}^{N_i} \mathrm{E}_{\mathrm{m}}\{\mathrm{E}_{\mathrm{m}}(y_{ij}|v_i)|\bar{y}_{iw}\}$$

minimizes the MSPE among the class of all predictors that depend on the data only through $\bar{y}_{iw}$. Because an explicit model for the unobserved units is not assumed, $\mathrm{E}_{\mathrm{m}}\{\mathrm{E}_{\mathrm{m}}(y_{ij}|v_i)|\bar{y}_{iw}\}$ is unknown for any unobserved unit. It follows that $\hat{\bar{Y}}_i^{\mathrm{BP}}$ is not computable. However, we may treat the latter as an unknown finite population mean, which can be estimated unbiasedly with respect to the sampling design by

$$\hat{\bar{Y}}_i^{\mathrm{EBP}} = \sum_{j=1}^{n_i} w_{ij} \mathrm{E}_{\mathrm{m}}\{\mathrm{E}_{\mathrm{m}}(y_{ij}|v_i)|\bar{y}_{iw}\},$$

assuming that $\mathrm{E}_{\mathrm{m}}\{\mathrm{E}_{\mathrm{m}}(y_{ij}|v_i)|\bar{y}_{iw}\}$ is fully specified for any observed unit y_{ij}, $i = 1, \ldots, m$, $j = 1, \ldots, n_i$. Some alternative expressions of $\hat{\bar{Y}}_i^{\mathrm{EBP}}$ are the following. We state the results as a theorem.

Theorem 3.1. We have $\hat{\bar{Y}}_i^{\mathrm{EBP}} = \tilde{\zeta}$, where $\tilde{\zeta} = \mathrm{E}_{\mathrm{m}}(\zeta|\bar{y}_{iw})$ with $\zeta = \mathrm{E}_{\mathrm{m}}(\bar{y}_{iw}|v_i)$. Furthermore, we have

$$\tilde{\zeta} = \frac{\int \zeta_i(v) f(\bar{y}_{iw}, v) f(v) dv}{\int f(\bar{y}_{iw}, v) f(v) dv},$$

where $f(\bar{y}_{iw}, v)$, $f(v)$ are nonnegative functions such that, under the assumed model, $f(\bar{y}_{iw}, v) = f(\bar{y}_{iw}|v_i)|_{v_i=v}$, the conditional density of $\bar{y}_{iw}$ given $v_i = v$, and $f(v)$ is the density of v_i.

Note that the EBP defined above depends on $\bar{y}_{iw}$ and often on θ, a vector of unknown parameters, that is, $\tilde{\zeta}_i = u_i(\bar{y}_{iw}, \theta)$ for some function $u_i(\cdot, \cdot)$.

When θ is unknown, it is replaced by $\hat{\theta}$, a model-consistent estimator (i.e., an estimator that is consistent under the assumed model). This gives the following model-assisted EBP of $\bar{Y}_i$ based on $\bar{y}_{iw}$:

$$\hat{\zeta}_i = u_i(\bar{y}_{iw}, \hat{\theta}). \tag{3.54}$$

One important property of the model-assisted EBP is that it is design consistent. More specifically, Jiang and Lahiri (2005a) showed that, under some regularity conditions, $\tilde{\zeta}_i$ agrees asymptotically with $\bar{y}_{iw}$ regardless of the model and θ, as long as n_i is large. Here, the asymptotic agreement is in the sense that $\tilde{\zeta}_i - \bar{y}_{iw} \to 0$ in $\mathrm{P_d}$, the probability measure with respect to the design. A similar result also holds with respect to the assumed model. In other words, the proposed model-assisted EBP is design consistent as well as model consistent. Such a property is not possessed by the EBP discussed earlier (see Section 3.6.2.1). See Example 3.9 below.

Similar to the MSE of the EBP discussed in Section 3.6.2.1, an estimator of the MSPE of the model-assisted EBP can be obtained such that it is second-order correct: that is, the bias of the MSPE estimator is $o(m^{-1})$, where m is the number of domains. See Section 3.7.4 for details.

3.6.3 A Simulated Example

We conclude this section with a simulated example to study the finite sample performance of the model-assisted EBP introduced in Section 3.6.2.2 as well as its MSPE estimator. We investigate the randomization-based properties of different estimators using a limited Monte Carlo simulation experiment. In other words, the computations of biases and MSEs in this section do not depend on the model used to generate the fixed finite population; they are all based on the sampling design. The first part of the simulation study focuses on the evaluation of the model-assisted EBP proposed in the previous section (JL) compared to a direct estimator (DIR) and an EBLUP. The second part evaluates two different MSPE estimators of the model-assisted EBP. The first MSPE estimator does not include the term $2g_i/m$ needed to achieve the second-order efficiency [see (3.67)]. The second MSPE estimator includes this term and is second-order correct.

We consider an EBLUP (same as the empirical Bayes estimator considered by Ghosh and Meeden 1986) of the finite population small area means $\bar{Y}_i = N^{-1}\sum_{j=1}^N y_{ij}$, $i = 1, \dots, m$ using the following nested error model for the finite population,

$$y_{ij} = \mu + v_i + e_{ij},$$

where v_is and e_{ij}s are independent with $v_i \sim N(0, \sigma_v^2)$ and $e_{ij} \sim N(0, \sigma_e^2)$, $i = 1, \dots, m$, $j = 1, \dots, k$. The model-assisted EBP is developed under the assumption that the above model holds for the sampled units and the existence of the random effects v_i for the unobserved units (but otherwise making no additional assumption on the unobserved units of the finite population). Both the model-assisted EBP and EBLUP use REML estimators of the variance components σ_v^2 and σ_e^2. Throughout the simulation we take $m = 30$, $k = 20$, $N = 200$, $\mu = 50$, and $\sigma_e^2 = 1$. For each finite population unit, a size measure (x) is generated from an exponential distribution with scale parameter 200.

Table 3.1. Randomization-based average bias, MSPE and relative bias of MSPE estimators

	Average Bias			MSPE (% Improvement)			RB	
σ_v^2	DIR	EB	JL	DIR	EB	JL	NSOC	SOC
.4	−.001	−.010	−.006	.111	.040	.057	36.5	−2.0
					(64.0)	49.0		
.6	−.001	−.011	−.005	.111	.047	.077	19.5	12.5
					(57.3)	(30.3)		
1	−.001	−.011	−.003	.111	.053	.096	10.2	8.7
					(52.5)	(13.7)		
6	−.001	−.011	−.001	.111	.057	.111	7.0	6.9
					(48.9)	(.43)		

We first investigate the performances of the model-assisted EBP and EBLUP and the MSPE estimators of the model-assisted EBP for four different values of σ_v^2. The finite population is generated from the above nested error model which is most favorable to the EBLUP. For a particular simulation run, we draw a sample of size $k = 20$ from each of the $m = 30$ small areas using a probability proportional to size (PPS) with replacement sampling design. Table 3.1 displays the randomization-based biases and MSPEs of DIR, JL, and EB (or EBLUP) and the percentage of relative biases of the two different MSPE estimators. All the results are based on 50,000 simulation runs.

The percentage improvement is defined to be the relative gain in MSPE over DIR expressed in percentage. DIR is the best in terms of the average bias, followed by JL and EB. The average bias of DIR remains more or less the same for different choices of σ_v. The same is true for the average MSPE. The bias of the EBLUP does not change for a variation of σ_v but the MSPE decreases with the increase of σ_v. The purpose of comparing the two MSPE estimators is to understand the effect of the additional term involving $g_i(\theta)$. For each small area $i = 1, \ldots, 30$, we calculate the relative bias (RB) of each MSPE estimator as follows,

$$\text{RB} = \frac{\text{E}(\text{mspe}_i) - \text{MSPE}_i}{\text{MSPE}_i},$$

where mspe_i is an arbitrary estimator of the true MSPE_i for area i. We then average these RBs for all the small areas and report the average relative bias (ARB) in Table 3.1. The performances of both MSPE estimators improve when σ_v^2 increases. The contribution from the term involving $g_i(\theta)$ is significant especially for small σ_v. For example, this additional term brings the percentage of relative bias of the first estimator from 36% to -2% when $\sigma_v = .4$, quite a remarkable contribution! When σ_v is large, the effect of this additional term diminishes.

Table 3.2. Randomization-based average bias and MSPE

	Average Bias			MSPE (% Improvement)		
(v, e)	DIR	EB	JL	DIR	EB	JL
(N,N)	.027	.239	.058	.118	.115 (2.2)	.103 (12.9)
(DE,N)	.027	.239	.051	.118	.119 (−.7)	.106 (10.4)
(N,DE)	.029	.254	.063	.124	.125 (−1.2)	.108 (12.6)
(DE,DE)	.030	.254	.057	.124	.127 (−2.6)	.110 (10.8)
(EXP,N)	.034	.270	.057	.124	.136 (−9.8)	.111 (10.3)
(EXP,EXP)	.030	.243	.051	.125	.118 (5.2)	.113 (9.7)

In Table 3.2, we compare the robustness properties of our proposed estimator with the EBLUP. We considered several finite populations, each generated from a model different from the above nested error model. We generated the finite populations from a nested error regression model (i.e., μ is replaced by $x_{ij}\mu$) using combinations of distributions to generate the random effect v_i and the pure error e_{ij}. We considered various combinations of normal, shifted exponential and shifted double exponentials to generated (v_i, e_{ij}). In each case, the means of v_i and e_{ij} are zeros and variances $\sigma_v^2 = 1 = \sigma_e^2$. In terms of the MSPE, the proposed model-assisted EBP is a clear winner. We note that in some situations the model-based EBLUP performs worse than the direct estimator.

3.7 Further Results and Technical Notes

3.7.1 More on NLGSA

In this section, we give more details about NLGSA introduced in Section 3.6.1.2. Note that as long as the αs are solved from (3.33) as functions of the βs, (3.32) can be fairly easily solved, because the number of fixed effects is usually not very large. Thus, we focus on solving (3.33) given β.

Suppose that the random effects are independent (and normal). This means that G, the covariance matrix of $\alpha = (\alpha_k)_{1\leq k\leq m}$, is diagonal, say, $G = \text{diag}(d_1, \ldots, d_m)$. Furthermore, assume a canonical link function $\xi_i = \eta_i$. Write $z_i = (z_{ik})_{1\leq k\leq m}$. Then Equations (3.33) are equivalent to

$$\frac{\alpha_k}{d_k} + \sum_{i=1}^{n} \frac{z_{ik}}{a_i(\phi)} b'\left(x_i'\beta + \sum_{l=1}^{m} x_{il}\alpha_l\right) = \sum_{i=1}^{n} \frac{z_{ik}}{a_i(\phi)} y_i, \qquad 1 \leq k \leq m.$$

Let $f_k(\alpha_1, \ldots, \alpha_{k-1}, \alpha_{k+1}, \ldots, \alpha_m)$ denote the unique solution λ to the following equation

$$\frac{\lambda}{d_k} + \sum_{i=1}^{n} \frac{z_{ik}}{a_i(\phi)} b'\left(x_i'\beta + z_{ik}\lambda + \sum_{l\neq k} z_{il}\alpha_l\right) = \sum_{i=1}^{n} \frac{z_{ik}}{a_i(\phi)} y_i.$$

A recursive algorithm is characterized by

$$\alpha_k^{(t)} = f_k(\alpha_1^{(t)}, \ldots, \alpha_{k-1}^{(t)}, \alpha_{k+1}^{(t-1)}, \ldots, \alpha_m^{(t-1)}), \qquad 1 \leq k \leq m$$

for $t = 1, 2, \ldots$, or, equivalently,

$$\frac{\alpha_k^{(t)}}{d_k} + \sum_{i=1}^{n} \frac{z_{ik}}{a_i(\phi)} b'\left(x_i'\beta + \sum_{l=1}^{k} z_{il}\alpha_l^{(t)} + \sum_{l=k+1}^{m} z_{il}\alpha_l^{(t-1)}\right)$$
$$= \sum_{i=1}^{n} \frac{z_{ik}}{a_i(\phi)} y_i, \qquad 1 \leq k \leq m.$$

Jiang (2000b) proved the following theorem regarding the convergence of the above NLGSA.

Theorem 3.2 (Global convergence of NLGSA). For any fixed β and arbitrary initial values, the NLGSA converges to the unique solution $\tilde{\alpha} = \tilde{\alpha}(\beta)$ to (3.3).

The proof used the global convergence theorem of Luenberger (1984). Note that it is easy to show that, given β, (3.3) has a unique solution $\tilde{\alpha} = \tilde{\alpha}(\beta)$. See Jiang (2000b) for detail.

3.7.2 Asymptotic Properties of PQWLS Estimators

In some ways, asymptotic theory regarding random effects is different from that about fixed parameters. First, the individual random effects are typically not identifiable [see the discussion below (3.38)]. Therefore, any asymptotic theory must take care, in particular, of the identifiability problem. Second, the number of random effects m should be allowed to increase with the sample size n. Asymptotic properties of estimators of fixed parameters when the number of parameters increases with n have been studied by Portnoy in a series of papers (e.g., Portnoy 1984).

To explore the asymptotic behavior of PGWLS estimators, we need to assume that m increases at a slower rate than n; that is, $m/n \to 0$. The case that m/n does not go to zero is discussed in the next section. First, we explain how the matrix P_A is chosen for the penalty term in (3.40). Note that the first term in (3.40), that is,

$$l_{\mathrm{C}}(\gamma) = \sum_{i=1}^{n} w_i\{y_i\eta_i - b_i(\eta_i)\},$$

depends on $\gamma = (\beta', \alpha')'$ only through $\eta = X\beta + Z\alpha$. However, γ cannot be identified by η, so there may be many vectors γ corresponding to the same η. The idea is therefore to consider a restricted space $S = \{\gamma : P_A\alpha = 0\}$, such that within this subspace γ is uniquely determined by η. Here we define a map $T : \gamma \to \tilde{\gamma} = (\tilde{\beta}, \tilde{\alpha})'$ as follows: $\tilde{\alpha} = P_{A^\perp}\alpha$, $\tilde{\beta} = \beta + (X'X)^{-1}X'ZP_A\alpha$. Obviously, T does not depend on the choice of A. Because $X\tilde{\beta} = X\beta + Z\alpha - P_{X^\perp}ZP_A\alpha = X\beta + Z\alpha$, we have $l_{\mathrm{C}}(\gamma) = l_{\mathrm{C}}(\tilde{\gamma})$. Let

$$G_A = \begin{pmatrix} X & Z \\ 0 & A' \end{pmatrix}.$$

The proofs of the following lemmas and theorem can be found in Jiang (1999).

Lemma 3.1. $\mathrm{rank}(G_A) = p + m$, where p is the dimension of β.

Corollary 3.1. Suppose that $b_i''(\cdot) > 0$, $1 \le i \le n$. Then, there can be only one maximizer of l_{P}.

Let B be a matrix, v a vector, and V a vector space. Define $\lambda_{\min}(B)|_V = \inf_{v \in V\setminus\{0\}}(v'Bv/v'v)$. Also, let $H = (X\ Z)'(X\ Z)$.

Lemma 3.2. For any positive numbers b_j, $1 \le j \le p$ and a_k, $1 \le k \le m$, we have

$$\begin{aligned}\lambda_{\min}(W^{-1}HW^{-1})|_{WS} &\ge \frac{\lambda_{\min}(G_A'G_A)}{(\max_{1\le j\le p} b_j^2) \vee (\max_{1\le k\le m} a_k^2)} \\ &> 0\end{aligned}$$

for any $W = \mathrm{diag}(b_1, \ldots, b_p, a_1, \ldots, a_m)$.

Let X_j (Z_k) denote the jth (kth) column of X (Z).

Theorem 3.3. Let $b_i''(\cdot)$ be continuous, $\max_{1\le i\le n}[w_i^2\mathrm{E}\{\mathrm{var}(y_i|\alpha)\}]$ be bounded, and

$$n^{-1}\left\{\left(\max_{1\le j\le p}|X_j|^2\right)\|(X'X)^{-1}X'Z\|^2 + \left(\max_{1\le k\le m}|Z_k|^2\right)\right\}|P_A\alpha|^2$$

converge to zero in probability. Let c_n, d_n be any sequences such that $\limsup(\max_{1\le j\le p}|\beta_j|/c_n) < 1$ and $\mathrm{P}(\max_{1\le k\le m}|\alpha_k|/d_n < 1) \to 1$. Also, let $M_i \geq c_n \sum_{j=1}^p |x_{ij}| + d_n \sum_{k=1}^m |z_{ik}|$, $1 \leq i \leq n$, and $\hat{\gamma}$ be the maximizer of l_{P} over $\Gamma(M) = \{\gamma : |\eta_i| \leq M_i, 1 \leq i \leq n\}$. Then,

$$n^{-1}\left\{\sum_{j=1}^p |X_j|^2(\hat{\beta}_j - \beta_j)^2 + \sum_{k=1}^m |Z_k|^2(\hat{\alpha}_k - \alpha_k)^2\right\} \longrightarrow 0$$

in probability, provided that

$$\frac{p+m}{n} = o(\omega^2),$$

where

$$\omega = \lambda_{\min}(W^{-1}HW^{-1})|_{WS} \min_{1\leq i\leq n}\{w_i \inf_{|u|\leq M_i} |b_i''(u)|\}$$

with $W = \mathrm{diag}(|X_1|, \ldots, |X_p|, |Z_1|, \ldots, |Z_m|)$.

Corollary 3.2. Suppose that the conditions of Theorem 3.3 hold and that $(p+m)/n = o(\omega^2)$.

(i) If p is fixed and $\liminf \lambda_{\min}(X'X)/n > 0$, $\hat{\beta}$ is consistent.

(ii) If $Z\alpha = Z_1\alpha_1 + \cdots + Z_q\alpha_q$, where each Z_u is a standard design matrix (see Section 2.4.2.1, Note 2), then, we have

$$\left(\sum_{v=1}^{m_u} n_{uv}\right)^{-1} \sum_{v=1}^{m_u} n_{uv}(\hat{\alpha}_{uv} - \alpha_{uv})^2 \longrightarrow 0$$

in probability, where $\alpha_u = (\alpha_{uv})_{1\leq v\leq m_u}$, $\hat{\alpha}_u = (\hat{\alpha}_{uv})_{1\leq v\leq m_u}$, and n_{uv} is the number of appearances of α_{uv} in the model.

The last result shows that, under suitable conditions, the PGWLS estimators of the fixed effects are consistent, and those of the random effects are consistent in some overall sense (but not necessarily for each individual random effect; Exercise 3.9).

Next, we consider a special class of GLMM, the so-called longitudinal GLMM, in which the responses are clustered in groups with each group associated with a single (possibly vector-valued) random effect. Suppose that the random effects $\alpha_1, \ldots, \alpha_m$ satisfy $\mathrm{E}(\alpha_i) = 0$. The responses are y_{ij}, $1 \leq i \leq m$, $1 \leq j \leq n_i$, such that, given the random effects, y_{ij}s are (conditionally) independent with $\mathrm{E}(y_{ij}|\alpha) = b_{ij}'(\eta_{ij})$, where $b_{ij}(\cdot)$ is differentiable. Furthermore, we have

$$\eta_{ij} = \mu + x_{ij}'\beta + z_i'\alpha_i,$$

where μ is an unknown intercept, $\beta = (\beta_j)_{1\leq j\leq s}$ (s is fixed) is an unknown vector of regression coefficients, and x_{ij} and z_i are known vectors. Such models are useful, for example, in the context of small-area estimation (e.g., Ghosh and Rao 1994), in which α_i represents a random effect associated with the ith small area. Here we are interested in the estimation of μ, β as well as $v_i = z_i'\alpha$, the so-called area-specific random effects. Therefore, without loss of generality, we may assume that

$$\eta_{ij} = \mu + x_{ij}'\beta + v_i,$$

where $v_1, \ldots, v_m$ are random effects with $\mathrm{E}(v_i) = 0$. It is clear that the model is a special case of GLMM. Following our earlier notation, it can be shown that, in this case, $A = 1_m \in \mathcal{B}(\mathcal{N}(P_{X^\perp}Z))$, $S = \{\gamma : v. = 0\}$, where $v. = \sum_{i=1}^m v_i$. Thus, (3.40) has a more explicit expression:

$$l_{\mathrm{P}}(\gamma) = \sum_{i=1}^m \sum_{j=1}^{n_i} w_{ij}\{y_{ij}\eta_{ij} - b_{ij}(\eta_{ij})\} - \frac{\lambda}{2} m\bar{v}^2,$$

where $\bar{v} = v./m$. Let

$$\delta_n = \min_{i,j}\{w_{ij} \inf_{|u|\le M_{ij}} b_{ij}''(u)\},$$

$$\lambda_n = \lambda_{\min}\left\{\sum_{i=1}^m \sum_{j=1}^{n_i} (x_{ij} - \bar{x}_i)(x_{ij} - \bar{x}_i)'\right\},$$

where $\bar{x}_i = n_i^{-1}\sum_{j=1}^{n_i} x_{ij}$. For the special longitudinal GLMM above, we have the following more explicit result (see Jiang 1999a).

Theorem 3.4. Suppose that $b_{ij}''(\cdot)$ is continuous. Furthermore, suppose that $w_{ij}^2\mathrm{E}\{\mathrm{var}(y_{ij}|v)\}$, $|x_{ij}|$ are bounded, $\liminf(\lambda_n/n) > 0$, and $\bar{v} \to 0$ in probability. Let c_n, d_n be such that $\limsup\{(|\mu| \vee |\beta|)/c_n\} < 1$ and $\mathrm{P}(\max_i |v_i|/d_n < 1) \to 1$. Let M_{ij} satisfy $M_{ij} \ge c_n(1 + |x_{ij}|) + d_n$. Finally, let $\hat{\gamma} = (\hat{\mu}, \hat{\beta}', \hat{v}')'$ be the maximizer of l_{P} over $\Gamma(M) = \{\gamma : |\eta_{ij}| \le M_{ij}, \forall i, j\}$. Then, we have $\hat{\beta} \to \beta$ in probability, and

$$\frac{1}{n}\sum_{i=1}^m n_i(\hat{a}_i - a_i)^2 \longrightarrow 0$$

in probability, where $a_i = \mu + v_i$ and $\hat{a}_i = \hat{\mu} + \hat{v}_i$, provided that $m/n = o(\delta_n^2)$. If the latter is strengthened to $(\min_{1\le i\le m} n_i)^{-1} = o(\delta_n^2)$, we have, in addition, $\hat{\mu} \to \mu$, $n^{-1}\sum_{i=1}^m n_i(\hat{v}_i - v_i)^2 \to 0$, and $m^{-1}\sum_{i=1}^m (\hat{v}_i - v_i)^2 \to 0$ in probability.

3.7.3 MSE of EBP

In this section, we give more details about the approximation and estimation of the MSE of EBP, introduced in Section 3.6.2.1. We assume, for simplicity, that the dispersion parameter ϕ is known, so that $b(\psi) = b(\theta)$ in (3.47).

First we have the following expression for $b(\theta)$,

$$\begin{aligned} b(\theta) &= \text{MSE}(\tilde{\zeta}) \\ &= \text{E}(\zeta^2) - \{\text{E}(\tilde{\zeta})\}^2 \\ &= \text{E}\{\zeta(\beta, \alpha_S)^2\} - [\text{E}\{u(y_S, \theta)\}]^2. \end{aligned} \tag{3.55}$$

Next, we use Taylor series expansion to approximate $\hat{\zeta} - \tilde{\zeta}$. Suppose that $|\hat{\theta} - \theta| = O(m^{-1/2})$ in a suitable sense (say, in probability). Then, we have, asymptotically,

$$\begin{aligned} \hat{\zeta} - \tilde{\zeta} &= u(y_S, \hat{\theta}) - u(y_S, \theta) \\ &= \left(\frac{\partial u}{\partial \theta'}\right)(\hat{\theta} - \theta) + o(m^{-1/2}). \end{aligned}$$

It follows that

$$\text{E}(\hat{\zeta} - \tilde{\zeta})^2 = m^{-1}\text{E}\left\{\left(\frac{\partial u}{\partial \theta'}\right)\sqrt{m}(\hat{\theta} - \theta)\right\}^2 + o(m^{-1}). \tag{3.56}$$

To obtain a further expression we use the following trick. First assume that $\hat{\theta}$ is an estimator based on y_{S-}. A consequence of this is that $\hat{\theta}$ is then independent of Y_S, and hence a further expression for the first term on the right side of (3.56) is easily obtained. We then argue that, if $\hat{\theta}$ is an estimator based on all the data, it only makes a difference of the order $o(m^{-1})$, and therefore the same approximation is still valid.

Suppose that $\hat{\theta} = \hat{\theta}_{S-}$, an estimator based on Y_{S-}. Then, by independence, we have

$$\begin{aligned} &\text{E}\left\{\left(\frac{\partial u}{\partial \theta'}\right)\sqrt{m}(\hat{\theta}_{S-} - \theta)\right\}^2 \\ &= \text{E}\left(\text{E}\left[\left\{\left(\frac{\partial u}{\partial \theta'}\right)\sqrt{m}(\hat{\theta}_{S-} - \theta)\right\}^2 \middle| y_S = w\right]\Bigg|_{w=y_S}\right) \\ &= \text{E}\left[\left\{\frac{\partial}{\partial \theta'}u(w, \theta)\right\} V_{S-}(\theta)\left\{\frac{\partial}{\partial \theta}u(w, \theta)\right\}\Bigg|_{w=y_S}\right] \\ &= \text{E}\left[\left\{\frac{\partial}{\partial \theta'}u(y_S, \theta)\right\} V_{S-}(\theta)\left\{\frac{\partial}{\partial \theta}u(y_S, \theta)\right\}\right] \\ &= e_{S-}(\theta), \end{aligned} \tag{3.57}$$

where $V_{S-}(\theta) = m\text{E}(\hat{\theta}_{S-} - \theta)(\hat{\theta}_{S-} - \theta)'$.

Let $\hat{\zeta}_1 = u(y_S, \hat{\theta}_{S-})$. Combining (3.47), (3.55), and (3.57), we obtain

$$\text{MSE}(\hat{\zeta}_1) = b(\theta) + m^{-1}e_{S-}(\theta) + o(m^{-1}). \tag{3.58}$$

Now, suppose that $\hat{\theta}$ is an estimator based on all data. We assume that $\hat{\theta}_{S-}$ satisfies $|\hat{\theta}_{S-} - \theta| = O(m^{-1/2})$ (in a suitable sense), and, in addition,

$|\hat{\theta}-\hat{\theta}_{S-}| = o(m^{-1/2})$. To see that this latter assumption is reasonable, consider a simple case in which one estimates the population mean μ by the sample mean $\hat{\mu}_m = m^{-1}\sum_{i=1}^m X_i$, where the X_is are i.i.d. observations. Then, we have, for example, $\hat{\mu}_m - \hat{\mu}_{m-1} = m^{-1}(X_m - \hat{\mu}_{m-1}) = O(m^{-1})$. Also note that $\hat{\mu}_m - \mu = O(m^{-1/2})$, $\hat{\mu}_{m-1} - \mu = O(m^{-1/2})$. (Here all the Os are in probability.) Note that

$$\mathrm{E}(\hat{\zeta} - \hat{\zeta}_1)(\hat{\zeta}_1 - \zeta) = \mathrm{E}(\hat{\zeta} - \hat{\zeta}_1)(\hat{\zeta}_1 - \tilde{\zeta}).$$

It follows, again by Taylor expansion and (3.58), that

$$\begin{aligned}\mathrm{MSE}(\hat{\zeta}) &= \mathrm{E}(\hat{\zeta} - \hat{\zeta}_1)^2 + 2\mathrm{E}(\hat{\zeta} - \hat{\zeta}_1)(\hat{\zeta}_1 - \tilde{\zeta}) + \mathrm{E}(\hat{\zeta}_1 - \zeta)^2 \\ &= \mathrm{MSE}(\hat{\zeta}_1) + o(m^{-1}) \\ &= b(\theta) + m^{-1}e(\theta) + o(m^{-1}),\end{aligned} \tag{3.59}$$

where $e(\theta)$ is $e_{S-}(\theta)$ with $V_{S-}(\theta)$ replaced by $V(\theta) = m\mathrm{E}(\hat{\theta} - \theta)(\hat{\theta} - \theta)'$.

Having obtained a second-order approximation to the MSE, we now consider the estimation of it. Note that, in (3.59), one may simply replace θ in $e(\theta)$ by $\hat{\theta}$, because this results in an error of the order $o(m^{-1})$ for the estimation of the MSE. However, one cannot do this to $b(\theta)$, because the bias may not be of the order $o(m^{-1})$. In fact, quite often we have $\mathrm{E}\{b(\hat{\theta}) - b(\theta)\} = O(m^{-1/2})$. However, if $|\hat{\theta} - \theta| = O(m^{-1/2})$ (in a suitable sense) and $\mathrm{E}(\hat{\theta} - \theta) = O(m^{-1})$, by Taylor expansion, we have

$$\begin{aligned}b(\hat{\theta}) &= b(\theta) + \left(\frac{\partial b}{\partial \theta'}\right)(\hat{\theta} - \theta) \\ &\quad + \frac{1}{2}(\hat{\theta} - \theta)'\left(\frac{\partial^2 b}{\partial\theta\partial\theta'}\right)(\hat{\theta} - \theta) + o(m^{-1}),\end{aligned}$$

and hence

$$\mathrm{E}\{b(\hat{\theta})\} = b(\theta) + m^{-1}B(\theta) + o(m^{-1}),$$

where

$$\begin{aligned}B(\theta) &= \left(\frac{\partial b}{\partial \theta'}\right) m\mathrm{E}(\hat{\theta} - \theta) \\ &\quad + \frac{1}{2}\mathrm{E}\left[\{\sqrt{m}(\hat{\theta} - \theta)\}'\left(\frac{\partial^2 b}{\partial\theta\partial\theta'}\right)\{\sqrt{m}(\hat{\theta} - \theta)\}\right].\end{aligned}$$

If we define $\widehat{\mathrm{MSE}}(\hat{\zeta})$ as (3.48), then it can be shown that (3.49) is satisfied.

Note that the arguments above are not a rigorous proof of (3.49) because, for example, $\mathrm{E}\{o_{\mathrm{P}}(m^{-1})\}$ is not necessarily $o(m^{-1})$. However, under regularity conditions a rigorous proof could be given. See, for example, Jiang and Lahiri (2001) for the case of binary responses.

Also note that the derivation above requires $\hat{\theta}$, $\hat{\theta}_{S-}$ satisfying certain conditions, namely,

$$|\hat{\theta} - \theta| = O_{\mathrm{P}}(m^{-1/2}), |\hat{\theta}_{S-} - \theta| = O_{\mathrm{P}}(m^{-1/2}),$$
$$|\hat{\theta} - \hat{\theta}_{S-}| = o_{\mathrm{P}}(m^{-1/2}), \mathrm{and} \mathrm{E}(\hat{\theta} - \theta) = O(m^{-1}). \tag{3.60}$$

Then the question is: are there such estimators? A class of estimators in GLMM that satisfy (3.60) are given by Jiang (1998). Also see Jiang and Zhang (2001). See section 4.2 for further discussion.

3.7.4 MSPE of the Model-Assisted EBP

Recall that the MSPE is defined as $\mathrm{MSPE}(\hat{\zeta}_i) = \mathrm{E}(\hat{\zeta}_i - \bar{Y}_i)^2$, where the expectation is taken with respect to both the sampling design and the assumed mixed model for the units in the sample. In this section, we assume that n_is are bounded for all i. Furthermore, we assume that the model holds for the sampled units so that (3.54) corresponds to the model-assisted EBP. Because the model is assumed to hold, we obtain an estimator of the MSPE whose bias is of the order $o(m^{-1})$ with respect to the assumed unit level mixed model. Under mild conditions, the bias is of the same order when an additional expectation is taken with respect to the sampling design. See Jiang and Lahiri (2005a) for further discussion.

So throughout the rest of this section, all expectations are with respect to the assumed model. We assume that $\zeta_i = \mathrm{E}(\bar{Y}_i|v_i)$, which holds, for example, under the normal linear mixed model of Prasad and Rao (1999) and the mixed logistic model, a special case of GLMM, of Jiang and Lahiri (2001). By this assumption and certain regularity conditions, it can be shown that $\bar{Y}_i - \zeta_i = O_{\mathrm{P}}(N_i^{-1/2})$. Thus, we have

$$(\hat{\zeta}_i - \bar{Y}_i)^2 = (\hat{\zeta}_i - \zeta_i)^2 + O_{\mathrm{P}}(N_i^{-1/2}).$$

Because of the above fact, we approximate $\mathrm{MSPE}(\hat{\zeta}_i)$ by $\mathrm{E}(\hat{\zeta}_i - \zeta_i)^2$. Note that here we assume that the population size N_i is much larger than m. To establish the results in the sequel rigorously one needs to show that the neglected terms are $o(m^{-1})$. Arguments to show that, for example, $\mathrm{E}\{o_{\mathrm{P}}(m^{-1})\} = o(m^{-1})$, are needed. Such results hold under suitable conditions that ensure uniform integrability. See Jiang and Lahiri (2001, Section 5). With this approximation, we have the following decomposition,

$$\begin{aligned}\mathrm{MSPE}(\hat{\zeta}_i) = {} & \mathrm{MSPE}(\tilde{\zeta}) + \mathrm{E}(\hat{\zeta}_i - \tilde{\zeta}_i)^2 \\ & + 2\mathrm{E}(\hat{\zeta}_i - \tilde{\zeta}_i)(\tilde{\zeta}_i - \zeta_i)^2 + o(m^{-1}).\end{aligned} \tag{3.61}$$

First, we have

$$
\begin{aligned}
\text{MSPE}(\tilde{\zeta}) &= \text{E}(\zeta_i^2) - \text{E}(\tilde{\zeta}_i^2) \\
&= \text{E}\left\{\sum_{j=1}^{n_i} w_{ij}\text{E}(y_{ij}|v_i)\right\}^2 + \text{E}\{u_i^2(\bar{y}_{iw}, \theta)\} \\
&\equiv b_i(\theta). \qquad (3.62)
\end{aligned}
$$

Second, by the same arguments as in the previous section, we have

$$
\text{E}(\hat{\zeta}_i - \tilde{\zeta}_i)^2 = e_i(\theta)m^{-1} + o(m^{-1}), \qquad (3.63)
$$

where

$$
e_i(\theta) = \text{E}\left\{\left(\frac{\partial u_i}{\partial \theta'}\right) V(\theta) \left(\frac{\partial u_i}{\partial \theta}\right)\right\}
$$

with $V(\theta) = m\text{E}(\hat{\theta} - \theta)(\hat{\theta} - \theta)'$.

Third, to obtain an approximation for the third term on the right side of (3.61), we make further assumptions on $\hat{\theta}$. Suppose that $\hat{\theta}$ is a solution to an estimating equation of the following type,

$$
M(\theta) = \frac{1}{m}\sum_{i=1}^{m} a_i(y_i, \theta) = 0, \qquad (3.64)
$$

where $y_i = (y_{ij})_{1 \le j \le n_i}$, $a_i(\cdot, \cdot)$ is vector valued such that $\text{E}a_i(y_i, \theta) = 0$ if θ is the true vector of parameters, $1 \le i \le m$. For example, it is easy to see that the maximum likelihood estimator of θ satisfies the above. It can be shown that if $\hat{\theta}$ satisfies (3.64), then

$$
\text{E}(\hat{\zeta}_i - \tilde{\zeta}_i)(\tilde{\zeta}_i - \zeta_i) = g_i(\theta)m^{-1} + o(m^{-1}), \qquad (3.65)
$$

where $g_i(\theta) = \text{E}[\omega_i(y, \theta)\{\text{E}(\zeta_i|\bar{y}_{iw}) - \text{E}(\zeta_i|y)\}]$ with

$$
\begin{aligned}
\omega_i(y, \theta) &= -\left(\frac{\partial u_i}{\partial \theta'}\right) A^{-1}a_i(y_i, \theta) + m\delta_i(y, \theta), \\
\delta_i(y, \theta) &= \left(\frac{\partial u_i}{\partial \theta'}\right) A^{-1}\left(\frac{\partial M}{\partial \theta'} - A\right) A^{-1}M(\theta) \\
&\quad + \frac{1}{2}\left(\frac{\partial u_i}{\partial \theta'}\right) A^{-1}\left\{M'(\theta)(A^{-1})'\text{E}\left(\frac{\partial^2 M_j}{\partial \theta \partial \theta'}\right) A^{-1}M(\theta)\right\} \\
&\quad + \frac{1}{2}M'(\theta)(A^{-1})'\frac{\partial^2 u_i}{\partial \theta \partial \theta'}A^{-1}M(\theta),
\end{aligned}
$$

$A = \text{E}(\partial M/\partial \theta')$. Here y represents the vector of all the data; that is, $y = (y_{ij})_{1 \le i \le m, 1 \le j \le n_i}$, M_j the jth component of M, and (b_j) a vector with components b_j.

Combining (3.61)–(3.63) and (3.65), we obtain the approximation

$$\text{MSPE}(\hat{\zeta}_i) = b_i(\theta) + \{e_i(\theta) + 2g_i(\theta)\}m^{-1} + o(m^{-1}). \tag{3.66}$$

Finally, if we define

$$\widehat{\text{MSPE}}(\hat{\zeta}_i) = b_i(\hat{\theta}) + \{\widehat{e_i(\theta)} + 2g_i(\hat{\theta}) - \widehat{B_i(\theta)}\}m^{-1}, \tag{3.67}$$

where

$$B_i(\theta) = m\left\{\left(\frac{\partial b_i}{\partial \theta'}\right)\text{E}(\hat{\theta}-\theta) + \frac{1}{2}\text{E}(\hat{\theta}-\theta)'\left(\frac{\partial^2 b_i}{\partial\theta\partial\theta'}\right)(\hat{\theta}-\theta)\right\},$$

and $\widehat{e_i(\theta)}$, $\widehat{B_i(\theta)}$ are estimators of $e_i(\theta)$, $B_i(\theta)$ given below, then, under suitable conditions, we have

$$\text{E}\{\widehat{\text{MSPE}}(\hat{\zeta}_i) - \text{MSPE}(\hat{\zeta}_i)\} = o(m^{-1}).$$

It remains to obtain estimators for $e_i(\theta)$ and $B_i(\theta)$. First, we have the following alternative expressions,

$$e_i(\theta) = \text{tr}\{V(\theta)G_1(\theta)\},$$
$$B_i(\theta) = \left(\frac{\partial b_i}{\partial\theta}\right)' v(\theta) + \frac{1}{2}\text{tr}\{V(\theta)G_2(\theta)\},$$

where $v(\theta) = m\text{E}(\hat{\theta}-\theta)$, $V(\theta) = m\text{E}(\hat{\theta}-\theta)(\hat{\theta}-\theta)'$,

$$G_1(\theta) = \text{E}\left(\frac{\partial u_i}{\partial\theta}\right)\left(\frac{\partial u_i}{\partial\theta}\right)' \quad \text{and} \quad G_2(\theta) = \frac{\partial^2 d_i}{\partial\theta\partial\theta'}.$$

$G_j(\theta)$ can be estimated by a plug-in estimator; that is, $G_j(\hat{\theta})$, $j = 1, 2$. As for $v(\theta)$ and $V(\theta)$, we propose to use the following sandwich-type estimators,

$$\widehat{V(\theta)} = m\left(\sum_{i=1}^m \left.\frac{\partial a_i}{\partial\theta'}\right|_{\theta=\hat{\theta}}\right)^{-1}\left(\sum_{i=1}^m \hat{a}_i\hat{a}_i'\right)\left(\sum_{i=1}^m \left.\frac{\partial a_i'}{\partial\theta}\right|_{\theta=\hat{\theta}}\right)^{-1}, \tag{3.68}$$

$$\begin{aligned}\widehat{v(\theta)} &= \frac{1}{m}\sum_{i=1}^m \hat{A}^{-1}\left(\left.\frac{\partial a_i}{\partial\theta'}\right|_{\theta=\hat{\theta}}\right)\hat{A}^{-1}\hat{a}_i \\ &\quad -\frac{1}{2}\hat{A}^{-1}\left\{\frac{1}{m}\sum_{i=1}^m \hat{a}_i'(\hat{A}^{-1})'\hat{H}_j\hat{A}^{-1}\hat{a}_i\right\},\end{aligned} \tag{3.69}$$

where $\hat{a}_i = a_i(y_i, \hat{\theta})$,

$$\hat{A} = \frac{1}{m}\sum_{i=1}^m \left.\frac{\partial a_i}{\partial\theta'}\right|_{\theta=\hat{\theta}},$$

$$\hat{H}_j = \frac{1}{m}\sum_{i=1}^m \left.\frac{\partial^2 a_{i,j}}{\partial\theta\partial\theta'}\right|_{\theta=\hat{\theta}},$$

and, as before, (b_j) represents a vector whose jth component is b_j. The derivations of (3.68) and (3.69) are given in Jiang and Lahiri (2005a).

3.8 Exercises

3.1. Show that, under the assumption made in the first paragraph of Section 3.2, the vector of observations, $y = (y_1, \dots, y_n)'$, has the same distribution as the Gaussian linear mixed model (1.1), where $\alpha \sim N(0, G)$, $\epsilon \sim N(0, \tau^2 I)$, and α and ϵ are independent.

3.2. Suppose that, in Example 3.3, the conditional distribution of y_i given α is binomial(k_i, p_i) instead of Bernoulli, where k_i is a known positive integer, and p_i is the same as in Example 3.3. Show that, with a suitable link function, this is a special case of GLMM. What is the dispersion parameter ϕ in this case? And what is $a_i(\phi)$?

3.3. Show that the likelihood function under the GLMM in Example 3.5 is given by (3.4).

3.4. Derive the Laplace approximation (3.6). What is the constant c in (3.7)? Please explain.

3.5. Verify (3.13) and obtain an expression for r. Show that r has expectation zero.

3.6. Consider the following simple mixed logistic model, which is a special case of GLMM. Suppose that, given the random effects $\alpha_1, \dots, \alpha_m$, binary responses y_{ij}, $i = 1, \dots, m$, $j = 1, \dots, n$ are conditionally independent with conditional probability $p_{ij} = \mathrm{P}(y_{ij} = 1|\alpha)$, where $\alpha = (\alpha_i)_{1 \le i \le m}$, such that $\mathrm{logit}(p_{ij}) = \mu + \alpha_i$, where $\mathrm{logit}(p) = \log\{p/(1-p)\}$ and μ is an unknown parameter. Furthermore, suppose that the random effects $\alpha_1, \dots, \alpha_m$ are independent and distributed as $N(0, \sigma^2)$, where σ^2 is an unknown variance. Show that, when $\sigma^2 = 0$, the approximation (3.15) is identical to the exact log-likelihood function under this model. What about the approximation (3.16), that is, the penalized quasi log-likelihood? Is it identical to the exact log-likelihood when $\sigma^2 = 0$?

3.7. Consider the Poisson-gamma HGLM of Example 3.6. Show that in this case the HMLE for β is the same as the (marginal) MLE for β.

3.8. Show that, in the case of Example 3.5, the equations (3.33) reduce to (3.34) and (3.35).

3.9. Consider the following simple mixed logistic model. Given the random effects $\alpha_1, \dots, \alpha_m$, responses y_{ij}, $i = 1, \dots, m$, $j = 1, \dots, k$, are conditionally independent with $\mathrm{logit}(p_{ij}) = \mu + \alpha_i$, where $p_{ij} = \mathrm{P}(y_{ij} = 1|\alpha)$. Specify the conditions of Corollary 3.2 as well as the conclusion.

3.10. For the same example in Exercise 3.9, specify the conditions of Theorem 3.4 as well as the conclusion.

3.11. Verify that the mixed logistic model of Example 3.8 is a special case of GLMM.

3.12. Consider the behavior of EBP in Example 3.8 (Continued) in section 3.6.2.1. Show that, with $x_{ij} = x_i$, one has, as $\sigma \to \infty$, $u_i(y_{i\cdot}, \theta) \approx \mathrm{logit}(\bar{y}_{i\cdot}) - x_i'\beta$.

3.13. Consider estimation of the MSE of $\hat{\alpha}_i$ in Example 3.8 (Continued) in section 3.6.2.1. Show that, in this case, the term $b(\theta)$ in (3.48) can be expressed as

$$b(\theta) = \sigma^2 - \sum_{k=0}^{n_i} u_i^2(k, \theta) p_i(k, \theta),$$

where

$$\begin{aligned} p_i(k, \theta) &= \mathrm{P}(y_{i\cdot} = k) \\ &= \sum_{z \in S(n_i, k)} \exp\left(\sum_{j=1}^{n_i} z_j x'_{ij} \beta \right) \mathrm{E}[\exp\{s_i(z_\cdot, \sigma\xi, \beta)\}] \end{aligned}$$

with $S(l, k) = \{z = (z_1, \ldots, z_l) \in \{0, 1\}^l : z_\cdot = z_1 + \cdots + z_l = k\}$.

3.13. Show that the numerator on the right side of (3.51) is $O(m^{-1})$.

4

Generalized Linear Mixed Models: Part II

4.1 Likelihood-Based Inference

As mentioned in Section 3.4, the likelihood function under a GLMM typically involves integrals with no analytic expressions, and therefore is difficult to evaluate. For relatively simple models, the likelihood function may be evaluated by numerical integration techniques. See, for example, Hinde (1982), and Crouch and Spiegelman (1990). Such a technique is tractable if the integrals involved are low-dimensional. The following is an example.

Example 4.1. Suppose that, given the random effects $\alpha_1, \dots, \alpha_m$, binary responses y_{ij}, $i = 1, \dots, m$, $j = 1, \dots, k$ are conditionally independent such that $\text{logit}(p_{ij}) = x'_{ij}\beta + \alpha_i$, where β is a vector of unknown regression coefficients, and $p_{ij} = \text{P}(y_{ij} = 1|\alpha)$. Furthermore, the random effects α_i, $1 \le i \le m$ are independent and distributed as $N(0, \sigma^2)$. It can be shown (Exercise 4.1) that the log-likelihood function under this model can be expressed as

$$l(\beta, \sigma^2) = \sum_{i=1}^{m} \int \frac{1}{\sqrt{2\pi\sigma^2}} \exp\left\{\sum_{j=1}^{k} s_{ij}(y_i, v, \beta) - \frac{v^2}{2\sigma^2}\right\} dv, \tag{4.1}$$

where $y_i = (y_{ij})_{1 \le j \le k}$ and

$$s_{ij}(y_i, v, \beta) = y_{ij}(x'_{ij}\beta + v) - \log\{1 + \exp(x'_{ij}\beta + v)\}.$$

It is clear that only one-dimensional integrals are involved in the log-likelihood function. Such integrals may be evaluated numerically. More specifically, suppose that $f(x)$ is a univariate function and one wishes to numerically evaluate the integral

$$I = \int f(x) dx.$$

The integral may be approximated by $\sum_{l=1}^{L} f(x_l)\Delta_l$, where $A = x_0 < x_1 < \cdots < x_L = B$, $\Delta_l = x_l - x_{l-1}$, and $A < 0$, $B > 0$ such that the absolute values of A, B are sufficiently large and those of Δ_ls are sufficiently small. This is, perhaps, the simplest numerical integration algorithm, but it by no mean is the most efficient (Exercise 4.2). In fact, some more efficient algorithms have been developed to numerically evaluate a one-dimensional integral. For example, one of the standard approaches in numerical integration is called Gaussian quadrature. Consider an approximation to an integral as follows,

$$\int_a^b w(x)f(x)dx \approx \sum_{j=1}^N w_j f(x_j).$$

Here the w_js and x_js are called weights and abscissas, respectively. A feature of Gaussian quadrature is that it chooses the weights and abscissas such that the above approximation is exact if $f(x)$ is a polynomial. It is easy to understand that such a choice will be dependent on the function $w(x)$. For example, for $w(x) = 1$, $N = 10$, the weights and abscissas are determined below for a ten-point Gauss–Legendre integration. Here the abscissas are symmetric around the midpoint of the range of integration, $x^* = (a+b)/2$, expressed as x^* and $x^* \pm d \cdot u_j$, $j = 1, \dots, 5$, where $d = (b-a)/2$ and u_j is given by 0.148874, 0.433395, 0.679410, 0.865063, 0.973907 for $j = 1, 2, 3, 4, 5$ up to the sixth decimal. Similarly, the weights are equal for symmetric abscissas, and for x^* and $x^* \pm d \cdot u_j$, $j = 1, \dots, 5$ the corresponding weights are 0, 0.295524, 0.269267, 0.219086, 0.149451, 0.066671. Other functions $w(x)$ that are commonly used include $w(x) = 1/\sqrt{1-x^2}$, $-1 < x < 1$ (Gauss–Chebyshev), $w(x) = e^{-x^2}$, $-\infty < x < \infty$ (Gauss–Hermite), and $w(x) = x^\alpha e^{-x}$, $0 < x < \infty$ (Gauss–Laguerre), where α is a positive constant. See, for example, Press et al. (1997) for more details. Numerical integration routines such as Gaussian quadrature have been implemented in SAS (NLMIXED), Stata, and MIXOR.

However, numerical integration is generally intractable if the dimension of integrals involved is greater than two. Alternatively, the integrals may be evaluated by Monte Carlo methods. It should be pointed out that, for problems involving irreducibly high-dimensional integrals, naive Monte Carlo usually does not work. For example, the high-dimensional integral in Example 3.5 cannot be evaluated by a naive Monte Carlo method. This is because a product of, say, 1600 terms with each term less than one is numerically zero. Therefore, an i.i.d. sum of such terms will not yield anything but zero without a huge simulation sample size! In the following sections, we introduce methods developed by researchers using advanced Monte Carlo techniques for computing the maximum likelihood estimators in GLMM.

4.1.1 A Monte Carlo EM Algorithm for Binary Data

McCulloch (1994) considered a threshold model, in which the response u_i ($i = 1, \dots, n$) represents an unobserved continuous variable and one only observes $y_i = 1_{(u_i > 0)}$, that is, whether or not u_i exceeds a threshold which, without loss of generality, is set to 0. Furthermore, it is assumed that

$$u = X\beta + Z_1\alpha_1 + \cdots + Z_s\alpha_s + \epsilon, \tag{4.2}$$

where β is a vector of unknown fixed effects, $\alpha_1, \dots, \alpha_s$ are independent vectors of random effects such that $\alpha_r \sim N(0, \sigma_r^2 I_{m_r})$, $1 \leq r \leq s$, and X, $Z_1, \dots, Z_s$ are known matrices. As usual, here ϵ represents a vector of errors that is independent of the random effects and distributed as $N(0, \tau^2 I_n)$. It is easy to show that the model is a special case of GLMM with binary responses (Exercise 4.3). The problem of interest is to estimate the fixed parameters β, σ_r^2, $1 \leq r \leq s$ as well as to predict the realized values of the random effects. McCulloch proposed to use an EM algorithm for inference about the model. Before introducing the method, we give a brief overview of the EM algorithm and its application in linear mixed models.

1. The EM algorithm. A key element in the EM algorithm is the so-called "complete data". Usually, these consist of the observed data, denoted by y, and some unobserved random variables, denoted by ξ. For example, ξ may be a vector of missing observations or a vector of random effects. The idea is to choose ξ appropriately so that maximum likelihood becomes trivial for the complete data. Let $w = (y, \xi)$ denote the complete data, which are assumed to have a probability density $f(w|\theta)$ depending on a vector θ of unknown parameters. In the E-step of the algorithm, one computes the conditional expectation

$$Q\{\theta|\theta^{(k)}\} = \mathrm{E}\left\{\log f(w|\theta)|y, \theta^{(k)}\right\},$$

where $\theta^{(k)}$ is the estimated θ at step k (the current step). Note that Q is a function of θ. Then, in the M-step, one maximizes $Q\{\theta|\theta^{(k)}\}$ with respect to θ to obtain the next step estimator $\theta^{(k+1)}$. The process is iterated until convergence. For more details, see, for example, Lange (1999). Laird and Ware (1982) applied the EM algorithm to estimation of the variance components in a Gaussian mixed model. Suppose that, in (4.2), u is replaced by y; that is, the observed data follow a Gaussian mixed model. The complete data then consist of y, $\alpha_1, \dots, \alpha_s$. The log-likelihood based on the complete data has the following expression,

$$\begin{aligned} l = c - \frac{1}{2}\Bigg\{ & n\log(\tau^2) + \sum_{r=1}^{s} m_r \log(\sigma_r^2) + \sum_{r=1}^{r} \frac{\alpha_r'\alpha_r}{\sigma_r^2} \\ & + \frac{1}{\tau^2}\left(y - X\beta - \sum_{r=1}^{s} Z_r\alpha_r\right)'\left(y - X\beta - \sum_{r=1}^{s} Z_r\alpha_r\right)\Bigg\}, \end{aligned}$$

where c does not depend on the data or parameters. Thus, to complete the E-step, one needs expressions for $\mathrm{E}(\alpha_r|y)$ and $\mathrm{E}(\alpha_r'\alpha_r|y)$, $1 \leq r \leq s$. By the theory of multivariate normal distribution, it is easy to show (Exercise 4.4)

$$\mathrm{E}(\alpha_r|y) = \sigma_r^2 Z_r' V^{-1}(y - X\beta), \tag{4.3}$$

$$\mathrm{E}(\alpha_r'\alpha_r|y) = \sigma_r^4(y - X\beta)'V^{-1}Z_rZ_r'V^{-1}(y - X\beta) + \sigma_r^2 m_r - \sigma_r^4 \mathrm{tr}(Z_r'V^{-1}Z_r), 1 \le r \le s, \quad (4.4)$$

where $V = \mathrm{Var}(y) = \tau^2 I_n + \sum_{r=1}^{s} \sigma_r^2 Z_r Z_r'$. Once the E-step is completed, the M-step is straightforward because

$$(\sigma_r^2)^{(k+1)} = m_r^{-1}\mathrm{E}(\alpha_i'\alpha_i|y)|_{\beta=\beta^{(k)},\sigma^2=(\sigma^2)^{(k)}}, 1 \le r \le s, \quad (4.5)$$

$$\beta^{(k+1)} = (X'X)^{-1}X'\left\{y - \sum_{q=1}^{s} Z_q \mathrm{E}(\alpha_q|y)|_{\beta=\beta^{(k)},\sigma^2=(\sigma^2)^{(k)}}\right\}, \quad (4.6)$$

where $\sigma^2 = (\sigma_j^2)_{1\le j\le s}$.

vspace2mm

2. Monte Carlo EM by Gibbs sampler. To apply the EM algorithm to the threshold model, the complete data consist of u, $\alpha_1, \dots, \alpha_s$. The procedure is very similar to that for Gaussian mixed models with y replaced by u. However, there is one major difference: the observed data are y, not u, which is not normal. As a result, expressions (4.3) and (4.4) no longer hold. In fact, no analytic expressions exist for the left sides of (4.3) and (4.4) in this case. In some simple cases, the conditional expectations may be evaluated by numerical integration, as discussed earlier. However, for more complicated random effects structure (e.g., crossed random effects), numerical integration may be intractable. McCulloch (1994) proposed using a Gibbs sampling approach to approximate the conditional expectations. The Gibbs sampler is a special case of the Metropolis–Hastings algorithm, a device for constructing a Markov chain with a prescribed equilibrium distribution π on a given state space. At each step of the algorithm, say, state i, a new destination state j is proposed; that is, sampled, according to a probability density $p_{ij} = p(j|i)$. Then, a random number is drawn uniformly from the interval $[0, 1]$ to determine if the proposed stage is acceptable. Namely, if the random number is less than

$$r = \min\left(\frac{\pi_j p_{ji}}{\pi_i p_{ij}}, 1\right),$$

the proposed stage is accepted. Otherwise, the proposed step is declined, and the chain remains in place. See, for example, Gelman et al. (2003) for more detail. Here r is called the acceptance probability. In the case of the Gibbs sampler, one component of the sample point will be updated at each step. If the component to be updated is chosen at random, then it can be shown that the acceptance probability is one (e.g., Lange 1999, pp. 332). However, in practice, the updates of the components are typically done according to the natural order of the components. The following outlines how the Gibbs sampler is used to generate a sample from the conditional distribution of $u|y$ (see McCulloch 1994). For each $1 \le i \le n$,

1. Compute $\sigma_i^2 = \mathrm{var}(u_i|u_j, j \neq i)$, $c_i = \mathrm{Cov}(u_i, u_{-i})$, where $u_{-i} = (u_j)_{j \neq i}$, and $\mu_i = \mathrm{E}(u_i|u_j, j \neq i) = x_i'\beta + c_i'(u_{-i} - X_{-i}\beta)$, where X_{-i} is X with its ith row x_i' removed. Note that c_i is an $(n-1) \times 1$ vector. Then,

2. Simulate u_i from a truncated normal distribution with mean μ_i and standard deviation σ_i. If $y_i = 1$, simulate u_i truncated above 0; if $y_i = 0$, simulate u_i truncated below 0.

Using the algorithm, McCulloch (1994) analyzed the salamander mating data (McCullagh and Nelder 1989, §14.5) and obtained the MLE of the parameters under a GLMM.

4.1.2 Extensions

1. MCEM with Metropolis–Hastings algorithm. The Monte Carlo EM method (MCEM method) described above has one limitation: that is, it applies only to the case of binary responses with a probit link and normal random effects. McCulloch (1997) extended the method in several ways. First, he used the Metropolis–Hastings algorithm instead of Gibbs to generate random samples for the MCEM. This allowed him to relax the dependency of his earlier results on the normal (mixed model) theory. More specifically, he considered a GLMM with a canonical link function such that

$$f(y|\alpha, \beta, \phi) = \exp\left\{\frac{y_i\eta_i - b(\eta_i)}{a(\phi)} + c(y_i, \phi)\right\}, \tag{4.7}$$

where $\eta_i = x_i'\beta + z_i'\alpha$ with x_i and z_i known, and $a(\cdot)$, $b(\cdot)$, and $c(\cdot,\cdot)$ are known functions. Furthermore, it is assumed that $\alpha \sim f(\alpha|\theta)$. Here, as usual, y and α denote the vectors of responses and random effects, respectively, and β, θ and ϕ the vectors of fixed effects, variance components, and additional dispersion parameter. Then, the main computational issue is to evaluate the conditional expectations $\mathrm{E}[\log\{f(y|\alpha, \beta, \phi)\}|y]$ and $\mathrm{E}[\log\{f(\alpha|\theta)\}|y]$. It can be shown that, at the kth step of any cycle, the acceptance probability of the Metropolis–Hastings algorithm is given by

$$r_k = \frac{\prod_{i=1}^n f(y_i|\alpha^*, \beta, \phi)}{\prod_{i=1}^n f(y_i|\alpha, \beta, \phi)},$$

where α denotes the previous draw from the conditional distribution of $\alpha|y$, and α^* is a candidate draw whose k component is u_k^*, a generated new value, and the other components are the same as those of α. An advantage of this expression is that it does not depend on the distribution of α; only the conditional distribution of $y|\alpha$ is involved. Thus, for example, the normality assumption is no longer important here.

2. Monte Carlo Newton–Raphson procedure. As the Newton–Raphson procedure is also widely used in maximum likelihood estimation, a Monte Carlo analogue of the Newton–Raphson algorithm (MCNR) was also developed. It can be shown (Exercise 4.5) that the ML equations for estimating β, θ, and ϕ may be expressed as

$$\mathrm{E}\left[\frac{\partial}{\partial\beta}\log\{f(y|\alpha,\beta,\phi)\}\bigg|y\right] = 0, \tag{4.8}$$

$$\mathrm{E}\left[\frac{\partial}{\partial\phi}\log\{f(y|\alpha,\beta,\phi)\}\bigg|y\right] = 0, \tag{4.9}$$

$$\mathrm{E}\left[\frac{\partial}{\partial\theta}\log\{f(\alpha|\theta)\}\bigg|y\right] = 0. \tag{4.10}$$

Equation (4.10) is often fairly easy to solve. For example, if the random effects are normally distributed, the left side of (4.10) has an analytic expression. On the other hand, the left sides of (4.8) and (4.9) typically involve conditional expectations of functions of α given y. To evaluate this expression, McCulloch (1997) used a scoring technique. First Taylor-expand the left side of (4.7) as a function of β around the true β and ϕ, just as in the derivation of the Newton–Raphson algorithm. Then, the conditional expectation is taken. By a similar derivation as in McCullagh and Nelder (1989, pp. 42), one obtains an iterative equation of the form

$$\beta^{(m+1)} = \beta^{(m)} + [\mathrm{E}\{X'W^{(m)}X|y\}]^{-1}X'\mathrm{E}[W^{(m)}G^{(m)}\{y-\mu^{(m)}\}|y],$$

where $G = \partial\eta/\partial\mu = \mathrm{diag}(\partial\eta_i/\partial\mu_i)$ [see (4.7)],

$$W^{-1} = \mathrm{diag}\left\{\left(\frac{\partial\eta_i}{\partial\mu_i}\right)^2 \mathrm{var}(y_i|\alpha)\right\},$$

and the superscript (m) means evaluation at $\beta^{(m)}$ and $\phi^{(m)}$.

3. Simulated ML. Finally, a method of simulated maximum likelihood (SML) was proposed. As noted earlier, a naive Monte-Carlo method often does not work here (see the discussion below Example 4.1). McCulloch (1997) proposed using a method called *importance sampling.* The idea is the following. Suppose that one needs to evaluate an integral of the form

$$I(f) = \int f(x)dx$$

for some function $f(x) \geq 0$. Note that the integral may be expressed as

$$\begin{aligned} I(f) &= \int \frac{f(x)}{h(x)}h(x)dx \\ &= \mathrm{E}\left\{\frac{f(\xi)}{h(\xi)}\right\}, \end{aligned}$$

where h is a pdf such that $h(x) > 0$ if $f(x) > 0$, and ξ is a random variable with pdf h. Thus, if one can generate a sequence of i.i.d. samples $\xi_1, \ldots, \xi_K$ with the pdf h, one can approximate the integral by

$$\mathrm{E}_p(f) \approx \frac{1}{K}\sum_{k=1}^{K}\frac{f(\xi_k)}{h(\xi_k)}.$$

See Gelman et al. (2003) for more details.

In the current situation, the likelihood function can be written as

$$\begin{aligned} L(\beta, \phi, \theta|y) &= \int f(y|\alpha, \beta, \phi) f(\alpha|\theta) d\alpha \\ &= \int \frac{f(y|\alpha, \beta, \phi) f(\alpha|\theta)}{g(\alpha)} g(\alpha) d\alpha \\ &\approx \frac{1}{K}\sum_{k=1}^{K}\frac{f(y|\alpha^{(k)}, \beta, \phi) f(\alpha^{(k)}|\theta)}{g(\alpha^{(k)})}, \end{aligned}$$

where $\alpha^{(k)}$, $k = 1, \dots, K$ are generated i. i. d. random vectors with (joint) pdf h. Note that this gives an unbiased estimator of the likelihood function regardless of g. The question then is how to choose g, which is called the importance sampling distribution. We use an example to illustrate the above methods, including the choice of g in SML.

Example 4.2. McCulloch (1997) used the following example to illustrate the MCEM, MCNR and SML methods. It is a special case of Example 4.1 with $x_{ij}'\beta = \beta x_{ij}$; that is, there is a single covariate x_{ij} and β is a scalar. The likelihood function can be expressed as

$$L(\beta, \sigma^2|y) = \prod_{i=1}^{m}\int\left[\prod_{j=1}^{k}\frac{\exp\{y_{ij}(\beta x_{ij} + u)\}}{1 + \exp(\beta x_{ij} + u)}\right]\frac{\exp(-u^2/2\sigma^2)}{(2\pi\sigma^2)^{1/2}} du.$$

For MCEM and MCNR, the acceptance probability for the Metropolis–Hastings algorithm is given by

$$r_k = \min\left[1, \prod_{i=1}^{m}\exp\{y_{i\cdot}(\alpha_i^* - \alpha_i)\}\prod_{j=1}^{n}\frac{1 + \exp(\beta x_{ij} + \alpha_i)}{1 + \exp(\beta x_{ij} + \alpha_i^*)}\right],$$

where $y_{i\cdot} = \sum_{j=1}^{k} y_{ij}$. Furthermore, for the MCNR iterations, one has $\mu = (\mu_{ij})_{1\le i\le n, 1\le j\le k}$ with $\mu_{ij} = h(\beta x_{ij} + \alpha_i)$ and $h(x) = e^x/(1 + e^x)$, and $W = \mathrm{diag}(W_{ij}, 1 \le i \le n, 1 \le j \le k)$ with $W_{ij} = \mu_{ij}(1 - \mu_{ij})$. As for SML, there is a question of what to use for g, the importance sampling distribution. In a simulation study, McCulloch used the distribution $N(0, \sigma^2)$, which is the same as the distribution of the random effects, as g. In other words, g is chosen as the pdf of the true distribution of the random effects. Even given such an advantage, SML still seems to perform poorly in this simple example.

On the other hand, both MCEM and MCNR performed reasonably well in two simulated examples, including the one discussed above (McCulloch 1997,

Section 6). In the concluding remark of the paper, the author further noted that MCEM and MCNR may be followed by a round of SML, which "usually refines the estimates and also gives accurate estimates of the maximum value of the likelihood." We recall this interesting observation later. The simulation results also show that the PQL estimator (Breslow and Clayton 1993) may perform poorly compared to MCEM or MCNR, which is not surprising given the inconsistency of the PQL estimator (see Section 3.5.2.4).

4.1.3 MCEM with I.I.D. Sampling

In this and the next sections, we introduce MCEM methods developed by Booth and Hobert (1999). Unlike McCulloch (1994, 1997), who used Markov chains to generate Monte Carlo samples, Booth and Hobert used i.i.d. sampling to construct Monte Carlo approximations at the E-step. More specifically, the authors used two methods to generate the Monte Carlo samples. The first is importance sampling; the second is rejection sampling. Furthermore, they suggested a rule for automatically increasing the Monte Carlo sample size as the algorithm proceeds, whenever necessary. This latter method is known as *automation.* We first introduce the two methods of generating i.i.d. Monte Carlo samples, and leave the automated method to the next section.

1. Importance sampling. The importance sampling was introduced earlier with the SML method, where we pointed out that an important issue for importance sampling is how to choose g, the importance sampling distribution. In McCulloch (1997), the author did not give a general suggestion on what g to use, but in a simulated example he used the true distribution of the random effects as g. Of course, such a choice would not be possible in practice, but it did not work well anyway, as the simulation results showed. Booth and Hobert suggested a multivariate-t distribution that matches (approximately) the mode and curvature of the conditional distribution of the random effects given the data. They noted that the E-step is all about the calculation of $Q\{\psi|\psi^{(l)}\} = \mathrm{E}[\log\{f(y,\alpha|\psi)\}|y;\psi^{(l)}]$, where $\psi = (\beta', \phi, \theta')'$ and l represents the current step. The expected value is computed under the conditional distribution of α given y, which has density

$$f(\alpha|y;\psi) \propto f(y|\alpha;\beta,\phi)f(\alpha|\theta). \tag{4.11}$$

There is a normalizing constant involved in the above expression, which is the (marginal) density function $f\{y|\psi^{(l)}\}$. (This is why $\propto$ is used instead of $=$.) However, as the authors pointed out, the constant does not play a role in the next M-step, because it depends only on $\psi^{(l)}$, whereas the next-step maximization is over ψ. Let $\alpha_1^*, \dots, \alpha_K^*$ be an i.i.d. sample generated from g, the importance sampling distribution. Then, we have the approximation

$$Q\{\psi|\psi^{(l)}\} \approx \frac{1}{K}\sum_{k=1}^{K} w_{kl}\log\{f(y,\alpha_k^*|\psi)\}, \tag{4.12}$$

where $w_{kl} = f\{\alpha_k^*|y;\psi^{(l)}\}/g(\alpha_k^*)$, known as the *importance weights*. The right side of (4.12) is then maximized with respect to ϕ in the M-step to obtain $\psi^{(l+1)}$. Note that the right side of (4.12) is not a completely known function (of ϕ), but subject to an unknown constant which is $f\{y|\psi^{(l)}\}$. However, as noted earlier, this constant makes no difference in the M-step, therefore we simply ignore it. In other words, the function that is actually maximized is the right side of (4.12) with $f\{y|\psi^{(l)}\}$ replaced by 1.

As for the importance sampling distribution g, Booth and Hobert proposed using a multivariate t-distribution whose mean and covariance matrix match the Laplace approximations of the mean and covariance matrix of $f(\alpha|y;\psi)$. An m-multivariate t-distribution with mean vector μ, covariance matrix Σ, and d degrees of freedom has the joint pdf

$$g(x) = \frac{\Gamma\{(m+d)/2\}}{(\pi d)^{m/2}\Gamma(d/2)}|\Sigma|^{-1/2}\left\{1 + \frac{1}{d}(x-\mu)'\Sigma^{-1}(x-\mu)\right\}^{-(m+d)/2},$$

$-\infty < x < \infty$. It remains to determine μ, Σ, and d. To this end, write $f = f(\alpha|y;\psi) = c\exp\{l(\alpha)\}$, where c is the unknown normalizing constant. Then, the mode of f, $\tilde{\alpha}$, is the solution to $l^{(1)}(\alpha) = 0$, where $l^{(1)}$ represents the vector of first derivatives. This is the Laplace approximation to the mean (e.g., de Bruijn 1981, §4). Similarly, the Laplace approximation to the covariance matrix is $-l^{(2)}(\tilde{\alpha})^{-1}$, where $l^{(2)}$ represents the matrix of second derivatives. However, Booth and Hobert did not offer a guideline for choosing the degree of freedom d and noted that the optimal choice of d would be a topic of further investigation. In the simulation studies that the authors conducted, $d = 40$ was used for the degrees of freedom.

2. Rejection sampling. Alternatively, i.i.d. samples may be generated from $f(\alpha|y;\psi)$ by multivariate rejection sampling as follows (Geweke 1996, §3.2). Write the conditional density as $f = cf_1 f_2$, where c is the normalizing constant, and f_1, f_2 are the two factors on the right side of (4.11). (i) First draw α from f_2 and, independently, u from the Uniform$[0,1]$ distribution. (ii) If $u \le f_1(\alpha)/\tau$, where $\tau = \sup_\alpha f_1(\alpha)$, accept α. Otherwise, return to (i). Note that f_1 corresponds to a likelihood function under a GLM. Therefore, τ can be found using the iterative WLS procedure for fitting the GLMs (McCullagh and Nelder 1989, pp. 206). Furthermore, it can be shown that τ need not change at each step of the MCEM algorithm (Booth and Hobert 1999, pp. 271–272).

4.1.4 Automation

One question in using the Monte Carlo methods is to determine the Monte Carlo sample size (MC size). This is particularly important in MCEM, because the process is often time consuming. As noted by Wei and Tanner (1990), it is inefficient to start with a large MC size when the current approximation is far from the truth. On the other hand, at some point, one may need to increase the MC size, if there is evidence that the approximation error is overwhelmed by the Monte Carlo error. It is clear that the key is to evaluate the relative size of the Monte Carlo error with respect to the approximation one, so that one knows when is the right time to increase the MC size and by how much.

Booth and Hobert (1999) propose an automated method that at each iteration of the MCEM determines the appropriate MC size. They first obtain an approximation to the variance of the estimator at the lth iteration (see below). Then, at the $(l+1)$th iteration, they construct an approximate $100(1-a)\%$ confidence region $(0 < a < 1)$ for $\psi_*^{(l+1)}$, which maximizes $Q\{\psi|\psi^{(l)}\}$. If the previous estimator $\psi^{(l)}$ lies within the region, the EM step is swamped by Monte Carlo errors. This means that the MC size K needs to increase. The proposed amount of increase in K is K/r, where r is a positive integer chosen by the researcher. The authors claim that they had been successfully using the method by choosing $r \in \{3, 4, 5\}$ with $a = 0.25$. Again, the optimal choice of a and r is subject to further investigation.

We now explain how to construct an approximate confidence region for $\psi_*^{(l+1)}$ given $\psi^{(l)}$. Denote the right side of (4.12) by $Q_{\rm m}\{\psi|\psi^{(l)}\}$. Because $\psi^{(l+1)}$ maximizes $Q_{\rm m}$, under regularity conditions, one has $Q_{\rm m}^{(1)}\{\psi^{(l+1)}|\psi^{(l)}\} = 0$, where $f^{(j)}$ denotes the jth derivative (vector or matrix), $j = 1, 2$. Thus, by Taylor expansion, we have

$$0 \approx Q_{\rm m}^{(1)}\{\psi_*^{(l+1)}|\psi^{(l)}\} + Q_{\rm m}^{(2)}\{\psi_*^{(l+1)}|\psi^{(l)}\}\{\psi^{(l+1)} - \psi_*^{(l+1)}\}.$$

This gives an approximation

$$\psi^{(l+1)} \approx \psi_*^{(l+1)} - [Q_{\rm m}^{(2)}\{\psi_*^{(l+1)}|\psi^{(l)}\}]^{-1} Q_{\rm m}^{(1)}\{\psi_*^{(l+1)}|\psi^{(l)}\},$$

which suggests that, given $\psi^{(l)}$, $\psi^{(l+1)}$ is approximately normally distributed with mean $\psi_*^{(l+1)}$ and covariance matrix

$$\left[Q_{\rm m}^{(2)}\{\psi_*^{(l+1)}|\psi^{(l)}\}\right]^{-1} \mathrm{Var}\left[Q_{\rm m}^{(1)}\{\psi_*^{(l+1)}|\psi^{(l)}\}\right] \times \left[Q_{\rm m}^{(2)}\{\psi_*^{(l+1)}|\psi^{(l)}\}\right]^{-1}. \tag{4.13}$$

(Note that, under regularity conditions, the matrix of second derivatives is symmetric.) An estimate of the covariance matrix is obtained by replacing $\psi_*^{(l+1)}$ in (4.13) by $\psi^{(l+1)}$, and approximating the middle term in (4.13) by

$$\frac{1}{K^2}\sum_{k=1}^{K} w_{kl}^2 \left[\frac{\partial}{\partial\psi}\log\{f(y, \alpha_k|\psi^{(l+1)})\}\right]\left[\frac{\partial}{\partial\psi}\log\{f(y, \alpha_k|\psi^{(l+1)})\}\right]'.$$

The MCEM methods using i.i.d. samples have the following advantages over those using Markov chains (McCulloch 1994, 1997). First, the assessment of the Monte Carlo errors is straightforward. Note that such an assessment is critical to the automated method. A related theoretical advantage is that conditions for the central limit theorem used in the normal approximation in the i.i.d. case is much easier to verify than under a Markov chain. As for the comparison of the two methods, Booth and Hobert found that rejection sampling is more efficient in a small sample (i.e., data) situation, whereas importance sampling works better in a large sample case. In terms of computational speed, the authors showed that for the same example that was considered by McCulloch (1997), the rejection sampling and importance sampling methods are about 2.5 times and 30 times faster, respectively, than the Metropolis–Hastings sampling method of McCulloch (1997).

It is interesting to note that McCulloch (1997) also used importance sampling in his SML method, but reported poor performance nevertheless. One possible explanation is that the choice of the importance sampling distribution, g, made a difference. In the simulation study of McCulloch (1997), the author used the (marginal) distribution of the random effects as g, whereas in Booth and Hobert (1999), the authors used a multivariate-t distribution that approximately matched the mean and covariance matrix of the conditional distribution of α given y. Note that the latter is the distribution from which one wishes to sample. It is possible that the multivariate-t had provided a better approximation than the marginal distribution of α. One may also note that Booth and Hobert used importance sampling between the iterations of EM algorithm, whereas McCulloch used it in SML as a one-time solver to the ML problem. In fact, McCulloch also reported better performance of SML when the latter was used as a follow-up to his MCEM or MCNR, which somehow is inline with Booth and Hobert's findings.

In conclusion, there have been some good advances in computing the maximum likelihood estimators in GLMM using Monte Carlo EM algorithms. On the other hand, the procedures are still computationally intensive as compared to the approximate inference methods introduced in the previous chapter. However, with the fast developments of computer hardware and technology, it is very probable that computation of the exact MLE in GLMM will eventually become a routine operation. On the other hand, some important theoretical problems regarding the MLE remain unsolved. For example, the salamander mating data has been analyzed by many authors, whereas some others used the same model and data structure for simulation studies. See, for example, McCullagh and Nelder (1989), Karim and Zeger (1992), Drum and McCullagh (1993), Lin and Breslow (1996), Lin (1997), Jiang (1998a), and Jiang and Zhang (2001). In particular, McCulloch (1994) and Booth and Hobert (1999) have used MCEM algorithms to compute the MLE for this dataset. However, one fundamental question has yet to be answered: suppose that the numbers of salamanders involved in the experiments increase, and that the assumed GLMM is correct. Is the MLE consistent? The answer is not obvious at all.

4.1.5 Maximization by Parts

Although MCEM methods have been a main approach in likelihood-based inference about GLMM, alternative procedures have also been proposed. In this section, we introduce a method proposed by Song et al. (2005), which they called maximization by parts (MBP). Again, the objective was to overcome some of the computational difficulties in maximum likelihood estimation. One of these difficulties is the computation of the second derivatives of the log-likelihood function. For example, the Newton–Raphson procedure requires calculations of both the first and second derivatives. If the likelihood function is complicated, the derivation and calculation of its derivatives, especially the second derivatives, can be both analytically and computationally challenging. The following is an example.

Example 4.3 (The Gaussian copula). A d-variate Gaussian copula distribution is defined as a d-variate distribution whose cdf is given by

$$C(u_1,\dots,u_k|\Gamma)=\Phi_{d,\Gamma}\{\Phi^{-1}(u_1),\dots,\Phi^{-1}(u_d)\},$$

$0<u_1,\dots,u_d<1$, where Γ is a $(d\times d)$ correlation matrix, $\Phi_{d,\Gamma}$ and Φ denote the cdfs of $N_d(0,\Gamma)$ and $N(0,1)$, respectively. It follows that the (joint) pdf of the Gaussian copula is given by

$$c(u_1,\dots,u_d|\Gamma)=|\Gamma|^{-1/2}\exp\left\{\frac{1}{2}w'(I-\Gamma^{-1})w)\right\}, \tag{4.14}$$

where $z=(w_j)_{1\le j\le d}$ with $w_j=\Phi^{-1}(u_j)$ (Exercise 4.6). Suppose that one observes d-dimensional independent vectors $y_1,\dots,y_n$ such that $y_i=(y_{ij})_{1\le j\le d}$ follows a d-variate Gaussian copula distribution with cdf

$$F(y_i|\theta)=C\{F_1(y_1|\theta_1),\dots,F_d(y_d|\theta_d)|\Gamma\},$$

where $F_j(\cdot|\theta_j)$ is a univariate cdf and θ_j is an unknown vector of parameters associated with F_j, $1\le j\le d$ (e.g., Song 2000). Here θ represents the vector of all the different parameters involved in θ_j, $1\le j\le d$, and Γ. Then (Exercise 4.6), the joint pdf of y_i is given by

$$f(y_i|\theta)=c\{F_1(y_{i1}|\theta_1),\dots,F_d(y_{id}|\theta_d)|\Gamma\}\prod_{j=1}^d f_j(y_{ij}|\theta_j), \tag{4.15}$$

where $f_j(\cdot|\theta_j)=(\partial/\partial y_{ij})F_j(y_{ij}|\theta_j)$. Furthermore, the marginal cdf of y_{ij} is $F_j(\cdot|\theta_j)$ and $f_j(\cdot|\theta_j)$, respectively (Exercise 4.6). Thus, the likelihood function under the assumed model can be expressed as

$$L(\theta)=\prod_{i=1}^n\left[c\{F_1(y_{i1}|\theta_1),\dots,F_d(y_{id}|\theta_d)|\Gamma\}\prod_{j=1}^d f_j(y_{ij}|\theta_j)\right].$$

As noted by Song et al. (2005), it is fairly straightforward to compute the first derivatives of the log-likelihood $l(\theta) = \log\{L(\theta)\}$, but it is much harder to derive analytically the second derivatives of l.

The idea of MBP is very easy to illustrate. Write the log-likelihood function as

$$l(\theta) = l_w(\theta) + l_e(\theta). \tag{4.16}$$

Let $\dot{l}$ denote the vector of first (partial) derivatives. Then, the likelihood equation

$$\dot{l}(\theta) = 0 \tag{4.17}$$

can be written as

$$\dot{l}_w(\theta) = -\dot{l}_e(\theta). \tag{4.18}$$

Here the θs on both sides of (4.18) are supposed to be the same, but they do not have to be so in an iterative equation, and this is the idea of MBP. The initial estimator $\hat{\theta}^{(1)}$ is a solution to $\dot{l}_w(\theta) = 0$. Then, use the equation

$$\dot{l}_w\{\hat{\theta}^{(2)}\} = -\dot{l}_e\{\hat{\theta}^{(1)}\} \tag{4.19}$$

to update to get the next step estimator $\hat{\theta}_2$, and so on.

It is easy to see that, if the sequence

$$\hat{\theta}^{(l)}, \quad l = 1, 2, \dots \tag{4.20}$$

converges, the limit, say, θ^*, satisfies (4.17). Furthermore, Jiang (2005b) observed the following. The left side of (4.17), evaluated at the sequence (4.20), has absolute values

$$|\dot{l}_e(\hat{\theta}_1) - \dot{l}_e(\hat{\theta}_0)|, \ |\dot{l}_e(\hat{\theta}_2) - \dot{l}_e(\hat{\theta}_1)|, \ |\dot{l}_e(\hat{\theta}_3) - \dot{l}_e(\hat{\theta}_2)|, \dots . \tag{4.21}$$

Suppose that the function $l_e(\cdot)$ is at least locally uniformly continuous. Consider the distances between consecutive points in (4.20):

$$|\hat{\theta}_1 - \hat{\theta}_0|, \ |\hat{\theta}_2 - \hat{\theta}_1|, \ |\hat{\theta}_3 - \hat{\theta}_2|, \dots \tag{4.22}$$

(here $\hat{\theta}_0$ serves as a "starting point"). If (4.22) shows sign of decreasing, which will be the case if the sequence is indeed convergent, (4.21) is expected to do the same. This means that the left side of (4.17) decreases in absolute value along the sequence (4.20). Note that, because the MLE satisfies (4.17), the absolute value of the left side of (4.17), evaluated at an estimator, may be used as a measure of "closeness" of the estimator to the MLE; and the efficiency of the estimator is expected to increase as it gets closer to the MLE.

From a practical standpoint, the most important issue regarding MBP seems to be the decomposition (4.16). We now use an example to illustrate.

Example 4.3 (Continued). For the Gaussian copula model, one decomposition of the log-likelihood has

$$l_w(\theta) = \sum_{i=1}^{n}\sum_{j=1}^{d} \log\{f_j(y_{ij}|\theta_j)\},$$

$$l_e(\theta) = \frac{1}{2}\left\{\sum_{i=1}^{n} z_i(\theta)'(I_d - \Gamma^{-1})z_i(\theta)\right\},$$

where the jth component of $z_i(\theta)$ is $\Phi^{-1}\{F_j(y_{ij}|\theta_j)\}$, $1 \leq j \leq d$. It is clear that l_w corresponds to the log-likelihood under a model with independent observations, and l_e is the difference between the real log-likelihood and the "working" independent log-likelihood.

In general, a condition for a good choice of l_w is the so-called *information dominance*. In other words, $\ddot{l}_w$ needs to be larger than $\ddot{l}_e$ in a certain sense (Song et al. 2005, Theorem 2). However, because $\ddot{l}$ is difficult to evaluate, this condition is not easy to verify. On the other hand, the argument above suggests a potentially practical procedure to verify that one has had a good choice of l_w, that is, to let the procedure run for a few steps. If the sequence (4.22) shows sign of decreasing, even if not after every step, it is an indication that a good choice has been made. This is because the same argument then shows that the left side of (4.17) decreases in absolute value along the sequence (4.20), hopefully to zero.

Another condition for choosing l_w is that $\dot{l}_w(\theta) = 0$ is an unbiased estimating equation, or, alternatively, that $\hat{\theta}_1$ is a consistent estimator. This condition is satisfied in the Gaussian copula model (Exercise 4.7).

The MBP method is potentially applicable to at least some class of GLMMs. It is suggested that the hierarchical log-likelihood of Lee and Nelder (1996; see our earlier discussion in Section 3.5.4) may be used as l_w. However, if the random effects in the GLMM are normally distributed, this will lead to a biased estimating equation. In fact, the solution to such an equation may not be consistent (Clayton 1996; Jiang 1999c). The choice of l_w or the performance of MBP with the proposed hierarchical log-likelihood l_w remain unclear to date.

Assuming that MBP is applicable to GLMM, the next question is how much does MBP help. As noted earlier, the procedure has a computational advantage in situations where $\ddot{l}$ is much more difficult to deal with (numerically or analytically) than $\dot{l}$. The Gaussian copula model provides a good example. Now let us consider a GLMM example.

Example 4.4. Suppose that, given the random effects u_i, $1 \leq i \leq a$ and v_j, $1 \leq j \leq b$, y_{ij} are conditionally independent such that

$$\text{logit}\{\mathrm{P}(y_{ij} = 1|u, v)\} = \beta_0 + \beta_1 x_{ij} + u_i + v_j,$$

where $u = (u_i)_{1 \le i \le a}$, $v = (v_j)_{1 \le j \le b}$, x_{ij} is a known covariate, and β_0 and β_1 are unknown coefficients. Furthermore, suppose that the random effects are independent with $u_i \sim N(0, \sigma^2)$, $v_j \sim N(0, \tau^2)$. It is more convenient to use the following expressions: $u_i = \sigma\xi_i$, $v_j = \tau\eta_j$, where $\xi_1, \dots, \xi_a$ and $\eta_1, \dots, \eta_b$ are i. i. d. $N(0, 1)$ random variables. Then, the log-likelihood function under this GLMM has the following expression,

$$
\begin{aligned}
l = c + \log \Bigg[\int \cdots \int \exp \Bigg\{ & \sum_{i=1}^{a} \sum_{j=1}^{b} \phi_{ij}(y_{ij}, \beta, \sigma\xi_i, \tau\eta_j) \\
& - \frac{1}{2} \sum_{i=1}^{a} \xi_i^2 - \frac{1}{2} \sum_{j=1}^{b} \eta_j^2 \Bigg\} d\xi_1 \cdots d\xi_a d\eta_1 \cdots d\eta_b \Bigg],
\end{aligned}
$$

where c is a constant, and

$$
\begin{aligned}
\phi_{ij}(y_{ij}, \beta, u_i, v_j) &= y_{ij}(\beta_0 + \beta_1 x_{ij} + u_i + v_j) \\
&\quad - \log\{1 + \exp(\beta_0 + \beta_1 x_{ij} + u_i + v_j)\}.
\end{aligned}
$$

For simplicity, let us assume that the variance components σ and τ are known, so that β_0 and β_1 are the only unknown parameters. It can be shown that the first and second derivatives of l have the following forms (Exercise 4.8):

$$
\begin{aligned}
\frac{\partial l}{\partial \beta_s} &= \frac{\int \cdots \int \exp\{\cdots\} \psi_s d\xi d\eta}{\int \cdots \int \exp\{\cdots\} d\xi d\eta}, \\
\frac{\partial^2 l}{\partial \beta_s \partial \beta_t} &= \frac{\int \cdots \int \exp\{\cdots\} \psi_{s,t} d\xi d\eta}{\int \cdots \int \exp\{\cdots\} d\xi d\eta} \\
&\quad - \left[\frac{\int \cdots \int \exp\{\cdots\} \psi_s d\xi d\eta}{\int \cdots \int \exp\{\cdots\} d\xi d\eta} \right] \\
&\quad \times \left[\frac{\int \cdots \int \exp\{\cdots\} \psi_t d\xi d\eta}{\int \cdots \int \exp\{\cdots\} d\xi d\eta} \right],
\end{aligned}
$$

$s, t = 0, 1$, where $d\xi = d\xi_1 \cdots d\xi_a$, $d\eta = d\eta_1 \cdots d\eta_b$,

$$
\begin{aligned}
\psi_s &= \sum_{i=1}^{a} \sum_{j=1}^{b} \frac{\partial \phi_{ij}}{\partial \beta_s}, \\
\psi_{s,t} &= \left(\sum_{i=1}^{a} \sum_{j=1}^{b} \frac{\partial \phi_{ij}}{\partial \beta_s} \right) \left(\sum_{i=1}^{a} \sum_{j=1}^{b} \frac{\partial \phi_{ij}}{\partial \beta_t} \right) \\
&\quad + \sum_{i=1}^{a} \sum_{j=1}^{b} \frac{\partial^2 \phi_{ij}}{\partial \beta_s \partial \beta_t}.
\end{aligned}
$$

Now the fact is, the integrals involved in $\dot{l}$ are equally difficult to evaluate as those involved in $\ddot{l}$. (Note that these are $(a + b)$-dimensional integrals.)

Nevertheless, new integrals do emerge in $\ddot{l}$; that is, there are three different integrals in $\dot{l}$, and six different ones in $\ddot{l}$. In general, if there are p unknown parameters, there may be as many as $p+1$ different integrals in $\dot{l}$, and as many as $p+1+p(p+1)/2 = (1/2)(p+1)(p+2)$ different integrals in $\ddot{l}$. If p is large, it is quite a saving in computation, provided that any single one of the integrals (involved in $\dot{l}$) can be evaluated.

4.1.6 Bayesian Inference

GLMM can be naturally formulated in a Bayesian framework, and thus analyzed using the Bayesian methods. The main difference in the model is that a (joint) prior is assumed for β and G, the covariance matrix of α. For example, a flat prior is sometimes used; that is, $\pi(\beta, G) \propto$ constant. The main objective of Bayesian inference is to obtain the posteriors for β, G, and α. In the following, we first describe the method for a special class of GLMM, the so-called longitudinal GLMMs.

Suppose that there are m independent clusters such that, within the ith cluster, the responses y_{ij}, $j = 1, \dots, n_i$ are conditionally independent given a d-dimensional vector α_i of random effects with conditional density

$$f(y_{ij}|\alpha_i) = \exp\left\{\frac{y_{ij}\theta_{ij} - b(\theta_{ij})}{\phi} + c(y_{ij}, \phi)\right\}, \tag{4.23}$$

where ϕ is a dispersion parameter, and the functions $b(\cdot)$ and $c(\cdot, \cdot)$ are the same as before. Furthermore, let $\mu_{ij} = \mathrm{E}(y_{ij}|\alpha_i)$ and assume that

$$g(\mu_{ij}) = x_{ij}'\beta + z_{ij}'\alpha_i, \tag{4.24}$$

where g is the link function, and x_{ij} and z_{ij} are known vectors. The random effects α_i is assumed to be distributed as $N(0, G)$. So far, no Bayesian modeling has come into play.

The model is completed by assuming that (β, G) has a joint prior density $\pi(\beta, G)$. The joint posterior for β and G is then given by (Exercise 4.9)

$$f(\beta, G|y) = \frac{\prod_{i=1}^m \int f(y_i|\beta, \alpha_i) f(\alpha_i|G)\pi(\beta, G) d\alpha_i}{\int \prod_{i=1}^m \int f(y_i|\beta, \alpha_i) f(\alpha_i|G)\pi(\beta, G) d\alpha_i d\beta dG}, \tag{4.25}$$

where $y_i = (y_{ij})_{1 \le j \le n_i}$, $f(y_i|\beta, \alpha_i)$ is the conditional density of y_i given β and α_i, and

$$f(\alpha_i|G) = \frac{1}{(2\pi)^{d/2}|G|^{1/2}} \exp\left(-\frac{1}{2}\alpha_i' G^{-1}\alpha_i\right).$$

It is easy to see that, if $\pi(\beta, G)$ is a flat prior (i.e., constant), the numerator in (4.25) is simply the likelihood function. Similarly, the posterior for α_i is given by

$$f(\alpha_i|y) = \frac{\int f(y_i|\alpha_i,\beta)f(\alpha_i|G)\pi(\beta,G)d\beta dG}{\int f(y_i|\alpha_i,\beta)f(\alpha_i|G)\pi(\beta,G)d\alpha_i d\beta dG}. \tag{4.26}$$

The posteriors (4.25) and (4.26) are typically numerically intractable, especially when the dimension of α_i, d, is greater than one. Therefore, Monte Carlo methods were proposed to handle the computation. For example, Zeger and Karim (1991) used the Gibbs sampler to evaluate the posteriors. The Gibbs sampler was introduced in section 4.1.1.2. In this case, the procedure calls for drawing samples from the following conditional distributions: $[\beta|\alpha, y]$, $[G|\alpha]$, and $[\alpha|\beta, G, y]$. The first conditional distribution can be approximated by a multivariate normal distribution, if the sample size is large. The mean of the multivariate normal distribution is the MLE obtained by fitting a GLM of y_{ij} on x_{ij} using $z_{ij}'\alpha_i$ as offsets (e.g., McCullagh and Nelder 1989). The covariance matrix of the multivariate distribution is the inverse of the Fisher information matrix, also obtained by fitting the GLM. However, in small samples the normal approximation may not be adequate. In such a case rejection sampling (see section 4.1.3.2) was used by Zeger and Karim to generate random samples. As for the second conditional distribution, it is known that if $\pi(G) \propto |G|^{-(d+1)/2}$, the posterior for G^{-1} is a Wishart distribution with $m - d + 1$ degrees of freedom and parameters $S = \sum_{i=1}^m \alpha_i\alpha_i'$ (e.g., Box and Tiao 1973). Thus, a random sample from the posterior of G can be drawn by first generating a standard Wishart random matrix with $m - d + 1$ degrees of freedom (e.g., Odell and Feiveson 1996), say, W, and then computing $G = (T'WT)^{-1}$, where T is the Choleski decomposition of S^{-1} satisfying $S^{-1} = T'T$ (see Appendix B). Finally, the third conditional distribution is the most difficult to generate. Zeger and Karim (1991) again used rejection sampling for this step. They used the idea of matching the mode and curvature of a multivariate Gaussian distribution. Note that a similar method was used by Booth and Hobert in their importance sampling procedure for MCEM (see Section 4.1.3.1).

In a related work, Karim and Zeger (1992) applied the Gibbs sampler to analyze the salamander mating data. Note that the GLMM for the salamander data is different from the above longitudinal GLMM in that the random effects are crossed rather than within clusters. See Section 4.4.3 for more details.

An important issue in Bayesian analysis is the propriety of the posterior (e.g., Hobert and Casella 1996). The issue was not addressed in Zeger and Karim (1991) nor in Karim and Zeger (1992). Ghosh et al. (1998) considered a class of longitudinal GLMMs useful in small area estimation, and provided sufficient conditions that ensure the propriety of the posterior. As in Zeger and Karim (1991), it is assumed that the observations come from m strata or local areas, and there are n_i observations, y_{ij}, $j = 1, \dots, n_i$ from the ith stratum. Again, the y_{ij} are conditionally independent with conditional density

$$f(y_{ij}|\theta_{ij}, \phi_{ij}) = \exp\left\{\frac{y_{ij}\theta_{ij} - b(\theta_{ij})}{\phi_{ij}} + c(y_{ij}, \phi_{ij})\right\}, \tag{4.27}$$

where ϕ_{ij} is known. Furthermore, the natural parameters θ_{ij}s satisfy

$$h(\theta_{ij}) = x'_{ij}\beta + \alpha_i + \epsilon_{ij}, \tag{4.28}$$

where h is a known function, α_is are the random effects and ϵ_{ij}s are errors. It is assumed that the α_is and ϵ_{ij}s are independent with $\alpha_i \sim N(0, \sigma_1^2)$ and $\epsilon_{ij} \sim N(0, \sigma_0^2)$. It is easy to see that (4.28) is more restrictive than (4.24). On the other hand, unlike (4.23), (4.27) allows ϕ_{ij} to be dependent on i and j, which may incorporate weights in the observations (e.g., McCullagh and Nelder 1989, pp. 29). As for the prior, Ghosh et al. (1998) assumed that β, σ_1^2, and σ_0^2 are mutually independent with $\beta \sim \text{Uniform}(R^p)$ $(p < m)$, $\sigma_1^{-2} \sim \text{Gamma}(a/2, b/2)$, and $\sigma_0^{-2} \sim \text{Gamma}(c/2, d/2)$, where a $\text{Gamma}(\lambda, \nu)$ distribution has pdf $f(u) \propto u^{\nu-1}e^{-\lambda u}$, $u > 0$. The following theorem was proved by Ghosh et al. (1998).

Theorem 4.1. Suppose that $a, c > 0$, $n-p+d > 0$, and $m+b > 0$, where $n = \sum_{i=1}^m n_i$ is the total sample size. If

$$\int_{\bar{\theta}_{ij}}^{\underline{\theta}_{ij}} \exp\left\{\frac{\theta y_{ij} - b(\theta)}{\phi_{ij}}\right\} h'(\theta)d\theta < \infty$$

for all y_{ij} and $\phi_{ij} > 0$, where $(\bar{\theta}_{ij}, \underline{\theta}_{ij})$ is the support of θ_{ij}, the (joint) posterior of the θ_{ij}s given y is proper.

The same authors also considered the so-called spatial GLMM with a Bayesian formulation. A spatial GLMM is such that the random effects corresponding to contiguous areas would have stronger (positive) correlation than noncontiguous areas. Sufficient conditions were also given that ensure propriety of the posterior under a spatial GLMM. See Ghosh et al. (1998) for details.

We conclude this section with an example of small area estimation using hierarchical Bayesian GLMM.

Example 4.5 (The National Health Interview Survey). Malec et al. (1997) published a study involving small area estimation using data from the National Health Interview Survey (NHIS). The NHIS is a multistage interview survey conducted annually for the National Center for Health Statistics to provide health and health care information for the civilian and noninstitutionalized population in the United States. The 1985–1994 NHIS sample involved about 200 primary sampling units (PSUs), selected from a stratified population of 1983 PSUs. Each PSU consists essentially of a single county or a group of contiguous counties. Within each sampled PSU, groups of households are aggregated into areal segments and sampled. Each year there is a new sample of approximately 50,000 households, or about 120,000 individuals. For more information about the design of the NHIS, see Massey et al. (1989).

Although the NHIS emphasizes national estimates, there is also a need for estimates for small geographical areas or subpopulations. For example, Lieu *et al.* (1993) used data from the 1988 NHIS child health supplement to compare access to health care and doctors for different races of children aged 10–17. Such problems are known as small area estimation because usually the sample sizes for the small geographical areas or subpopulations are fairly small. Clearly, the usual design-based estimator, which uses only the sample survey data for the particular small area of interest, is unreliable due to relatively small samples that are available from the area. Many methods exist for inference about small areas based on the idea of "borrowing strength." See Rao (2003) for an overview.

Most of the variables in the NHIS are binary. In this particular study (Malec et al. 1997), the binary variable Y indicates whether the individual has made at least one visit to a physician within the past year ($Y = 1$), or otherwise ($Y = 0$). Other available data include, for each sampled individual, demographic variables such as age, race, and sex; socioeconomic variables such as highest education level attained and presence of a telephone; and location of residence. The main interest of this study is to provide an estimate of a population proportion for a small geographical area or subpopulation. Such an estimate is directly associated with an estimate of the total. For example, to estimate the proportion of males in Iowa who have made at least one visit to a doctor within the past year, one estimates the total Θ of male Iowans who have made such visits and divide Θ by the total number of male Iowans, which is assumed to be known.

It is assumed that each individual in the population belongs to one of K mutually exclusive and exhaustive classes based on the individual's socioeconomic/demographic status. Let Y_{ijk} denote a binary random variable for individual j in cluster i, class k, where $i = 1, \dots, L$, $k = 1, \dots, K$, and $j = 1, \dots, N_{ik}$. Furthermore, given p_{ik}, the Y_{ijk}s are independent Bernoulli with $\mathrm{P}(Y_{ijk} = 1|p_{ik}) = p_{ik}$. A vector of M covariates, $X_k = (X_{k1}, \dots, X_{kM})'$, is assumed to be the same for each individual j in cluster i, class k such that

$$\text{logit}(p_{ik}) = X_k' \beta_i,$$

where $\beta_i = (\beta_{i1}, \dots, \beta_{iM})'$ is a vector of regression coefficients. Moreover, it is assumed that, conditional on η and Γ, the β_is are independent with

$$\beta_i \sim N(G_i \eta, \Gamma),$$

where each row of G_i is a subset of the cluster-level covariates $(Z_{i1}, \dots, Z_{ic})$ not necessarily related to X_k, η is a vector of regression coefficients, and Γ an $M \times M$ positive definite matrix. Finally, a reference prior distribution π is assigned to η and Γ such that

$$\pi(\eta, \Gamma) \propto \text{constant}.$$

Specifically, for the NHIS problem considered by Malec et al. (1997), the authors proposed the following model in which

$$X_k'\beta_i = \alpha + \beta_{i1}X_{0k} + \beta_{i2}X_{15,k} + \beta_{i3}X_{25,k} + \beta_{i4}X_{55,k}$$
$$+\beta_{i5}Y_kX_{15,k} + \beta_{i6}Y_kX_{25,k} + \beta_{i7}Z_k,$$

where Y_k and Z_k are indicator variables corresponding to gender and race, such that $Y_k = 1$ if class k corresponds to male and $Z_k = 1$ if class k corresponds to white; $X_{a,k} = \max(0, A_k - a)$ with A_k being the midpoint of the ages within class k [e.g., if class k corresponds to black females ages 40–45, then $X_{15,k} = \max(0, 42.5 - 15)$]. The authors indicated that the model is developed based on visual displays of the relationships between the log-odds of the presence/absence of at least one doctor visit within the past year and age, for each race by sex class, using the SAS forward stepwise logistic regression procedure PROC LOGISTIC.

The objective here is to make an inference about a finite population proportion P for a specified small area and subpopulation, expressed as

$$P = \frac{\sum_{i\in I}\sum_{k\in K}\sum_{j=1}^{N_{ik}} Y_{ijk}}{\sum_{i\in I}\sum_{k\in K} N_{ik}}.$$

Alternatively, one may consider the total of the small area or subpopulation,

$$\Theta = \left(\sum_{i\in I}\sum_{k\in K} N_{ik}\right) P$$
$$= \sum_{i\in I}\sum_{k\in K}\sum_{j=1}^{N_{ik}} Y_{ijk}.$$

Let y_{s} denote the vector of sampled observations. Then, because $\mathrm{E}(Y_{ijk}|p_{ik}) = p_{ik}$, the posterior mean of Θ can be expressed as

$$\mathrm{E}(\Theta|y_{\mathrm{s}}) = \sum_{i\in I}\sum_{k\in K}\sum_{j\in s_{ik}} y_{ijk} + \sum_{i\in I}\sum_{k\in K}\sum_{j\notin s_{ik}} \mathrm{E}(p_{ik}|y_{\mathrm{s}})$$
$$= \sum_{i\in I}\sum_{k\in K}\sum_{j\in s_{ik}} y_{ijk} + \sum_{i\in I}\sum_{k\in K}(N_{ik} - n_{ik})\mathrm{E}(p_{ik}|y_{\mathrm{s}}),$$

where s_{ik} denote the set of sampled individuals in cluster i and class k that has size n_{ik}, and

$$p_{ik} = \frac{\exp(X_k'\beta_i)}{1 + \exp(X_k'\beta_i)}.$$

Similarly, the posterior variance of Θ can be expressed as

$$\mathrm{var}(\Theta|y_{\mathrm{s}}) = \sum_{i\in I}\sum_{k\in K}(N_{ik} - n_{ik})\mathrm{E}\{p_{ik}(1 - p_{ik})|y_{\mathrm{s}}\}$$
$$+\mathrm{var}\left\{\sum_{i\in I}\sum_{k\in K}(N_{ik} - n_{ik})p_{ik}\middle| y_{\mathrm{s}}\right\}.$$

Note that the posteriors $f(\beta, \eta, \Gamma|y_s)$, where $\beta = (\beta_i)_{1\le i\le L}$, do not have simple closed-form expressions. Malec et al. used the Gibbs sampler (see Section 4.1.1.2) to evaluate the posterior means and variance above. See Malec et al. (1997; pp. 818) for more details.

4.2 Estimating Equations

The general framework of estimating functions was set up by V. P. Godambe some 30 years before that of generalized linear mixed models (Godambe 1960). In Godambe (1991), the author viewed the approach as an extension of the Gauss–Markov theorem. An estimating function is a function, possibly vector valued, that depends both on $y = (y_i)_{1\le i\le n}$, a vector of observations, and θ, a vector of parameters. Denoted by $g(y, \theta)$, the estimating function is required to satisfy

$$\mathrm{E}_\theta\{g(y,\theta)\} = 0 \tag{4.29}$$

for every θ. For simplicity, let us first consider the case that $y_1, \dots, y_n$ are independent with $\mathrm{E}(y_i) = \theta$, a scalar. Let $\mathcal{G}$ denote the class of estimating functions of the form

$$g(y,\theta) = \sum_{i=1}^{n} a_i(\theta)(y_i - \theta),$$

where $a_i(\theta)$ are differentiable functions with $\sum_i a_i(\theta) \neq 0$. Then, an extension of the Gauss–Markov theorem states the following (Godambe 1991).

Theorem 4.2. If $\mathrm{var}(y_i) = \sigma^2$, $1 \le i \le n$, $g^* = \sum_{i=1}^n (y_i - \theta)$ is an optimal estimating function within $\mathcal{G}$ and the equation $g^* = 0$ provides $\bar{y}$, the sample mean, as an estimator of θ.

The equation

$$g(y,\theta) = 0 \tag{4.30}$$

to be solved for θ is called an estimating equation. In Theorem 4.2, the optimality is in the following sense, which was also introduced by Godambe. Note that for (4.30) to be used as an estimating equation, the corresponding estimating function should be as close to zero as possible, if θ is the true parameter. In view of (4.29), this means that one needs to minimize $\mathrm{var}(g)$. On the other hand, in order to distinguish the true θ from a false one, it makes sense to maximize $\partial g/\partial\theta$, or the absolute value of its expected value. When both are put on the same scale, the two criteria for optimality can be combined by considering

$$\frac{\mathrm{var}(g)}{\{\mathrm{E}(\partial g/\partial\theta)\}^2} = \mathrm{var}(g_s), \tag{4.31}$$

where $g_s = g/\mathrm{E}(\partial g/\partial\theta)$ is a standardized version of g. Thus, the optimality in Theorem 4.2 is in the sense that

$$\mathrm{var}(g_s^*) \le \mathrm{var}(g_s) \quad \text{for any } g \in \mathcal{G}.$$

Now consider a multivariate version of the estimating function. Let y be a vector of responses that is associated with a vector x of explanatory variables. Here we allow x to be random as well. Suppose that the (conditional) mean of y given x is associated with θ, a vector of unknown parameters. For notational simplicity, write $\mu = \mathrm{E}_\theta(y|x) = \mu(x, \theta)$, and $V = \mathrm{Var}(y|x)$. Here Var represents the covariance matrix, and Var or E without subscript θ means to be taken at the true θ. Let $\dot{\mu}$ denote the matrix of partial derivatives; that is, $\dot{\mu} = \partial\mu/\partial\theta'$. Consider the following class of vector-valued estimating functions $\mathcal{H} = \{G = A(y - \mu)\}$, where $A = A(x, \theta)$, such that $\mathrm{E}(\dot{G})$ is nonsingular. The following theorem can be established.

Theorem 4.3. Suppose that V is known, and that $\mathrm{E}(\dot{\mu}'V^{-1}\dot{\mu})$ is nonsingular. Then, the optimal estimating function within $\mathcal{H}$ is given by $G^* = \dot{\mu}'V^{-1}(y - \mu)$, that is, with $A = A^* = \dot{\mu}'V^{-1}$.

Here the optimality is in a similar sense to the univariate case. Define the partial order of nonnegative definite matrices as $A \ge B$ if $A - B$ is nonnegative definite. Then, the optimality in Theorem 4.3 is in the sense that the estimating function G^* maximizes, in the partial order of nonnegative definite matrices, the generalized information criterion

$$\mathcal{I}(G) = \{\mathrm{E}(\dot{G})\}'\{\mathrm{E}(GG')\}^{-1}\{\mathrm{E}(\dot{G})\}, \tag{4.32}$$

where $\dot{G} = \partial G/\partial\theta'$. It is easy to see that (4.32) is, indeed, the Fisher information matrix when G is the score function corresponding to a likelihood – that is, $G = \partial \log(L)/\partial\theta$, where L is the likelihood function – and this provides another view of Godambe's criterion of optimality. Also, (4.32) is equal to the reciprocal of (4.31) in the univariate case, so that maximizing (4.32) is equivalent to minimizing (4.31). The proof of Theorem 4.3 is given in Section 4.6.1.

4.2.1 Generalized Estimating Equations (GEE)

In the case of longitudinal GLMM, the optimal estimating function according to Theorem 4.3 can be expressed as

$$G^* = \sum_{i=1}^m \dot{\mu}_i' V_i^{-1}(y_i - \mu_i),$$

where $y_i = (y_{ij})_{1\le j\le n_i}$, $\mu_i = \mathrm{E}(y_i)$, and $V_i = \mathrm{Var}(y_i)$. Here, as in the earlier sections, the covariates x_i are considered fixed rather than random. The corresponding estimating equation is known as the generalized estimating equation, or GEE (Liang and Zeger 1986), given by

$$\sum_{i=1}^{m} \dot{\mu}_i' V_i^{-1}(y_i - \mu_i) = 0. \tag{4.33}$$

In (4.33), it is assumed that V_i, $1 \leq i \leq m$ are known because, otherwise, the equation cannot be solved. However, the true V_is are unknown in practice. Note that, under a GLMM, the V_is may depend on a vector of variance components θ in addition to β; that is, $V_i = V_i(\beta, \theta)$, $1 \leq i \leq m$. If a GLMM is not assumed and neither is any other parametric model for the covariances, the V_is may be completely unknown. Liang and Zeger proposed to use "working" covariance matrices instead of the true V_is to obtain the GEE estimator. For example, one may use the identity matrices that correspond to a model assuming independent errors with equal variance. The method is justified in the following sense. As is shown by Liang and Zeger, under some regularity conditions, the resulting GEE estimator is consistent despite that the working covariances misspecify the true V_is. However, the estimator based on working V_is may be inefficient as compared to that based on the true V_is.

Alternatively, one may replace the V_is in (4.33) by their consistent estimators, say, $\hat{V}_i$s. For example, under a GLMM, if θ is replaced by a $\sqrt{m}$-consistent estimator, say, $\hat{\theta}$ [i.e., $\sqrt{m}(\hat{\theta} - \theta)$ is bounded in probability], the resulting GEE estimator is asymptotically as efficient as the GEE estimator based on the true V_is. (Of course, the latter is not an estimator unless the V_is are known.) This means that $\sqrt{m}(\hat{\beta} - \beta)$ has the same asymptotic covariance matrix as $\sqrt{m}(\tilde{\beta} - \beta)$, where $\hat{\beta}$ is the solution to (4.33) with θ replaced by $\hat{\theta}$, and $\tilde{\beta}$ is that with the true V_is (e.g., Liang and Zeger 1986). However, to find a $\sqrt{m}$-consistent estimator one typically needs to assume a parametric model for the V_is, which increases the risk of model misspecifications. Even under a parametric covariance model, the $\sqrt{m}$-consistent estimator may not be easy to compute, especially if the model is complicated. In the next section, we propose another alternative that offers a more robust and computationally attractive solution.

The GEE method has been used in the analysis of longitudinal data, in which β is often the problem of main interest. Although, under the GLMM assumption, β may be estimated by likelihood-based methods (see Section 4.1), there are some concerns about such methods. First, as discussed earlier, the likelihood based methods are computationally intensive, and therefore may be intractable for analysis involving variable selection (see Section 4.3). Second, the efficiency of the likelihood-based methods may be undermined in the case of model misspecification, which often occurs in the analysis of longitudinal data. For example, in longitudinal studies there often exists serial correlation among the repeated measures from the same subject. Such a serial correlation may not be taken into account by a GLMM. Note that, under the GLMM assumption, the repeated measures are conditionally independent given the random effects, which means that no (additional) serial correlation exists once the values of the random effects are specified. We consider an example.

Example 4.6. Consider the salamander mating example discussed earlier (see Section 3.3.1). McCullagh and Nelder (1989) proposed a GLMM for analyzing the data, in which random effects corresponding to the female/male animals were introduced. The dataset and model have been extensively studied. However, in most cases it was assumed that a different set of animals (20 for each sex) was used in each mating experiment, although, in reality, the same set of animals was repeatedly used in two of the experiments (McCullagh and Nelder 1989, §14.5). Furthermore, most of the GLMMs used in this context (with the exception of, perhaps, Jiang and Zhang 2001) assumed that no further correlation among the data exists given the random effects. However, the responses in this case should be considered longitudinal, because repeated measures were collected from the same subjects (once in the summer, and once in the autumn of 1986). Therefore, serial correlation may exist among the repeated responses even given the random effects (i.e., the animals). In other words, the true correlations among the data may not have been adequately addressed by the GLMMs.

The GEE method is computationally more attractive than the likelihood-based methods. More important, GEE does not require a full specification of the distribution of the data. In fact, consistency of the GEE estimator only requires correct specification of the mean functions; that is, μ_i, $1 \leq i \leq n$. Of course, for the estimator to maintain the (asymptotic) optimality (in the sense of Theorem 4.3), the covariance matrices V_i, $1 \leq i \leq n$ also need to be correctly specified (and consistently estimated), but such assumptions are still much weaker than the full specification of the distribution. For example, the GEE method is applicable to cases beyond the scope of GLMM, such as Example 4.6 above (see Jiang and Zhang 2001). See Section 4.2.4 below for more examples. On the other hand, so far the majority of the literature on (correct) specification of the V_is has been using parametric models for the variance–covariance structure of the data in order to obtain a $\sqrt{m}$-consistent estimator of θ (see earlier discussion). Such a method is sensitive to model misspecifications, and may be difficult to operate computationally, say, under a GLMM. Because of such concerns, a different approach is proposed in the next section.

4.2.2 Iterative Estimating Equations

As noted earlier, to obtain the optimal GEE estimator, one needs to know the true covariance matrices V_i in (4.33). In this section, we propose an iterative procedure, which allows one to obtain an estimator that is asymptotically as efficient as the optimal GEE estimator without knowing the true V_is. The method is an extension of the I-WLS method introduced in Section 1.4.3. Note that WLS is the special case of GEE in linear models (Exercise 4.10). We describe the extension below under the assumption of a semiparametric regression model and then discuss its application to longitudinal GLMMs.

We consider a follow-up study conducted over a set of prespecified visit times $t_1, \ldots, t_b$. Suppose that the responses are collected from subset i at the visit times t_j, $j \in J_i \subset J = \{1, \ldots, b\}$. Let $y_i = (y_{ij})_{j \in J_i}$. Here we allow the visit times to be dependent on the subject. This enables us to include some cases with missing responses, but not in an informative way. The latter case is considered in Section 4.5.3. Let $X_{ij} = (X_{ijl})_{1 \le l \le p}$ represent a vector of explanatory variables associated with y_{ij} so that $X_{ij1} = 1$. Write $X_i = (X_{ij})_{j \in J_i} = (X_{ijl})_{i \in J_i, 1 \le l \le p}$. Note that X_i may include both time-dependent and independent covariates so that, without loss of generality, it may be expressed as $X_i = (X_{i1}, X_{i2})$, where X_{i1} does not depend on j (i.e., time) whereas X_{i2} does. We assume that (X_i, Y_i), $i = 1, \ldots, m$ are independent. Furthermore, it is assumed that

$$\mathrm{E}(Y_{ij}|X_i) = g_j(X_i, \beta), \tag{4.34}$$

where β is a $p \times 1$ vector of unknown regression coefficients and $g_j(\cdot, \cdot)$ are fixed functions. We use the notation $\mu_{ij} = \mathrm{E}(Y_{ij}|X_i)$ and $\mu_i = (\mu_{ij})_{j \in J_i}$. Note that $\mu_i = \mathrm{E}(Y_i|X_i)$. In addition, denote the (conditional) covariance matrix of Y_i given X_i as

$$V_i = \mathrm{Var}(Y_i|X_i), \tag{4.35}$$

whose (j,k)th element is $v_{ijk} = \mathrm{cov}(Y_{ij}, Y_{ik}|X_i) = \mathrm{E}\{(Y_{ij} - \mu_{ij})(Y_{ik} - \mu_{ik})|X_i\}$, $j, k \in J_i$. Note that the dimension of V_i may depend on i. Let $D = \{(j,k) : j, k \in J_i \text{ for some } 1 \le i \le n\}$.

Our main interest is to estimate β, the vector of regression coefficients. According to the earlier discussion, if the V_is are known, β may be estimated by the GEE (4.33). On the other hand, if β is known, the covariance matrices V_i can be estimated by the method of moments as follows. Note that for any $(j,k) \in D$, some of the v_{ijk}s may be the same, either by the nature of the data or by the assumptions. Let L_{jk} denote the number of different v_{ijk}s. Suppose that $v_{ijk} = v(j,k,l)$, $i \in I(j,k,l)$, where $I(j,k,l)$ is a subset of $\{1, \ldots, m\}$, $1 \le l \le L_{jk}$. For any $(j,k) \in D$, $1 \le l \le L_{jk}$, define

$$\hat{v}(j,k,l) = \frac{1}{n(j,k,l)} \sum_{i \in I(j,k,l)} \{Y_{ij} - g_j(X_i, \beta)\}\{Y_{ik} - g_k(X_i, \beta)\}, \tag{4.36}$$

where $n(j,k,l) = |I(j,k,l)|$, the cardinality. Then, define $\hat{V}_i = (\hat{v}_{ijk})_{j,k \in J_i}$, where $\hat{v}_{ijk} = \hat{v}(j,k,l)$, if $i \in I(j,k,l)$.

The main points of the previous paragraph may be summarized as follows. If the V_is were known, one could estimate β by the GEE; on the other hand, if β were known, one could estimate the V_is by the method of moments. It is clear that there is a cycle here, which motivates the following iterative procedure. Starting with an initial estimator of β, use (4.36), with β replaced by the initial estimator, to obtain the estimators of the V_is; then use (4.33) to update the estimator of β and repeat the process. We call such a procedure iterative estimating equations, or IEE. If the procedure converges, the limiting estimator is called the IEE estimator, or IEEE. It is easy to see that IEE is an extension of I-WLS discussed in Section 1.4.3.

In practice, the initial estimate of β may be obtained as the solution to (4.33) with $V_i = I$, the identity matrix (with the suitable dimension).

As in the case of I-WLS, one may conjecture about linear convergence of IEE as well as asymptotic efficiency of IEEE in the sense that the latter is asymptotically as efficient as the optimal GEE estimator obtained by solving (4.33) with the true V_is. In Section 4.5.2 we give conditions under which these conjectures are indeed true.

To apply IEE to a longitudinal GLMM, we denote the responses by y_{ij}, $i = 1, \dots, m$, $j = 1, \dots, n_i$, and let $y_i = (y_{ij})_{1 \le j \le n_i}$. We assume that each y_i is associated with a vector of random effects α_i that has dimension d such that (4.24) holds. Furthermore, we assume that the responses from different clusters $y_1, \dots, y_m$ are independent. Finally, suppose that

$$\alpha_i \sim f(u|\theta), \tag{4.37}$$

where $f(\cdot|\theta)$ is a d-variate pdf known up to a vector of dispersion parameters θ such that $\mathrm{E}_\theta(\alpha_i) = 0$. Let $\psi = (\beta', \theta')'$. Then, we have

$$\begin{aligned} \mathrm{E}(y_{ij}) &= \mathrm{E}\{\mathrm{E}(y_{ij}|\alpha_i)\} \\ &= \mathrm{E}\{h(x_{ij}'\beta + z_{ij}'\alpha_i)\} \\ &= \int h(x_{ij}'\beta + z_{ij}'u) f(u|\theta) du, \end{aligned}$$

where $h = g^{-1}$. Let $W_i = (X_i \ Z_i)$, where $X_i = (x_{ij}')_{1 \le j \le n_i}$, $Z_i = (z_{ij}')_{1 \le j \le n_i}$. For any vectors $a \in R^p$, $b \in R^d$, define

$$\mu_1(a, b, \psi) = \int h(a'\beta + b'u) f(u|\theta) du.$$

Furthermore, for any $n_i \times p$ matrix A and $n_i \times d$ matrix B, let $C = (A \ B)$, and $g_j(C, \psi) = \mu_1(a_j, b_j, \psi)$, where a_j' and b_j' are the jth rows of A and B, respectively. Then, it is easy to see that

$$\mathrm{E}(y_{ij}) = g_j(W_i, \psi). \tag{4.38}$$

It is clear that (4.38) is simply (4.34) with X_i replaced by W_i, and β replaced by ψ. Note that here, because W_i is a fixed matrix of covariates, we have $\mathrm{E}(y_i|W_{ij}) = \mathrm{E}(y_{ij})$. In other words, the longitudinal GLMM satisfies the semiparametric regression model introduced above, hence IEE applies.

The IEE approach is marginal in that it does not make use of an explicit model involving the random effects; only the expression (4.38) is needed. The advantage of this approach is robustness. For example, it does not require that the random effects $\alpha_1, \dots, \alpha_m$ are independent, neither does it require a parametric expression, such as (4.38), for the variance and covariance. A method relying on fewer assumptions is usually more robust to model misspecifications. Also note that the independence assumption regarding $y_1, \dots, y_m$ is easier to check than the same assumption about the random effects $\alpha_1, \dots, \alpha_m$, because the y_is are observed. The disadvantage of the marginal approach is that it does not provide estimates of the random effects, which are of interest in some cases. For example, in small area estimation (e.g., Rao 2003), the random effects are associated with the small area means, which are often of main interest. See Lee and Nelder (2004) for an overview with discussions on the use of random effects models and marginal models. We consider an example.

Example 4.7. Consider a random-intercept model with binary responses. Let y_{ij} be the response for subject i collected at time t_j. We assume that given a subject-specific random effect (the random-intercept) α_i, binary responses y_{ij}, $j = 1, \dots, k$ are conditionally independent with conditional probability $p_{ij} = \mathrm{P}(y_{ij} = 1|\alpha_i)$, which satisfies $\mathrm{logit}(p_{ij}) = \beta_0 + \beta_1 t_j + \alpha_i$, where β_0, β_1 are unknown coefficients. Furthermore, we assume that $\alpha_i \sim N(0, \sigma^2)$, where $\sigma > 0$ and is unknown. Let $y_i = (y_{ij})_{1\le j\le k}$. It is assumed that $y_1, \dots, y_m$ are independent, where m is the number of subjects.

It is easy to show that, under the assumed model, one has

$$\begin{aligned} \mathrm{E}(y_{ij}) &= \int_{-\infty}^{\infty} h(\beta_0 + \beta_1 t_j + \sigma u) f(u) du \\ &\equiv \mu(t_j, \psi), \end{aligned}$$

where $h(x) = e^x/(1+e^x)$, $f(u) = (1/\sqrt{2\pi})e^{-u^2/2}$, and $\psi = (\beta_0, \beta_1, \sigma)'$. Write $\mu_j = \mu(t_j, \psi)$, and $\mu = (\mu_j)_{1\le j\le k}$. We have

$$\begin{aligned} \frac{\partial \mu_j}{\partial \beta_0} &= \int_{-\infty}^{\infty} h'(\beta_0 + \beta_1 t_j + \sigma u) f(u) du, \\ \frac{\partial \mu_j}{\partial \beta_1} &= t_j \int_{-\infty}^{\infty} h'(\beta_0 + \beta_1 t_j + \sigma u) f(u) du, \\ \frac{\partial \mu_j}{\partial \sigma} &= \int_{-\infty}^{\infty} h'(\beta_0 + \beta_1 t_j + \sigma u) u f(u) du. \end{aligned}$$

Also, it is easy to see that the y_is have the same (joint) distribution, hence $V_i = \mathrm{Var}(y_i) = V_0$, an unspecified $k \times k$ covariance matrix, $1 \le i \le m$. Thus, the GEE equation for estimating ψ is given by

$$\sum_{i=1}^{m} \dot{\mu}' V_0^{-1}(y_i - \mu) = 0,$$

provided that V_0 is known. On the other hand, if ψ is known, V_0 can be estimated by the method of moments as follows,

$$\hat{V}_0 = \frac{1}{m} \sum_{i=1}^{m} (y_i - \mu)(y_i - \mu)'.$$

The IEE procedure then iterates between the two steps when both V_0 and ψ are unknown, starting with $V_0 = I$, the k-dimensional identity matrix.

The mean function μ_j above involves a one-dimensional integral, which can be approximated by a simple Monte Carlo method, namely,

$$\mu_j \approx \frac{1}{L}\sum_{l=1}^{L} h(\beta_0 + \beta_1 t_j + \sigma\xi_l),$$

where ξ_l, $l = 1, \dots, L$ are independent $N(0,1)$ random variables generated by a computer. Similar approximations can be obtained for the derivatives. The Monte Carlo method is further explored in the next section.

The GEE (or IEE) method considered so far applies only to the situation where the responses are independently clustered. In other words, the covariance matrix of the data is blockdiagonal. However, such a block-diagonal covariance structure may not always exist. For example, when the GLMM involves crossed random effects, such as in the salamander mating example, the data cannot be independently clustered. In the following sections we discuss some GEE-type estimators that apply to GLMMs in general, that is, without requiring a blockdiagonal covariance structure for the data.

4.2.3 Method of Simulated Moments

The method of simulated moments, or MSM, has been known to econometricians since the late 1980s. See, for example, McFadden (1989) and Lee (1992). The method applies to cases where the moments cannot be expressed as analytic functions of the parameters, and therefore direct computation of the method of moments (MM) estimators is not possible. These moments are then approximated by Monte Carlo methods, and this is the only difference between MSM and MM. To develop a MSM for GLMMs, let us first consider a simple example.

Example 4.8. Let y_{ij} be a binary outcome with $\text{logit}\{\text{P}(y_{ij} = 1|\alpha)\} = \mu + \alpha_i$, $1 \le i \le m$, $1 \le j \le k$, where $\alpha_1, \dots, \alpha_m$ are i.i.d. random effects with $\alpha_i \sim N(0, \sigma^2)$, $\alpha = (\alpha_i)_{1 \le i \le m}$, and μ, σ are unknown parameters with $\sigma \ge 0$. It is more convenient to use the following expression: $\alpha_i = \sigma u_i$, $1 \le i \le m$, where $u_1, \dots, u_m$ are i.i.d. $N(0,1)$ random variables. It is easy to show (Exercise 4.11) that a set of sufficient statistics for μ and σ are $y_{i\cdot} = \sum_{j=1}^{k} y_{ij}$, $1 \le i \le m$. Thus, we consider the following MM estimating equations based on the sufficient statistics,

$$\frac{1}{m}\sum_{i=1}^{m} y_{i\cdot} = \text{E}(y_{1\cdot}),$$

$$\frac{1}{m}\sum_{i=1}^{m} y_{i\cdot}^2 = \text{E}(y_{1\cdot}^2).$$

It is easy to show (Exercise 4.11) that $\mathrm{E}(y_{1\cdot}) = k\mathrm{E}\{h_\theta(\zeta)\}$ and $\mathrm{E}\{y_{1\cdot}^2\} = k\mathrm{E}\{h_\theta(\zeta)\} + k(k-1)\mathrm{E}\{h_\theta^2(\zeta)\}$, where $h_\theta(x) = \exp(\mu+\sigma x)/\{1+\exp(\mu+\sigma x)\}$ and $\zeta \sim N(0,1)$. It is more convenient to consider the following equivalent equations.

$$\frac{y_{\cdot\cdot}}{mk} = \mathrm{E}\{h_\theta(\zeta)\}, \tag{4.39}$$

$$\frac{1}{mk(k-1)}\sum_{i=1}^{m}(y_{i\cdot}^2 - y_{i\cdot}) = \mathrm{E}\{h_\theta^2(\zeta)\}, \tag{4.40}$$

where $y_{\cdot\cdot} = \sum_{i=1}^m y_{i\cdot}$. Let $u_1, \ldots, u_L$ be a sequence of $N(0,1)$ random variables generated by a computer. Then, the right sides of (4.39) and (4.40) may be approximated by $L^{-1}\sum_{l=1}^L h_\theta(u_l)$ and $L^{-1}\sum_{l=1}^L h_\theta^2(u_l)$, respectively. The equations then get solved to obtain the MSM estimators of μ and σ.

To see how the MSM estimators perform, a small simulation study is carried out with $m = 20$ or 80 and $k = 2$ or 6. The true parameters are $\mu = 0.2$ and $\sigma^2 = 1.0$. The results in Table 4.1 are based on 1000 simulations, where the estimator of σ^2 is the square of the estimator of σ.

Table 4.1. Simulated mean and standard error

		Estimator of μ		Estimator of σ^2	
m	k	Mean	SE	Mean	SE
20	2	0.31	0.52	2.90	3.42
20	6	0.24	0.30	1.12	0.84
80	2	0.18	0.22	1.08	0.83
80	6	0.18	0.14	1.03	0.34

To describe the general procedure for MSM, we assume that the conditional density of y_i given the vector of random effects α has the following form,

$$f(y_i|\alpha) = \exp[(w_i/\phi)\{y_i\xi_i - b(\xi_i)\} + c_i(y_i, \phi)], \tag{4.41}$$

where ϕ is a dispersion parameter, and w_is are known weights. Typically, $w_i = 1$ for ungrouped data; $w_i = n_i$ for grouped data if the response is an average, where n_i is the group size; and $w_i = 1/n_i$ if the response is a group sum. Here $b(\cdot)$ and $c_i(\cdot,\cdot)$ are the same as in the definition of GLMM. As for ξ_i, we assume a canonical link, that is, (3.2) with $\eta_i = \xi_i$. Furthermore, we assume that $\alpha = (\alpha_1', \ldots, \alpha_q')'$, where α_r is a random vector whose components are independent and distributed as $N(0, \sigma_r^2)$, $1 \le r \le q$. Furthermore, $Z = (Z_1, \ldots, Z_q)$, so that $Z\alpha = Z_1\alpha_1 + \cdots + Z_q\alpha_q$. The following expression of α is sometimes more convenient,

$$\alpha = Du, \tag{4.42}$$

where D is blockdiagonal with the diagonal blocks $\sigma_r I_{m_r}$, $1 \le r \le q$, and $u \sim N(0, I_m)$ with $m = m_1 + \cdots + m_q$. First assume that ϕ is known. Let $\theta = (\beta', \sigma_1, \ldots, \sigma_q)'$. Consider an unrestricted parameter space $\theta \in \Theta = R^{p+q}$. This allows computational convenience for using MSM because, otherwise, there will be constraints on the parameter space. Of course, this raises the issue of identifiability, because both $(\beta', \sigma_1, \ldots, \sigma_q)'$ and $(\beta', -\sigma_1, \ldots, -\sigma_q)$ correspond to the same model. Nevertheless, it is enough to make sure that β and $\sigma^2 = (\sigma_1^2, \ldots, \sigma_q^2)'$ are identifiable. In fact, in Section 4.5.4, we show that, under suitable conditions, the MSM estimators of β and σ^2 are consistent, therefore, the conditions also ensure identifiability of β and σ^2.

We first derive a set of sufficient statistics for θ. It can be shown (Exercise 4.12) that the marginal density of y can be expressed as

$$L = \int \exp\left\{c + a(y, \phi) + \frac{b(u, \theta)}{\phi} - \frac{|u|^2}{2} + \left(\sum_{i=1}^{n} w_i x_i y_i\right)' \left(\frac{\beta}{\phi}\right) + \left(\sum_{i=1}^{n} w_i z_i y_i\right)' \left(\frac{D}{\phi}\right) u\right\} du, \tag{4.43}$$

where c is a constant, $a(y, \phi)$ depends only on y and ϕ, and $b(u, \theta)$ depends only on u and θ. It follows that a set of sufficient statistics for θ is given by

$$\begin{aligned} S_j &= \textstyle\sum_{i=1}^n w_i x_{ij} y_i, \quad 1 \le j \le p, \\ S_{p+l} &= \textstyle\sum_{i=1}^n w_i z_{i1l} y_i, \quad 1 \le l \le m_1, \\ &\vdots \\ S_{p+m_1+\cdots+m_{q-1}+l} &= \textstyle\sum_{i=1}^n w_i z_{iql} y_i, \quad 1 \le l \le m_q, \end{aligned}$$

where $Z_r = (z_{irl})_{1 \le i \le n, 1 \le l \le m_r}$, $1 \le r \le q$. Thus, a natural set of MM equations can be formulated as

$$\sum_{i=1}^{n} w_i x_{ij} y_i = \sum_{i=1}^{n} w_i x_{ij} \mathrm{E}_\theta(y_i), \quad 1 \le j \le p, \tag{4.44}$$

$$\sum_{l=1}^{m_r} \left(\sum_{i=1}^{n} w_i z_{irl} y_i\right)^2 = \sum_{l=1}^{m_r} \mathrm{E}_\theta \left(\sum_{i=1}^{n} w_i z_{irl} y_i\right)^2, \quad 1 \le r \le q. \tag{4.45}$$

Although the S_js are sufficient statistics for the model parameters only when ϕ is known (which, of course, includes the special cases of binomial and Poisson distributions), one may still use Equations (4.44) and (4.45) to estimate θ even if ϕ is unknown, provided that the right-hand sides of these equations do not involve ϕ. Note that the number of equations in (4.44) and (4.45) is identical to the dimension of θ.

However, for the right sides of (4.44) and (4.45) not to depend on ϕ, some changes have to be made. For simplicity, in the following we assume that Z_r, $1 \le r \le q$ are standard design matrices in the sense that each Z_r consists only of 0s and 1s, and there is exactly one 1 in each row and at least one 1 in each column. Then, if we denote the ith row of Z_r by $z'_{ir} = (z_{irl})'_{1\le l\le m_r}$, we have $|z_{ir}|^2 = 1$ and, for $s \neq t$, $z'_{sr}z_{tr} = 0$ or 1. Let $I_r = \{(s,t) : 1 \le s \neq t \le n, z'_{sr}z_{tr} = 1\} = \{(s,t) : 1 \le s \neq t \le n, z_{sr} = z_{tr}\}$. Then, it can be shown (Exercise 4.13) that

$$\sum_{l=1}^{m_r} \mathrm{E}_\theta \left(\sum_{i=1}^n w_i z_{irl} y_i\right)^2 = \sum_{i=1}^n w_i^2 \mathrm{E}_\theta(y_i^2) + \sum_{(s,t)\in I_r} w_s w_t \mathrm{E}(y_s y_t). \tag{4.46}$$

It is seen that the first term on the right side of (4.46) depends on ϕ, and the second term does not depend on ϕ (Exercise 4.13). Therefore, a simple modification of the earlier MM equations that will eliminate ϕ would be to replace (4.45) by the following equations,

$$\sum_{(s,t)\in I_r} w_s w_t y_s y_t = \sum_{(s,t)\in I_r} w_s w_t \mathrm{E}_\theta(y_s y_t), \qquad 1 \le r \le q. \tag{4.47}$$

Furthermore, write $u = (u'_1, \ldots, u'_q)'$ with $u_r = (u_{rl})_{1\le l\le m_r}$. Note that $u_r \sim N(0, I_{m_r})$. Then, the right side of (4.44) can be expressed as

$$X'_j W \mathrm{E}\{e(\theta, u)\},$$

where X_j is the jth column of X, $W = \mathrm{diag}(w_i, 1 \le i \le n)$, and $e(\theta, u) = \{b'(\xi_i)\}_{1\le i\le n}$ with $\xi_i = \sum_{j=1}^p x_{ij}\beta_j + \sum_{r=1}^q \sigma_r z'_{ir} u_r$ (Exercise 4.14). Similarly, the right side of (4.47) can be expressed as

$$\mathrm{E}\{e(\theta, u)' W H_r W e(\theta, u)\},$$

where H_r is the $n \times n$ symmetric matrix whose (s,t) entry is $1_{\{(s,t)\in I_r\}}$. Thus, the final MM equations that do not involve ϕ are given by

$$\begin{aligned} \sum_{i=1}^n w_i x_{ij} y_i &= X'_j W \mathrm{E}\{e(\theta, u)\}, \qquad 1 \le j \le p, \\ \sum_{(s,t)\in I_r} w_s w_t y_s y_t &= \mathrm{E}\{e(\theta, u)' W H_r W e(\theta, u)\}, \qquad 1 \le r \le q, \end{aligned} \tag{4.48}$$

where the expectations on the right-hand sides are with respect to $u \sim N(0, I_m)$. In order to solve these equations, we approximate the right-hand sides by a simple Monte Carlo method. Let $u^{(1)}, \ldots, u^{(L)}$ be generated i.i.d. copies of u. Then, the right sides of (4.48) can be approximated by the Monte Carlo averages; that is,

$$X_j'W\left[\frac{1}{L}\sum_{l=1}^{L}e\{\theta,u^{(l)}\}\right], \qquad 1\le j\le p,$$

$$\frac{1}{L}\sum_{l=1}^{L}e\{\theta,u^{(l)}\}'WH_rWe\{\theta,u^{(l)}\}, \qquad 1\le r\le q. \tag{4.49}$$

In conclusion, (4.48) with the right sides approximated by (4.49) are the MSM equations for estimating θ. Note that, quite often, the expressions inside the expectations on the right sides of (4.48) only involve some components of u. This means that one does not need to generate the entire vector u, and thus reduce the computation. We consider another example.

Example 4.9. The following example was considered by McGilchrist (1994) and Kuk (1995) in their simulation studies. Suppose that, given the random effects $u_1,\dots,u_{15}$, which are independent and distributed as $N(0,1)$, responses y_{ij}, $i=1,\dots,15$, $j=1,2$ are conditionally independent such that $y_{ij}|u\sim\text{binomial}(6,\pi_{ij})$, where $u=(u_i)_{1\le i\le 15}$, $\text{logit}(\pi_{ij})=\beta_0+\beta_1x_{ij}+\sigma u_i$ with $x_{i1}=2i-16$ and $x_{i2}=2i-15$. The MSM equations for estimating β_0, β_1, and σ take the following form (Exercise 4.15),

$$\begin{aligned}\sum_{i=1}^{15}(y_{i1}+y_{i2}) &= \frac{6}{L}\sum_{l=1}^{L}\sum_{i=1}^{15}(\pi_{i1l}+\pi_{i2l}),\\ \sum_{i=1}^{15}(x_{i1}y_{i1}+x_{i2}y_{i2}) &= \frac{6}{L}\sum_{l=1}^{L}\sum_{i=1}^{15}(x_{i1}\pi_{i1l}+x_{i2}\pi_{i2l}),\\ \sum_{i=1}^{15}y_{i1}y_{i2} &= \frac{36}{L}\sum_{l=1}^{L}\sum_{i=1}^{15}\pi_{i1l}\pi_{i2l},\end{aligned} \tag{4.50}$$

where $\pi_{ijl}=h(\beta_0+\beta_1x_{ij}+\sigma u_{il})$ with $h(x)=e^x/(1+e^x)$, and u_{il}, $1\le i\le 15$, $1\le l\le L$ are random variables generated independently from an $N(0,1)$ distribution.

Finally, we discuss how to estimate the standard errors of the MSM estimators. Define $\hat{\psi}=(\hat{\beta}',|\hat{\sigma}_1|,\dots,|\hat{\sigma}_q|)'$, where $\hat{\theta}=(\hat{\beta}',\hat{\sigma}_1,\dots,\hat{\sigma}_q)'$ is the MSM estimator of θ. Write the MSM equations as $\hat{M}=\tilde{M}(\hat{\theta})$, where $\tilde{M}$ is the vector of simulated moments. Similarly, let $M(\theta)$ denote the vector of moments. We assume, without loss of generality, that $\sigma_r\ge 0$, $1\le r\le q$ for the true θ. Because for large m and L, the simulated moments approximate the corresponding moments, and $\hat{\psi}$ is a consistent estimator of θ, we have, by Taylor expansion, $\hat{M}=\tilde{M}(\hat{\theta})\approx M(\hat{\theta})=M(\hat{\psi})\approx M(\theta)+\dot{M}(\theta)(\hat{\psi}-\theta)\approx M(\theta)+\dot{M}(\theta)J^{-1}(\theta)(\hat{\varphi}-\varphi)$, where $\dot{M}(\cdot)$ is the matrix of first derivatives, $\varphi=(\beta',\sigma_1^2,\dots,\sigma_q^2)'$, $\hat{\varphi}$ is the corresponding MSM estimator of φ, and $J(\theta)=\text{diag}(1,\dots,1,2\sigma_1,\dots,2\sigma_q)$. Thus, an approximate covariance matrix of $\hat{\varphi}$ is given by

$$\text{Var}(\hat{\varphi})\approx J(\theta)\{\dot{M}(\theta)^{-1}\}\text{Var}(\hat{M})\{\dot{M}(\theta)^{-1}\}'J(\theta). \tag{4.51}$$

In practice, $J(\theta)$ can be estimated by $J(\hat{\psi})$, and $\dot{M}(\theta)$ can be estimated by first replacing θ by $\hat{\psi}$ and then approximating the moments by simulated moments, as we did earlier. As for $\mathrm{Var}(\hat{M})$, although one could derive its parametric form, the latter is likely to involve ϕ, the dispersion parameter which is sometimes unknown. Alternatively, the covariance matrix of $\hat{M}$ can be estimated using a parametric bootstrap method as follows. First generate data from the GLMM, treating $\hat{\varphi}$ as the true θ. The generated data are a bootstrap sample, denoted by $y^*_{i,k}$, $1 \le i \le n$, $1 \le k \le K$. Then, compute the vector of sample moments based on the bootstrap sample, say, $\hat{M}^*_k$, $1 \le k \le K$. A bootstrap estimate of $\mathrm{Var}(\hat{M})$ is then given by $\widehat{\mathrm{Var}}(\hat{M}) = (K-1)^{-1}\sum_{k=1}^K \left(\hat{M}^*_k - \bar{\hat{M}}^*\right)\left(\hat{M}^*_k - \bar{\hat{M}}^*\right)'$, where $\bar{\hat{M}}^* = K^{-1}\sum_{k=1}^K \hat{M}^*_k$. To see how the method works, we revisit Example 4.9.

Example 4.9 (Continued). A simulation study was carried out for the model considered here with $L = 100$. Two sets of true parameters were considered: (i) $\sigma^2 = 1.0$ and (ii) $\sigma^2 = 2.0$, and in both cases $\beta_0 = 0.2$, $\beta_1 = 1.0$. The results based on 1000 simulations are summarized in Table 4.2 and compared with the approximate restricted maximum likelihood (AREML) estimator of McGilchrist (1994) and the iterative bias correction (IBC) estimator of Kuk (1995). The numbers in parentheses are averages of the estimated SEs. The AREML method is similar to the PQL of Breslow and Clayton (1993) discussed in Section 3.5.2. For the most part, the method is based on a link between BLUP and REML (e.g., Speed 1991) and a quadratic approximation to the conditional log-density of the responses given the random effects. The IBC method iteratively corrects the bias of PQL, which results in an asymptotically unbiased estimator. However, the latter method may be computationally intensive.

It appears that MSM is doing quite well in terms of the bias, especially compared with AREML. On the other hand, the standard errors of MSM estimators seem larger than those of AREML and IBC estimators. Finally, the estimated SEs are very close to the simulated ones, an indication of good performance of the above method of standard error estimation.

Table 4.2. Comparison of estimators

	Average of Estimators			SE of Estimator		
True Parameter	MSM	AREML	IBC	MSM	AREML	IBC
$\beta_0 = .2$	.20	.25	.19	.32 (.31)	.31	.26
$\beta_1 = .1$	.10	.10	.10	.04 (.04)	.04	.04
$\sigma^2 = 1.0$	.93	.91	.99	.63 (.65)	.54	.60
$\beta_0 = .2$	.21	.11	.20	.42 (.37)	.35	.36
$\beta_1 = .1$	.10	.10	.10	.05 (.05)	.05	.05
$\sigma^2 = 2.0$	1.83	1.68	1.90	1.19 (1.04)	.80	.96

4.2.4 Robust Estimation in GLMM

Although the MSM estimators are consistent, simulation results suggested that these estimators may be inefficient in the sense that the variances of the estimators are relatively large. In this section, we propose an improvement of the MSM.

We first give an extension of GLMM. Recall that in a GLM (McCullagh and Nelder, 1989), it is assumed that the distribution of the response is a member of a known exponential family. Thus, for a linear model to fit within the generalized linear models, one may have to assume that the distribution of the response is normal. However, the definition of a linear model does not require normality, and many of the techniques developed in linear models do not require the normality assumption. Therefore, GLMs, as defined, do not extend linear models in a full sense.

In view of this, we consider a broader class of models than the GLMM, in which the form of the conditional distribution, such as the exponential family, is not required. The method can be described under an even broader framework. Let θ be a vector of parameters under an assumed model. Suppose that there is a vector of base statistics, say, S, which typically is of higher dimension than θ. We assume that the following conditions are satisfied.

(i) The mean of S is a known function of θ.

(ii) The covariance matrix of S is a known function of θ, or at least is consistently estimable.

(iii) Certain smoothness and regularity conditions hold.

Let the dimension of θ and S be r and N, respectively. If only (i) is assumed, an estimator of θ may be obtained by solving the following equation,

$$BS = Bu(\theta), \tag{4.52}$$

where B is a $r \times N$ matrix and $u(\theta) = \mathrm{E}_\theta(S)$. This is called the first-step estimator, in which the choice of B is arbitrary. It can be shown (see Section 4.6.5) that, under suitable conditions, the first-step estimator is consistent, although it may not be efficient. To improve the efficiency, we further require (ii). Denote the first-step estimator by $\tilde{\theta}$, and consider a Taylor expansion around the true θ. We have $\tilde{\theta} - \theta \approx (BU)^{-1}Q(\theta)$, where $U = \partial u/\partial\theta'$ and $Q(\theta) = B\{S - u(\theta)\}$. Note that $Q(\tilde{\theta}) = 0$. Denote the covariance matrix of S by V. Then, we have

$$\mathrm{Var}(\tilde{\theta}) \approx \{(BU)^{-1}\}BVB'\{(BU)^{-1}\}'.$$

By Theorem 4.3, the optimal B is $U'V^{-1}$. Unfortunately, this optimal B depends on θ, which is exactly what we wish to estimate. Our approach is to replace θ in the optimal B by $\tilde{\theta}$, the first-step estimator. This leads to what we call the second-step estimator, denoted $\hat{\theta}$, and obtained by solving

$$\tilde{B}S = \tilde{B}u(\theta), \tag{4.53}$$

where $\tilde{B} = U'V^{-1}|_{\theta=\tilde{\theta}}$. It can be shown that, under suitable conditions, the second-step estimator is consistent and asymptotically efficient in the sense that its asymptotic covariance matrix is the same as that of the solution to the optimal estimating equation, that is, (4.52) with $B = U'V^{-1}$.

Note. It might appear that one could do better by allowing B in (4.52) to depend on θ, that is, $B = B(\theta)$. However, Theorem 4.3 shows that the asymptotic covariance matrix of the estimator corresponding to the optimal $B(\theta)$ is the same as that corresponding to the optimal B (which is a constant matrix). Therefore, the complication does not result in a real gain.

We now consider an extended version of GLMMs. Suppose that, given a vector $\alpha = (\alpha_k)_{1\leq k\leq m}$ of random effects, responses $y_1, \dots, y_n$ are conditionally independent such that

$$\mathrm{E}(y_i|\alpha) = h(\xi_i), \tag{4.54}$$

$$\mathrm{var}(y_i|\alpha) = a_i(\phi)v(\eta_i), \tag{4.55}$$

where $h(\cdot)$, $v(\cdot)$, and $a_i(\cdot)$ are known functions, ϕ is a dispersion parameter,

$$\xi_i = x_i'\beta + z_i'\alpha, \tag{4.56}$$

where β is a vector of unknown fixed effects, and $x_i = (x_{ij})_{1\leq j\leq p}$, $z_i = (z_{ik})_{1\leq k\leq m}$ are known vectors. Finally, we assume that

$$\alpha \sim F_\vartheta, \tag{4.57}$$

where F_ϑ is a multivariate distribution known up to a vector $\vartheta = (\vartheta_r)_{1\leq r\leq q}$ of dispersion parameters. Note that we do not require that the conditional density of y_i given α is a member of the exponential family, as in the original definition of GLMM (see Section 3.2). In fact, as shown, to obtain the first-step estimator, only (4.54) is needed.

To apply the method to the extended GLMMs, we need to first select the base statistics. By similar arguments as in the previous section, a natural choice may be the following,

$$\begin{aligned} S_j &= \sum_{i=1}^n w_i x_{ij} y_i, \qquad 1 \leq j \leq p, \\ S_{p+j} &= \sum_{s\neq t} w_s w_t z_{sk} z_{tk} y_s y_t, \qquad 1 \leq k \leq m. \end{aligned} \tag{4.58}$$

In fact, if $Z = (z_{ik})_{1\leq i\leq n, 1\leq k\leq m} = (Z_1 \cdots Z_q)$, where each Z_r is an $n \times n_r$ standard design matrix (see Section 4.2.3), $1 \leq r \leq q$, then, if one chooses $B = \mathrm{diag}(I_p, 1_{m_1}', \dots, 1_{m_q}')$, one obtains the MM eqnations of Jiang (1998a). Thus, the latter estimators are a special case of the first-step estimators. However, the following examples show that the second-step estimators can be considerably more efficient than the first-step ones.

Example 4.10 (Mixed logistic model). Consider the following mixed logistic model. Suppose that, given the random effects $\alpha_1, \dots, \alpha_m$, binary responses y_{ij}, $1 \le i \le m$, $1 \le j \le k_i$ are conditionally independent such that $\mathrm{logit}\{\mathrm{P}(y_{ij} = 1|\alpha)\} = \mu + \alpha_i$, where $\alpha = (\alpha_i)_{1 \le i \le m}$ and μ is an unknown parameter. Furthermore, suppose that the α_is are independent and distributed as $N(0, \sigma^2)$, where σ^2 is unknown.

It is easy to see that the base statistics (4.58) reduce to $y_{\cdot\cdot}$ and $y_{i\cdot}^2 - y_{i\cdot}$, $1 \le i \le m$, where $y_{i\cdot} = \sum_{j=1}^{n_i} y_{ij}$ and $y_{\cdot\cdot} = \sum_{i=1}^{m} y_{i\cdot}$.

A special case of this model is Example 4.8, in which $k_i = k$, $1 \le i \le m$, that is, the data are balanced. In fact, it can be shown that, in the latter case, the first-step estimators are the same as the second-step ones, and therefore are optimal (Exercise 4.17). However, when the data are unbalanced, the first-step estimator is no longer optimal. To see this, a simulation was carried out, in which $m = 100$, $n_i = 2$, $1 \le i \le 50$, and $n_i = 6$, $51 \le i \le 100$. The true parameters were chosen as $\mu = 0.2$ and $\sigma = 1.0$. The results based on 1000 simulations are summarized in Table 4.3, where SD represents the simulated standard deviation, and the overall MSE is the MSE of the estimator of μ plus that of the estimator of σ. There is about a 43% reduction of the overall MSE of the second-step estimator over the first-step one.

Table 4.3. Simulation results: mixed logistic model

Method of Estimation	Estimator of μ			Estimator of σ			Overall
	Mean	Bias	SD	Mean	Bias	SD	MSE
1st-step	.21	.01	.16	.98	−.02	.34	.15
2nd-step	.19	−.01	.16	.98	−.02	.24	.08

Because the first- and second-step estimators are developed under the assumption of the extended GLMM, the methods apply to some situations beyond (the classical) GLMM. The following is an example.

Example 4.11 (Beta-binomial). If $Y_1, \dots, Y_l$ are correlated Bernoulli random variables, the distribution of $Y = Y_1 + \cdots + Y_l$ is not binomial, and therefore does not belong to the exponential family. Here we consider a special case. Let p be a random variable with a beta$(\pi, 1-\pi)$ distribution, where $0 < \pi < 1$. Suppose that, given p, $Y_1, \dots, Y_l$ are independent Bernoulli(p) random variables, so that $Y|p \sim$ binomial(l, p). Then, it can be shown (Exercise 4.18) that the marginal distribution of Y is given by

$$\mathrm{P}(Y = k) = \frac{\Gamma(k+\pi)\Gamma(l-k+1-\pi)}{k!(l-k)!\Gamma(\pi)\Gamma(1-\pi)}, \qquad 1 \le k \le l. \tag{4.59}$$

This distribution is called beta-binomial(l, π). It follows that $\mathrm{E}(Y) = l\pi$ and $\mathrm{Var}(Y) = \phi l \pi(1-\pi)$, where $\phi = (l+1)/2$. It is seen that the mean function under beta-binomial(l, π) is the same as that of binomial(l, π), but the variance function is different. In other words, there is an overdispersion.

Now, suppose that, given the random effects $\alpha_1, \ldots, \alpha_m$, which are independent and distributed as $N(0, \sigma^2)$, responses y_{ij}, $1 \leq i \leq m$, $1 \leq j \leq n_i$ are independent and distributed as beta-binomial(l, π_i), where $\pi_i = h(\mu + \alpha_i)$ with $h(x) = e^x/(1 + e^x)$. Note that this is not a GLMM under the classical definition of Section 3.2, because the conditional distribution of y_{ij} is not a member of the exponential family. However, the model falls within the extended definition, because

$$\mathrm{E}(y_{ij}|\alpha) = l\pi_i, \tag{4.60}$$

$$\mathrm{var}(y_{ij}|\alpha) = \phi l \pi_i(1 - \pi_i). \tag{4.61}$$

If only (4.60) is assumed, one may obtain the first-step estimator of (μ, σ), for example, by choosing $B = \mathrm{diag}(1, 1'_m)$. If, in addition, (4.61) is assumed, one may obtain the second-step estimator. To see how much difference there is between the two, a simulation study was carried out with $m = 40$. Again, an unbalanced situation was considered: $n_i = 4$, $1 \leq i \leq 20$ and $n_i = 8$, $21 \leq i \leq 40$. We took $l = 2$, and the true parameters $\mu = 0.2$ and $\sigma = 1.0$. The results based on 1000 simulations are summarized in Table 4.4. Again, we see about 36% improvement of the second-step estimator over the first-step one.

Table 4.4. Simulation results: Beta-binomial

Method of	Estimation of μ			Estimation of σ			Overall
Estimation	Mean	Bias	SD	Mean	Bias	SD	MSE
1st-step	.25	.05	.25	1.13	.13	.37	.22
2nd-step	.25	.05	.26	1.09	.09	.25	.14

The improvements of the second-step estimators over the first-step ones in the precedent examples are not incidental. It can be shown that the second-step estimators are asymptotically optimal in the sense described earlier and, in particular, more efficient than the first-step estimators. See Section 4.6.5 for more details.

4.3 GLMM Selection

Recently, Jiang et al. (2006b) developed a general method, called *fence*, that potentially applies to a broad class of model selection problems, including GLMM selection problems.

4.3.1 A General Principle for Model Selection

The essential part of this procedure is a quantity $Q_M = Q_M(y, \theta_M)$, where M indicates the candidate model, y is an $n \times 1$ vector of observations, θ_M represents the vector of parameters under M, such that $\mathrm{E}(Q_M)$ is minimized when M is a true model and θ_M the true parameter vector under M. Here by true model we mean that M is a correct model but not necessarily the most efficient one. For example, in regression model selection, a true model is one that contains at least all the variables whose coefficients are nonzero, but the model remains true if an additional variable is added, whose coefficient is zero. We use the terms "true model" and "correct model" interchangeably. Below are some examples of Q_M.

1. Maximum likelihood (ML) model selection. If the model specifies the full distribution of y up to the parameter vector θ_M, an example of Q_M is the negative of the log-likelihood under M; that is,

$$Q_M = -\log\{f_M(y|\theta_M)\}, \tag{4.62}$$

where $f_M(\cdot|\theta_M)$ is the joint probability density function (pdf) of y with respect to a measure ν under M, given that θ_M is the true parameter vector. To see that $\mathrm{E}(Q_M)$ is minimized when M is a true model and θ_M the true parameter vector under M, let $f(y)$ denote the true pdf of y. Then, we have

$$\begin{aligned}
-\mathrm{E}(Q_M) &= \int \log\{f_M(y|\theta_M)\} f(y)\nu(dy) \\
&= \int \log\{f(y)\} f(y)\nu(dy) + \int \log\left\{\frac{f_M(y|\theta_M)}{f(y)}\right\} f(y)\nu(dy) \\
&\le \int \log\{f(y)\} f(y)\nu(dy) + \log\left\{\int \frac{f_M(y|\theta_M)}{f(y)} f(y)\nu(dy)\right\} \\
&= \int \log\{f(y)\} f(y)\nu(dy),
\end{aligned} \tag{4.63}$$

using the concave-function inequality. The lone term on the right side of (4.63) is equal to $-\mathrm{E}(Q_M)$ when M is a true model and θ_M the true parameter vector.

2. Mean and variance/covariance (MVC) model selection. If the model is only specified by the mean and covariance matrix of y, it is called a mean and variance/covariance model, or MVC model. In this case, we may consider

$$Q_M = |(T'V_M^{-1}T)^{-1}T'V_M^{-1}(y - \mu_M)|^2, \tag{4.64}$$

where μ_M and V_M are the mean vector and covariance matrix under M, and T is a given $n \times s$ matrix of full rank $s \le n$. To see that $\mathrm{E}(Q_M)$ is minimized when $\mu_M = \mu$, $V_M = V$, where μ and V denote the true mean vector and covariance matrix, note that

$$\mathrm{E}(Q_M) = \mathrm{tr}\{(T'V_M^{-1}T)^{-1}T'V_M^{-1}VV_M^{-1}T(T'V_M^{-1}T)^{-1}\} \\ +|(T'V_M^{-1}T)^{-1}T'V_M^{-1}(\mu_M - \mu)|^2. \tag{4.65}$$

The first term is the trace of the covariance matrix of the weighted least squares (WLS; see Section 1.4.3) estimator of β with the weight matrix $W = V_M^{-1}$ in the linear regression $y = T\beta + \epsilon$, where $\mathrm{E}(\epsilon) = 0$ and $\mathrm{Var}(\epsilon) = V$. Because the covariance matrix of the WLS estimator is minimized when $W = V^{-1}$ (i.e., $V_M = V$), the first term on the right side of (4.65) is minimized when $V_M = V$. On the other hand, the second term is zero when $\mu_M = \mu$.

3. Extended GLMM selection. Consider the problem of selecting an extended GLMM, introduced by Jiang and Zhang (2001), in which only the conditional mean of the response given the random effects is parametrically specified. It is assumed that, given a vector α of random effects, the responses $y_1, \dots, y_n$ are conditionally independent such that

$$\mathrm{E}(y_i|\alpha) = h(x_i'\beta + z_i'\alpha), \tag{4.66}$$

$1 \leq i \leq n$, where $h(\cdot)$ is a known function, β is a vector of unknown fixed effects, and x_i, z_i are known vectors. Furthermore, it is assumed that $\alpha \sim N(0, \Sigma)$, where the covariance matrix Σ depends on a vector ψ of variance components. Now the question is: how to select the function $h(\cdot)$, the fixed covariates (which are components of x_i), and the random effects factors (which correspond to subvectors of α). In other words, we have a problem of selecting a model for the conditional means.

For such a purpose, let β_M and ψ_M denote β and ψ under M, and $g_{M,i}(\beta_M, \psi_M) = \mathrm{E}\{h_M(x_i'\beta_M + z_i'\Sigma_M^{1/2}\xi)\}$, where h_M is the function h under M, Σ_M is the covariance matrix under M evaluated at ψ_M, and the expectation is taken with respect to $\xi \sim N(0, I_m)$ (which does not depend on M). Here m is the dimension of α and I_m the m-dimensional identity matrix. We consider the following,

$$Q_M = \sum_{i=1}^n \{y_i - g_{M,i}(\beta_M, \psi_M)\}^2. \tag{4.67}$$

It is easy to see that the Q_M given above satisfies the basic requirement: $\mathrm{E}(Q_M)$ is minimized when M is a true model and $\theta_M = (\beta_M', \psi_M')'$ is the true parameter vector under M. In fact, (4.67) corresponds to the Q_M in MVC model selection (see above) with $T = I$, the identity matrix. Note that, because V is not parametrically specified under the assumed model, it should not get involved in Q_M. Therefore, (4.67) is a natural choice for Q_M. Also note that, although (4.67) may be regarded as a residual sum of squares, the responses are correlated in the current situation.

An important issue in model selection is to control the dimensionality of the model, because otherwise the "bigger" model always wins. Here the dimension of a model M, $|M|$, is defined as the dimension of θ_M. A model is called *optimal* if it is a true model with the smallest dimension. Let $\hat{\theta}_M$ be defined as above. Let $\tilde{M} \in \mathcal{M}$ be such that $\hat{Q}_{\tilde{M}} = \min_{M \in \mathcal{M}} \hat{Q}_M$, where $\mathcal{M}$ represents the set of candidate models. We expect, at least in a large sample, that $\tilde{M}$ is a correct model. The question is: are there other correct models in $\mathcal{M}$ with smaller dimension than $\tilde{M}$?

To answer this question, we need to know what the difference $\hat{Q}_M - \hat{Q}_{\tilde{M}}$ is likely to be when M is a true model, and how different the difference might be when M is an incorrect model. Suppose that M^* is a correct model. If M is also a correct model, an appropriate measure of the difference $\hat{Q}_M - \hat{Q}_{M^*}$ is its standard deviation, denoted σ_{M,M^*}. On the other hand, if M is an incorrect model, the difference $\hat{Q}_M - \hat{Q}_{M^*}$ is expected to be much larger than σ_{M,M^*} (see arguments in Section 4.5.6). This leads to the following procedure. For simplicity, let us first consider the case that $\tilde{M}$ is unique.

1. Find $\tilde{M}$ such that $\hat{Q}_{\tilde{M}} = \min_{M \in \mathcal{M}} \hat{Q}_M$.
2. For each $M \in \mathcal{M}$ such that $|M| < |\tilde{M}|$, compute $\hat{\sigma}_{M,\tilde{M}}$, an estimator of $\sigma_{M,\tilde{M}}$. Then, M belongs to $\tilde{\mathcal{M}}_-$, the set of "true" models with $|M| < |\tilde{M}|$ if

$$\hat{Q}_M \le \hat{Q}_{\tilde{M}} + \hat{\sigma}_{M,\tilde{M}}. \tag{4.68}$$

3. Let $\tilde{\mathcal{M}} = \{\tilde{M}\} \cup \tilde{\mathcal{M}}_-$, $m_0 = \min_{M \in \tilde{\mathcal{M}}} |M|$, and $\mathcal{M}_0 = \{M \in \tilde{\mathcal{M}} : |M| = m_0\}$. Let M_0 be the model in $\mathcal{M}_0$ such that $\hat{Q}_{M_0} = \min_{M \in \mathcal{M}_0} \hat{Q}_M$. M_0 is the selected model.

The quantity $\hat{Q}_{\tilde{M}} + \hat{\sigma}_{M,\tilde{M}}$ serves as a "fence" to confine the true models (with dimensions smaller than $|\tilde{M}|$) and exclude the incorrect ones. For such a reason, the procedure is called *fence*. Note that the fence depends on M; that is, for different M the fence is different. The following outlines an effective algorithm for the fence. Let $d_1 < d_2 < \cdots < d_L$ be all the different dimensions of the models $M \in \mathcal{M}$.

The Fence Algorithm

(i) Find $\tilde{M}$.
(ii) Compute $\hat{\sigma}_{M,\tilde{M}}$ for all $M \in \mathcal{M}$ such that $|M| = d_1$; let $\mathcal{M}_1 = \{M \in \mathcal{M} : |M| = d_1 \text{ and (4.68) holds}\}$; if $\mathcal{M}_1 \neq \emptyset$, stop. Let M_0 be the model in $\mathcal{M}_1$ such that $\hat{Q}_{M_0} = \min_{M \in \mathcal{M}_1} \hat{Q}_M$; M_0 is the selected model.
(iii) If $\mathcal{M}_1 = \emptyset$, compute $\hat{\sigma}_{M,\tilde{M}}$ for all $M \in \mathcal{M}$ such that $|M| = d_2$; let $\mathcal{M}_2 = \{M \in \mathcal{M} : |M| = d_2 \text{ and (4.68) holds}\}$; if $\mathcal{M}_2 \neq \emptyset$, stop. Let M_0 be the model in $\mathcal{M}_2$ such that $\hat{Q}_{M_0} = \min_{M \in \mathcal{M}_2} \hat{Q}_M$; M_0 is the selected model.
(iv) Continue until the program stops (it will at some point).

In short, the algorithm may be described as follows. Check the candidate models, from the simplest to the most complex. Once one has discovered a model that falls within the fence and checked all the other models of the same simplicity (for membership within the fence), one stops.

In the case that $\tilde{M}$ is not unique, all one has to do is to redefine $\tilde{\mathcal{M}}$ in step 3 of the fence as $\tilde{\mathcal{M}} = \{M \in \mathcal{M} : |M| = |\tilde{M}|, \hat{Q}_M = \hat{Q}_{\tilde{M}}\} \cup \tilde{\mathcal{M}}_-$.

An extension of the fence that takes into account the issue of consistency is given by the same steps 1–3 above with (4.68) replaced by

$$\hat{Q}_M \leq \hat{Q}_{\tilde{M}} + c_n \hat{\sigma}_{M,\tilde{M}}, \tag{4.69}$$

where c_n is a sequence that $\to \infty$ slowly as $n \to \infty$. A similar effective algorithm can be outlined.

The key to the fence is the calculation of $\hat{\sigma}_{M,\tilde{M}}$ in step 2. Although for consistency (see Section 4.5.6) it is not required that $\hat{\sigma}_{M,M^*}$ be a consistent estimator of σ_{M,M^*}, as long as the former has the correct order, in practice, it is desirable to use a consistent estimator whenever possible. This is because, even if $\hat{\sigma}_{M,M^*}$ has the correct order, there is always a constant involved, which may be difficult to choose. A smaller constant is apparently to the benefit of larger models and thus results in overfitting; on the other hand, a larger constant would be in favor of smaller models, and hence prompts underfitting. Therefore, to balance the two sides, the best way would be to use a consistent estimator of σ_{M,M^*}, so that one need not worry about the constant. Here consistency is in the sense that $\hat{\sigma}_{M,M^*} = \sigma_{M,M^*} + o(\sigma_{M,M^*})$ or, equivalently, $\hat{\sigma}_{M,M^*}/\sigma_{M,M^*} \to 1$, in a suitable sense (e.g., in probability). In Section 4.5.6 we consider a special case, in which the data are clustered, and show how to obtain $\hat{\sigma}_{M,\tilde{M}}$.

4.3.2 A Simulated Example

We consider the following simulated example of GLMM selection. Suppose that three models are being considered.

Model I: Given the random effects $\alpha_1, \dots, \alpha_m$, binary responses y_{ij}, $i = 1, \dots, m$, $j = 1, \dots, k$ are conditionally independent such that

$$\operatorname{logit}(p_{ij}) = \beta_0 + \beta_1 x_i + \alpha_i,$$

where $p_{ij} = \mathrm{P}(y_{ij} = 1|\alpha)$; β_0, β_1 are fixed parameters; $x_i = 0$, $1 \leq i \leq [m/2]$ and $x_i = 1$, $[m/2] + 1 \leq i \leq m$ ($[x]$ represents the integer part of x). Furthermore, the random effects are independent and distributed as $N(0, \sigma^2)$.

Model II: Same as Model I except that $\beta_1 = 0$.

Model III: Same as Model I except that $\beta_0 = \beta_1 = 0$.

We first study the consistency of the MVC and ML model selection procedures in the situation where the data are generated from one of the candidate models. In other words, a true model belongs to the class of candidate models. Throughout the simulation studies, T was chosen as a block-diagonal matrix with $T_i = T_1$, $1 \leq i \leq m$, where T_1 is a $k \times l$ matrix with $l = [k/2]$, whose entries are generated from a Uniform$[0, 1]$ distribution, and then fixed throughout the simulations. The simulation results are summarized in Table 4.5. The columns for MVC and ML are probabilities of correct selection, reported as percentages estimated empirically from 100 realizations of the simulation. The numbers in parentheses are the percentages of selection of the other two models in order of increasing index of the model.

Table 4.5. Simulation results: consistency

True Model	m	k	l	β_0	β_1	σ	c_n	MVC	ML
I	100	4	2	$-.5$	1	1	1	82(5,13)	94(3,3)
I	200	4	2	$-.5$	1	1	1.1	97(1,2)	99(0,1)
II	100	4	2	$-.5$	NA	1	1	87(4,9)	88(5,7)
II	200	4	2	$-.5$	NA	1	1.1	93(4,3)	98(2,0)
III	100	4	2	NA	NA	1	1	87(3,10)	91(2,7)
III	200	4	2	NA	NA	1	1.1	96(0,4)	91(1,8)

We next study robustness of the MVC and ML fence procedures in the case where no true model (with respect to ML) is included in the candidate models. We consider one such case, in which the binary responses y_{ij} are generated as follows. Suppose that $(X_1, \dots, X_k)$ has a multivariate normal distribution such that $\mathrm{E}(X_j) = \mu$, $\mathrm{var}(X_j) = 1$, $1 \le j \le k$, and $\mathrm{cor}(X_s, X_t) = \rho$, $1 \le s \ne t \le k$. Then, let $Y_j = 1_{(X_j > 0)}$, $1 \le j \le k$. Denote the joint distribution of $(Y_1, \dots, Y_k)$ by $\mathrm{NB}(\mu, \rho)$ (here NB refers to "Normal-Bernoulli"). We then generate the data such that $y_1, \dots, y_m$ are independent, and the distribution of $y_i = (y_{ij})_{1 \le j \le k}$ follows one of the following models.

Model A: $y_i \sim \mathrm{NB}(\mu_1, \rho_1)$, $i = 1, \dots, [m/2]$, and $y_i \sim \mathrm{NB}(\mu_2, \rho_2)$, $i = [m/2] + 1, \dots, m$, where μ_j, ρ_j, $j = 1, 2$ are chosen to match the means, variances, and covariances under Model I. Note that one can do so because the means, variances and covariances under Model I depend only on three parameters, whereas there are four parameters under Model A.

Model B: $y_i \sim \mathrm{NB}(\mu, \rho)$, $i = 1, \dots, m$, where μ and ρ are chosen to match the mean, variance, and covariance under Model II. Note that, under Model II, the mean, variance, and covariance depend on two parameters.

Model C: Same as Model B except that μ and ρ are chosen to match the mean, variance, and covariance under Model III. Note that, under Model III, the mean is equal to $1/2$, the variance is $1/4$, and the covariance depends on a single parameter σ.

If the data are generated from Model A, Model I is a correct model with respect to MVC; similarly, if the data are generated from Model B, both Model I and II are correct with respect to MVC; and, if the data is generated from Model C, Models I–III are all correct in the sense of MVC. However, no model (I, II, or III) is correct from a ML standpoint. The simulation results are summarized in Table 4.6, in which β_0^*, β_1^*, and σ^* correspond to the parameters under the models in Table 4.5 with the matching mean(s), variance(s), and covariance(s). The columns for MVC and ML are probabilities of correct selection, reported as percentages estimated empirically from 100 realizations of the simulation. The numbers in parentheses are the percentages of selection of the other two models in order of increasing index of the model. β_0^*, β_1^* and σ^* are the matching parameters.

Table 4.6. Simulation results: robustness

True Model	m	k	l	β_0^*	β_1^*	σ^*	c_n	MVC	ML
A	100	4	2	−.5	1	1	1	83(7,10)	91(5,4)
A	200	4	2	−.5	1	1	1.1	97(2,1)	99(0,1)
B	100	4	2	−.5	NA	1	1	80(3,17)	91(4,5)
B	200	4	2	−.5	NA	1	1.1	95(3,2)	97(3,0)
C	100	4	2	NA	NA	1	1	83(8,9)	86(4,10)
C	200	4	2	NA	NA	1	1.1	91(1,8)	90(1,9)

Summary: It is seen in Table 4.5 that the numbers increase as m increases (and c_n slowly increases), a good indication of consistency. With the exception of one case (III/200), ML outperforms MVC, which is not surprising. What is a bit of surprise is that ML also seems quite robust in the situation where the true model is not one of the candidate models (therefore the objective is to select a model among the candidates that is closest to the reality). In fact, Table 4.6 shows that even in the latter case, ML still outperforms MVC (with the exception of one case: again, III/200). However, one has to keep in mind that there are many ways that a model can be misspecified, and here we only considered one of them (which misspecifies a NB as a GLMM). Furthermore, MVC has a computational advantage over ML, which is important in cases such as GLMM selection. Note that the computational burden usually increases with the sample size; on the other hand, the larger sample performance of MVC (i.e., $m = 200$) is quite close to that of ML.

A compromise would be to use MVC in cases of a large sample, and ML in cases of a small or moderate sample. Alternatively, one may use MVC for an initial round of model selection to narrow down the number of candidate models, and ML for a final round of model selection. For example, one may use MVC for steps 1 and 2 of the fence (see Section 4.3.1) to identify the subclass $\tilde{\mathcal{M}}$, and then apply ML (with steps 1–3) within $\tilde{\mathcal{M}}$ to identify the optimal model.

In Sections 4.4.2 and 4.4.3 we further illustrate the fence method using real-life data examples.

4.4 Real-Life Data Examples

4.4.1 Fetal Mortality in Mouse Litters

Brooks et al. (1997) presented six datasets recording fetal mortality in mouse litters. Here we consider the HS2 dataset from Table 4 of their paper, which reports the number of dead implants in 1328 litters of mice from untreated experimental animals.

The data may be considered as being summaries of the individual responses y_{ij}, $i = 1, \dots, 1328$, $j = 1, \dots, n_i$, where n_i is the size of the ith litter; $y_{ij} = 1$ if the jth implant in the ith litter is dead, and $y_{ij} = 0$ otherwise. The total number of responses is $N = \sum_{i=1}^{1328} n_i = 10,533$. For simplicity, the n_is are considered nonrandom.

Brooks et al. (1997) used a beta-binomial model to model the correlation among responses from the same litter. Here we consider an GLMM for the same purpose. Suppose that, given the random effects $\alpha_1, \dots, \alpha_m$, the binary responses y_{ij}, $1 \le i \le m$, $1 \le j \le n_i$ are conditionally independent such that

$$\text{logit}\{\text{pr}(y_{ij} = 1|\alpha)\} = \mu + \alpha_i \ ,$$

where μ is an unknown parameter. Furthermore, suppose that the α_i's are independent and distributed as $N(0, \sigma^2)$, where σ^2 is an unknown variance. Note that here $m = 1328$ and α_i is a random effect associated with the ith litter. The problem of interest then is to estimate the parameters μ and σ. Jiang and Zhang (2001) analyzed the data using the robust estimation method introduced in Section 4.2.4. Their first- and second-step estimates of μ are, with estimated standard errors in parentheses, -2.276 (0.047) and -2.296 (0.047), respectively. Both analyses have found the parameter μ highly significant with almost the same negative value. However, in this case, a parameter of greater interest is σ, the standard deviation of the litter effects. The first- and second-step estimates of σ are given by, with estimated standard errors in parentheses, 0.644 (0.059) and and 0.698 (0.057), respectively. Again, in both cases, the parameter was found highly significant, an indication of strong within-group correlations. The values of the first- and second-step estimates of σ differ slightly. Thus, we adopt the second-step estimate because it is supposed to be more efficient.

Furthermore, the within-group correlation between the binary responses can be estimated as follows. For any $j \neq k$, we have

$$\begin{aligned} \text{cov}(y_{ij}, y_{ik}) &= \text{E}(y_{ij}y_{ik}) - \text{E}(y_{ij})\text{E}(y_{ik}) \\ &= \text{E}\{h^2(\mu + \sigma\xi)\} - \{\text{E}h(\mu + \sigma\xi)\}^2, \end{aligned}$$

where $h(x) = e^x/(1 + e^x)$ and $\xi \sim N(0, 1)$. Thus, we have

$$\begin{aligned} \text{cov}(y_{ij}, y_{ik}) = &\int h^2(\mu + \sigma x)\frac{1}{\sqrt{2\pi}}e^{-x^2/2}dx \\ &-\left\{\int h(\mu + \sigma x)\frac{1}{\sqrt{2\pi}}e^{-x^2/2}dx\right\}^2 . \end{aligned} \tag{4.70}$$

The integrals involved in (4.70) can be evaluated by numerical integrations. Namely, if $|h(x)| \le 1$, then for any $\delta > 0$,

$$\int_{-\infty}^{\infty} h(x) \frac{1}{\sqrt{2\pi}} e^{-x^2/2} dx \approx \left(\frac{\delta}{\sqrt{\pi}}\right) \sum_{n=-N+1}^{N-1} \psi\left(\sqrt{2} n \delta\right) e^{-n^2\delta^2}.$$

See Goodwin (1949), who also evaluated the accuracy of this approximation. For example, for $h(x) = e^x/(1+e^x)$ with $\delta = 0.7$ and $N = 7$, the approximation error is less than 7.33×10^{-9}. The estimated covariance between y_{ij} and y_{ik} $(j \neq k)$ is then given by (4.70) with $\mu = -2.296$ and $\sigma = 0.698$.

4.4.2 Analysis of Gc Genotype Data: An Application of the Fence Method

Human group-specific component (Gc) is the plasma transport protein for Vitamin D. Polymorphic electrophoretic variants of Gc are found in all human populations. Daiger *et al.* (1984) presented data involving a series of monozygotic (MZ) and dizygotic (DZ) twins of known Gc genotypes in order to determine the heritability of quantitative variation in Gc. These included 31 MZ twin pairs, 13 DZ twin pairs, and 45 unrelated controls. For each individual, the concentration of Gc was available along with additional information about the sex, age, and Gc genotype of the individual. The genotypes are distinguishable at the Gc structural locus, and are classified as 1–1, 1–2, and 2–2.

Lange (2002) considered three statistical models for the Gc genotype data. Let y_{ij} represent the Gc concentration measured for the jth person who is one of the ith identical twin pair, $i = 1, \ldots, 31$, $j = 1, 2$. Furthermore, let y_{ij} represent the Gc concentration measured for the jth person who is one of the $(i-31)$th fraternal twin pairs, $i = 32, \ldots, 44$, $j = 1, 2$. Finally, Let y_i represent the Gc concentration for the $(i-44)$th person among the unrelated controls, $i = 45, \ldots, 89$. Then, the first model, Model I, can be expressed as

$$\begin{aligned} y_{ij} &= \mu_{1-1} 1_{(g_{ij}=1-1)} + \mu_{1-2} 1_{(g_{ij}=1-2)} + \mu_{2-2} 1_{(g_{ij}=2-2)} \\ &\quad + \mu_{\text{male}} 1_{(s_{ij}=\text{male})} + \mu_{\text{age}} a_{ij} + \epsilon_{ij}, \qquad i = 1, \ldots, 44, \qquad j = 1, 2, \end{aligned}$$

where g_{ij}, s_{ij}, and a_{ij} represent the genotype, sex, and age of the jth person in the i twin pair (identical or fraternal), and ϵ_{ij} is an error that is further specified later. If we let x_{ij} denote the vector whose components are $1_{(g_{ij}=1-1)}$, $1_{(g_{ij}=1-2)}$, $1_{(g_{ij}=2-2)}$, $1_{(s_{ij}=\text{male})}$, and a_{ij}, and β denote the vector whose components are μ_{1-1}, μ_{1-2}, μ_{2-2}, μ_{male} and μ_{age}, then the model can be expressed as

$$y_{ij} = x_{ij}'\beta + \epsilon_{ij}, \qquad i = 1, \ldots, 44, \qquad j = 1, 2. \tag{4.71}$$

Similarly, we have

$$\begin{aligned} y_i &= \mu_{1-1} 1_{(g_i=1-1)} + \mu_{1-2} 1_{(g_i=1-2)} + \mu_{2-2} 1_{(g_i=2-2)} \\ &\quad + \mu_{\text{male}} 1_{(s_i=\text{male})} + \mu_{\text{age}} a_i + \epsilon_i, \qquad i = 45, \ldots, 89, \end{aligned}$$

where g_i, s_i, and a_i are the genotype, sex, and age of the $(i-44)$th person in the unrelated control group, and ϵ_i is an error that is further specified. Let x_i denote the vector whose components are $1_{(g_i=1-1)}$, $1_{(g_i=1-2)}$, $1_{(g_i=2-2)}$, $1_{(s_i=\text{male})}$, and a_i, and β be the same as above; then we have

$$y_i = x_i'\beta + \epsilon_i, \qquad i = 45, \ldots, 89. \tag{4.72}$$

We now specify the distributions for the errors. Let $\epsilon_i = (\epsilon_{i1}, \epsilon_{i2})'$, $i = 1, \ldots, 44$. We assume that ϵ_i, $i = 1, \ldots, 89$ are independent. Furthermore, we assume that

$$\epsilon_i \sim N\left(\begin{bmatrix} 0 \\ 0 \end{bmatrix}, \sigma_{\text{tot}}^2 \begin{bmatrix} 1 & \rho_{\text{ident}} \\ \rho_{\text{ident}} & 1 \end{bmatrix}\right), \qquad i = 1, \ldots, 31,$$

where σ_{tot}^2 is the unknown total variance, and ρ_{ident} the unknown correlation coefficient between identical twins. Similarly, we assume

$$\epsilon_i \sim N\left(\begin{bmatrix} 0 \\ 0 \end{bmatrix}, \sigma_{\text{tot}}^2 \begin{bmatrix} 1 & \rho_{\text{frat}} \\ \rho_{\text{frat}} & 1 \end{bmatrix}\right), \qquad i = 32, \ldots, 44, \tag{4.73}$$

where ρ_{frat} is the unknown correlation coefficient between fraternal twins. Finally, we assume that

$$\epsilon_i \sim N(0, \sigma_{\text{tot}}^2), \qquad i = 45, \ldots, 89.$$

The second model, Model II, is the same as Model I except under the constraint $\rho_{\text{frat}} = \rho_{\text{ident}}/2$; that is, in (4.73) ρ_{frat} is replaced by $\rho_{\text{ident}}/2$.

The third model, Model III, is the same as Model I except under the constraints $\mu_{1-1} = \mu_{1-2} = \mu_{2-2}$; that is, in (4.71) and (4.72) we have

$$x_{ij}'\beta = \mu + \mu_{\text{male}} 1_{(s_{ij}=\text{male})} + \mu_{\text{age}} a_{ij} + \epsilon_{ij},$$
$$x_i'\beta = \mu + \mu_{\text{male}} 1_{(s_i=\text{male})} + \mu_{\text{age}} a_i + \epsilon_i.$$

Thus, under Model I, the parameters are

$$\theta_{\text{I}} = (\mu_{1-1}, \mu_{1-2}, \mu_{2-2}, \mu_{\text{male}}, \mu_{\text{age}}, \sigma_{\text{tot}}^2, \rho_{\text{ident}}, \rho_{\text{frat}})' \ (8 - \text{dimensional});$$

under Model II, the parameters are

$$\theta_{\text{II}} = (\mu_{1-1}, \mu_{1-2}, \mu_{2-2}, \mu_{\text{male}}, \mu_{\text{age}}, \sigma_{\text{tot}}^2, \rho_{\text{ident}})' \ (7 - \text{dimensional});$$

and, under Model III, the parameters are

$$\theta_{\text{III}} = (\mu, \mu_{\text{male}}, \mu_{\text{age}}, \sigma_{\text{tot}}^2, \rho_{\text{ident}}, \rho_{\text{frat}})' \ (6 - \text{dimensional});.$$

It is clear that all three models are Gaussian mixed models, which are special cases of GLMMs. We apply the fence method to this dataset to select an optimal model from the candidate models. More specifically, we consider

ML model selection (see Section 4.3.1.1). Note that, because Models II and III are submodels of Model I (in other words, Model I is the full model), we may take $\tilde{M}$ as Model I. The analysis resulted in the following values for $\hat{Q}_M$: $\hat{Q}_{\mathrm{I}} = 337.777$, $\hat{Q}_{\mathrm{II}} = 338.320$, and $\hat{Q}_{\mathrm{III}} = 352.471$. Furthermore, we obtained $\hat{\sigma}_{\mathrm{II,I}} = 1.367$ and $\hat{\sigma}_{\mathrm{III,I}} = 4.899$. Thus, Model II is in the fence and Model III is out. In conclusion, the analysis has selected Model II as the optimal model. This result is consistent with the finding of Lange (2002), who indicated that a "likelihood ratio test shows that there is virtually no evidence against the assumption $\rho_{\mathrm{frat}} = \rho_{\mathrm{ident}}/2$."

4.4.3 The Salamander-Mating Experiments: Various Applications of GLMM

Finally, we revisit the salamander-mating experiments discussed in Section 3.3.1. These data have been analyzed by numerous authors, and several different models have been proposed.

Lin and Breslow (1996) considered the following mixed logistic model, which is a special case of GLMM. Following the approach of Drum and McCullagh (1993), they assumed that a different group of animals (20 female and 20 male; 10 from each population) had been used in each experiment. Thus, the female random effects can be denoted by $\alpha_{\mathrm{f},1}, \dots, \alpha_{\mathrm{f},60}$, and the male random effects $\alpha_{\mathrm{m},1}, \dots, \alpha_{\mathrm{m},60}$. It was assumed that the $\alpha_{\mathrm{f},i}$s are independent with mean 0 and variance σ_{f}^2, the $\alpha_{\mathrm{m},j}$s are independent with mean 0 and variance σ_{m}^2, and the $\alpha_{\mathrm{f},i}$s and $\alpha_{\mathrm{m},j}$s are independent. Furthermore, let p_{ij} denote the conditional probability of successful mating given the effect of the ith female and jth male; that is, $p_{ij} = \mathrm{P}(y_{ij} = 1|\alpha_{\mathrm{f},i}, \alpha_{\mathrm{m},j})$. Lin and Breslow (1996) assumed that

$$\mathrm{logit}(p_{ij}) = x_{ij}'\beta + \alpha_{\mathrm{f},i} + \alpha_{\mathrm{m},j},$$

where x_{ij} is a vector of covariates consisting of the following components: an intercept, an indicator $\mathrm{WS_f}$ for WS female (1 for WS and 0 for RB), an indicator $\mathrm{WS_m}$ for WS male (1 for WS and 0 for RB), and the interaction $\mathrm{WS_f} \times \mathrm{WS_m}$; that is,

$$x_{ij}'\beta = \beta_0 + \beta_1\mathrm{WS_f} + \beta_2\mathrm{WS_m} + \beta_3\mathrm{WS_f} \times \mathrm{WS_m}. \tag{4.74}$$

Lin and Breslow fitted the model using three different methods. These are PQL (see Section 3.5.2) and its biased corrected versions (first and second order). The latter methods were developed to reduce the bias of PQL. See Breslow and Lin (1995) and Lin and Breslow (1996) for details. The results are summarized in Table 4.7, where $\mathrm{CPQL_1}$ and $\mathrm{CPQL_2}$ represent, respectively, the first-order and second-order bias-corrected PQL, and the numbers in parentheses are estimated standard errors.

Booth and Hobert (1999) used the MCEM method to obtained the maximum likelihood (ML) estimates of the parameters under an equivalent model.

Their covariate vector x_{ij} consisted of four indicators of different combinations of crossing: W/W for Whiteside female and Whiteside male, W/R for Whiteside female and Rough Butt male, R/W for Rough Butt female and Whiteside male, and R/R for Rough Butt female and Rough Butt male. Then, by noting the following simple relationships between the new indicators and the ones used by Lin and Breslow (1996),

$$\begin{aligned} \mathrm{W/W} &= \mathrm{WS_f} \times \mathrm{WS_m}, \\ \mathrm{W/R} &= \mathrm{WS_f} \times (1 - \mathrm{WS_m}), \\ \mathrm{R/W} &= (1 - \mathrm{WS_f}) \times \mathrm{WS_m}, \\ \mathrm{R/R} &= (1 - \mathrm{WS_f}) \times (1 - \mathrm{WS_m}), \end{aligned}$$

it is easy to obtain the association between the regression coefficients as $\beta_0 = \beta_{\mathrm{R/R}}$, $\beta_1 = \beta_{\mathrm{W/R}} - \beta_{\mathrm{R/R}}$, $\beta_2 = \beta_{\mathrm{R/W}} - \beta_{\mathrm{R/R}}$, and $\beta_3 = \beta_{\mathrm{R/R}} - \beta_{\mathrm{R/W}} - \beta_{\mathrm{W/R}} + \beta_{\mathrm{W/W}}$. The estimates of Booth and Hobert (1999) are included in Table 4.7 as comparison. The authors did not report the standard errors.

Also included in this table are the Bayes estimates obtained by Karim and Zeger (1992) using Gibbs sampling. The authors reported the median, and 5th and 95th percentiles of the posteriors of the parameters. The posterior medians are used as point estimates. The standard errors of the estimators of the regression coefficients are obtained using the following method. Because the posterior distributions for the βs are asymptotically normal, the interval between the 5th and 95th percentiles is approximately median plus/minus the standard error. This implies that the standard error is approximately the difference between the 95th and 5th percentiles divided by $2 \times 1.645 = 3.29$. The standard errors for the variance estimators are more complicated because the posteriors of the variances σ_{f}^2 and σ_{m}^2 are skewed.

Table 4.7. Estimates of parameters: PQL and bias-corrected PQL, ML and Bayes

Method	Intercept	$\mathrm{WS_f}$	$\mathrm{WS_m}$	$\mathrm{WS_f} \times \mathrm{WS_m}$	σ_f^2	σ_{m}^2
PQL	.79(.32)	−2.29(.43)	−.54(.39)	2.82(.50)	.72	.63
$\mathrm{CPQL_1}$	1.19(.37)	−3.39(.55)	−.82(.43)	4.19(.64)	.99	.91
$\mathrm{CPQL_2}$	.68(.37)	−2.16(.55)	−.49(.43)	2.65(.64)	—	—
ML	1.03	−2.98	−.71	3.65	1.40	1.25
Bayes	1.03((.43)	−3.01(.60)	−.69(.50)	3.74(.68)	1.50	1.36

It is seen that the interaction is highly significant regardless of the methods used. In fact, this can also be seen from a simple data summary. For example, for the summer experiment, the percentages of successful mating between the female and male animals from the two populations are 70.0% for WS–WS, 23.3% for WS–RB, 66.7% for RB–WS, and 73.3% for RB–RB. Thus, the percentage for WS-RB is much lower than all three other cases, which have similar percentages. Another factor that was found highly significant in all cases is $\mathrm{WS_f}$. Interestingly, its male counterpart, $\mathrm{WS_m}$ was found insignificant using all methods. It appeared that female animals played more significant roles than their male partners, and the (fixed) effects of male animals are mainly through their interactions with the females.

An assumption that has been used so far is that a different group of animals had been used in each experiment. Of course, this is not true in reality. However, the situation gets more complicated if this assumption is dropped. This is because there may be serial correlations not explained by the animal-specific random effects among the responses. See Example 4.6. Due to such considerations, Jiang and Zhang (2001) considered an extended version of GLMM for the pooled responses. More specifically, let y_{ij1} be the observed proportion of successful matings between the ith female and jth male in the summer and fall experiments that involved the same group of animals (so $y_{ij1} = 0$, 0.5 or 1), and y_{ij2} be the indicator of successful mating between the ith female and jth male in the last fall experiment that involved a new group of animals. It was assumed that, conditional on the random effects, $u_{k,i}$, $v_{k,j}$, $k = 1, 2$, i, $j = 1, \dots, 20$, which are independent and normally distributed with mean 0 and variances σ_{f}^2 and σ_{m}^2, respectively, the responses y_{ijk}, $(i, j) \in P$, $k = 1, 2$ are conditionally independent, where P represents the set of pairs (i, j) determined by the design, u, and v represent the female and male, respectively; $1, \dots, 10$ correspond to RB, and $11, \dots, 20$ to WS. Furthermore, it was assumed that the conditional mean of the response given the random effects satisfies one of the two models below: (i) (logit model) $\mathrm{E}(y_{ijk}|u, v) = h_1(x_{ij}'\beta + u_{k,i} + v_{k,j})$, $(i, j) \in P$, $k = 1, 2$, where $x_{ij}'\beta$ is given by (4.71), and $h_1(x) = e^x/(1 + e^x)$; (ii) (probit model) same as (i) with $h_1(x)$ replaced by $h_2(x) = \Phi(x)$, where $\Phi(\cdot)$ is the cdf of $N(0, 1)$. Note that it is not assumed that the conditional distribution of y_{ijk} given the random effects is a member of the exponential family. The authors then obtained the first-step estimators (see Section 4.2.4) of the parameters under both models. The results are given in table 4.8. The numbers in parentheses are the estimated standard errors, obtained from Theorem 4.9 in Section 4.5.5 under the assumption that the binomial conditional variance is correct. If the latter assumption fails, the standard error estimates are not reliable but the point estimates are still valid.

Table 4.8. First-step estimates with standard errors

Mean Function	β_0	β_1	β_2	β_3	σ_{f}	σ_{m}
Logit	0.95	−2.92	−0.69	3.62	0.99	1.28
	(0.55)	(0.87)	(0.60)	(1.02)	(0.59)	(0.57)
Probit	0.56	−1.70	−0.40	2.11	0.57	0.75
	(0.31)	(0.48)	(0.35)	(0.55)	(0.33)	(0.32)

Earlier, Karim and Zeger (1992) took an alternative approach by considering a GLMM with correlated random effects. In their Model B, the correlations among the responses are still solely due to the random effects (i.e., no serial correlations given the random effects), but the random effect for an animal is bivariate: a second random effect representing the season is added. This allows different but correlated effects for an animal used in two experiments. In addition, a season indicator is added to the fixed covariates. More specifically, their model can be expressed as

$$\text{logit}(p_{ijk}) = x_{ij}'\beta + \beta_4\text{FALL} + z_k'u_i + z_k'v_j,$$

where $p_{ijk} = \text{P}(y_{ijk} = 1|u_i, v_j)$, y_{ijk} is the indicator of successful mating between the ith female and jth male in the kth experiment, $x_{ij}'\beta$ is the same as (4.71), and FALL is the indicator of the fall season (0 for summer, 1 for fall). Here $k = 1$ (fall) or 2 (summer). Thus, FALL $= 1$ if $k = 1$ and FALL $= 0$ if $k = 2$. Note that the second group of animals was only used in the fall, for whom FALL is identical to 1. Furthermore, $z_k' = (1, \text{FALL})$, $u_i = (u_{i,1}, u_{i,2})'$, $v_j = (v_{j,1}, v_{j,2})'$, where u, v correspond to female and male, respectively. Thus, for the first group of animals, the random effect for the ith female is $u_{i,1}$ for the summer, and $u_{i,1} + u_{i,2}$ for the fall; whereas for the second group, the random effect for the ith female is $u_{i,1} + u_{i,2}$. We have similar expressions for the male random effects. Finally, it was assumed that the u_is and v_js are independent and bivariate normal such that $u_i \sim N(0, \Sigma_\text{f})$, $v_j \sim N(0, \Sigma_\text{m})$, where $\Sigma_\text{f} = (\sigma_{\text{f},rs})_{1\leq r,s\leq 2}$ and $\Sigma_\text{m} = (\sigma_{\text{m},rs})_{1\leq r,s\leq 2}$ are unknown covariance matrices. Karim and Zeger used Gibbs sampling to approximate the posterior distributions, assuming flat priors for both β and Σs. Although there was an issue of propriety of the posteriors when using flat priors (see discussion in Section 4.1.6), the problem did not seem to have occurred numerically in this case. The results of the posterior median, and 5th and 95th percentiles are given in Table 4.9. Note that the coefficient β_4 of the newly added seasonal

Table 4.9. Median and 5th and 95th percentiles of the posteriors

Parameter	β_0	β_1	β_2	β_3	β_4
Median	1.49	−3.13	−.76	3.90	−.62
Percentiles	.51,2.62	−4.26,−2.20	−1.82,.23	2.79,5.16	−1.51,.28
Parameter	$\sigma_{\text{f},11}$	$\sigma_{\text{f},12}$	$\sigma_{\text{f},22}$	—	—
Median	1.92	−2.17	3.79	—	—
Percentiles	.32,5.75	−6.46,−.26	1.03,9.53	—	—
Parameter	$\sigma_{\text{m},11}$	$\sigma_{\text{m},12}$	$\sigma_{\text{m},22}$	—	—
Median	1.25	.27	.23	—	—
Percentiles	.28,3.62	−.73,.88	.01,1.46	—	—

indicator is insignificant at the 5% level; the variance of $u_{i,2}$, the seasonal female random effect, is significant, and the variance of $v_{j,2}$, the seasonal male random effect, is barely significant at the 5% level.

It is remarkable that, although the models of Jiang and Zhang (2001) and Karim and Zeger (1992, Model B) are different from those of Lin and Breslow (1996), Booth and Hobert (1999), and Karim and Zeger (1992, Model A), the conclusions about the significance of the regression coefficients as well as their signs are essentially the same.

Finally, Jiang et al. (2006b) applied the fence method (see Section 4.3) to the salamander mating data. They considered the problem of selecting an extended GLMM (see Section 4.2.4) in this case. Following Jiang and Zhang (2001), we pool the data from the two experiments involving the same group of salamanders, so let y_{ij1} be the observed proportion of successful matings between the ith female and jth male in the two experiments. Let y_{ij2} be the indicator of successful mating between the ith female and jth male in the last experiment involving a new set of animals.

We assume that given the random effects, $u_{k,i}$, $v_{k,j}$, $k = 1, 2$, $i, j = 1, \dots, 20$, which are independent and normally distributed with mean 0 and variances σ^2 and τ^2, respectively, the responses y_{ijk}, $(i, j) \in P$, $k = 1, 2$ are conditionally independent, where P represents the set of pairs (i, j) determined by the design, which is partially crossed; u and v represent the female and male, respectively; $1, \dots, 10$ correspond to RB, and $11, \dots, 20$ correspond to WS. Furthermore, we consider the following models for the conditional means.

Model I: $\mathrm{E}(y_{ijk}|u, v) = h_1(\beta_0 + \beta_1 \mathrm{WS_f} + \beta_2 \mathrm{WS_m} + \beta_3 \mathrm{WS_f} \times \mathrm{WS_m} + u_{k,i} + v_{k,j})$, $(i, j) \in P$, $k = 1, 2$, where $h_1(x) = e^x/(1 + e^x)$; $\mathrm{WS_f}$ is an indicator for WS female (1 for WS and 0 for RB), $\mathrm{WS_m}$ is an indicator for WS male (1 for WS and 0 for RB) and $\mathrm{WS_f} \times \mathrm{WS_m}$ represents the interaction.

Model II: Same as Model I except dropping the interaction term.

Model III: Same as Model I with h_1 replaced by h_2, where $h_2(x) = \Phi(x)$, the cdf of $N(0, 1)$.

Model IV: Same as Model III except dropping the interaction term.

The models are special cases of the extended GLMMs introduced in Section 4.2.4. See Section 4.3.1 for a special application of the fence in this case. We apply the fence method (with $c_n = 1$) discussed in the latter section to this case. The analysis has yielded the following values of $\hat{Q}_M$ for M = I, II, III, and IV: $39.5292, 44.3782, 39.5292, 41.6190$. Therefore, we have $\tilde{M}$ = I or III. If we use $\tilde{M}$ = I, then $\hat{\sigma}_{M,\tilde{M}} = 1.7748$ for M = II and $\hat{\sigma}_{M,\tilde{M}} = 1.1525$ for M = IV. Therefore, neither M = II nor M = IV falls within the fence. If we use $\tilde{M}$ = III, then $\hat{\sigma}_{M,\tilde{M}} = 1.68$ for M = II and $\hat{\sigma}_{M,\tilde{M}} = 1.3795$ for M = IV. Thus, once again, neither M = II nor M = IV is inside the fence. In conclusion, the fence method has selected both Model I and Model III (either one) as the optimal model. Interestingly, these are exactly the models fitted by Jiang and Zhang (2001) using a different method, although the authors had not considered it a model selection problem. The eliminations of Model

II and Model IV are consistent with many of the previous studies (e.g., Karim and Zeger 1992; Breslow and Clayton 1993; Lin and Breslow 1996), which have found the interaction term significant, although the majority of these studies have focused on logit models. As by products of the fence procedure, the estimated regression coefficients and variance components under the two models selected by the fence are given in Table 4.10.

Table 4.10. Estimates of parameters for the salamander mating data

Model	β_0	β_1	β_2	β_3	σ	τ
I	1.00	−2.96	−0.71	3.62	0.97	1.24
III	0.90	−2.66	−0.64	3.25	1.08	1.49

4.5 Further Results and Technical Notes

4.5.1 Proof of Theorem 4.3

By Theorem 2.1 of Heyde (1997), to establish the optimality of G^*, it suffices to show that $\{\mathrm{E}(\dot{G})\}^{-1}\mathrm{E}(GG^{*'})$ is a constant matrix for all $G \in \mathcal{H}$. Let $G = A(y - \mu) \in \mathcal{H}$. We have

$$\begin{aligned}\mathrm{E}(GG^{*'}) &= \mathrm{E}\{A(y-\mu)(y-\mu)'V^{-1}\dot{\mu}\}\\ &= \mathrm{E}[A\mathrm{E}\{(y-\mu)(y-\mu)'|x\}V^{-1}\dot{\mu}]\\ &= \mathrm{E}(A\dot{\mu}).\end{aligned}$$

On the other hand, we have $\dot{G} = \dot{A}(y-\mu) - A\dot{\mu}$. Thus,

$$\begin{aligned}\mathrm{E}(\dot{G}) &= \mathrm{E}\{\dot{A}(y-\mu)\} - \mathrm{E}(A\dot{\mu})\\ &= \mathrm{E}\{\dot{A}\mathrm{E}(y-\mu|x)\} - \mathrm{E}(A\dot{\mu})\\ &= -\mathrm{E}(A\dot{\mu}).\end{aligned}$$

Therefore, $\{\mathrm{E}(\dot{G})\}^{-1}\mathrm{E}(GG^{*'}) = -I$, where I is the identity matrix, and this proves the theorem.

4.5.2 Linear Convergence and Asymptotic Properties of IEE

1. Linear convergence. We adapt a term from numerical analysis. An iterative algorithm that results in a sequence $x^{(m)}$, $m = 1, 2, \dots$ converges linearly to a limit x^*, if there is $0 < \rho < 1$ such that $\sup_{m\geq 1}\{|x^{(m)} - x^*|/\rho^m\} < \infty$ (e.g., Press et al. 1997).

Let $L_1 = \max_{1 \leq i \leq n} \max_{j \in J_i} s_{ij}$ with $s_{ij} = \sup_{|\tilde{\beta}-\beta| \leq \epsilon_1} |(\partial/\partial\beta) g_j(X_i, \tilde{\beta})|$, where β represents the true parameter vector, ϵ_1 is any positive constant, and $(\partial/\partial\beta) f(\tilde{\beta})$ means $(\partial f/\partial\beta)|_{\beta=\tilde{\beta}}$. Similarly, let $L_2 = \max_{1 \leq i \leq n} \max_{j \in J_i} w_{ij}$, where $w_{ij} = \sup_{|\tilde{\beta}-\beta| \leq \epsilon_1} \|(\partial^2/\partial\beta\partial\beta') g_j(X_i, \tilde{\beta})\|$. Also, let $\mathcal{V} = \{v : \lambda_{\min}(V_i) \geq \lambda_0, \lambda_{\max}(V_i) \leq M_0, 1 \leq i \leq n\}$, where $\lambda_{\min}$ and $\lambda_{\max}$ represent the smallest and largest eigenvalues, respectively, and δ_0 and M_0 are given positive constants. Note that $\mathcal{V}$ is a nonrandom set.

An array of nonnegative definite matrices $\{A_{n,i}\}$ is bounded from above if $\|A_{n,i}\| \leq c$ for some constant c; the array is bounded from below if $A_{n,i}^{-1}$ exists and $\|A_{n,i}^{-1}\| \leq c$ for some constant c. A sequence of random matrices is bounded in probability, denoted by $A_n = O_{\mathrm{P}}(1)$, if for any $\epsilon > 0$, there is $M > 0$ and $N \geq 1$ such that $\mathrm{P}(\|A_n\| \leq M) > 1 - \epsilon$, if $n \geq N$. The sequence is bounded away from zero in probability if $A_n^{-1} = O_{\mathrm{P}}(1)$. Note that the definition also applies to a sequence of random variables, considered as a special case of random matrices. Also, recall that p is the dimension of β and R the dimension of v. We make the following assumptions.

A1. For any $(j,k) \in D$, the number of different v_{ijk}s is bounded, that is, for each $(j,k) \in D$, there is a set of numbers $\mathcal{V}_{jk} = \{v(j,k,l), 1 \leq l \leq L_{jk}\}$, where L_{jk} is bounded, such that $v_{ijk} \in \mathcal{V}_{jk}$ for any $1 \leq i \leq n$ with $j, k \in J_i$.

A2. The functions $g_j(X_i, \beta)$ are twice continuously differentiable with respect to β; $\mathrm{E}(|Y_i|^4)$, $1 \leq i \leq n$ are bounded; and L_1, L_2, $\max_{1 \leq i \leq n}(\|V_i\| \vee \|V_i^{-1}\|)$ are $O_{\mathrm{P}}(1)$.

A3 (Consistency of GEE estimator). For any given V_i, $1 \leq i \leq n$ bounded from above and below, the GEE equation (4.33) has a unique solution $\hat{\beta}$ that is consistent.

A4 (Differentiability of GEE solution). For any v, the solution to (4.33), $\beta(v)$, is continuously differentiable with respect to v, and $\sup_{v \in \mathcal{V}} \|\partial\beta/\partial v\| = O_{\mathrm{P}}(1)$.

A5. $n(j,k,l) \to \infty$ for any $1 \leq l \leq L_{jk}$, $(j,k) \in D$, as $n \to \infty$.

The proof of the following theorem can be found in Jiang et al. (2006a).

Theorem 4.4. Under assumptions *A1–A5*, P(IEE converges) $\to 1$ as $n \to \infty$. Furthermore, we have $\mathrm{P}[\sup_{m \geq 1}\{|\hat{\beta}^{(m)} - \hat{\beta}^*|/(p\eta)^{m/2}\} < \infty] \to 1$, $\mathrm{P}[\sup_{m \geq 1}\{|\hat{v}^{(m)} - \hat{v}^*|/(R\eta)^{m/2}\} < \infty] \to 1$ as $n \to \infty$ for any $0 < \eta < (p \vee R)^{-1}$, where $(\hat{\beta}^*, \hat{v}^*)$ is the (limiting) IEEE.

Note 1. It is clear that the restriction $\eta < (p \vee R)^{-1}$ is unnecessary (because, for example, $(p\eta_1)^{-m/2} < (p\eta_2)^{-m/2}$ for any $\eta_1 \geq (p \vee R)^{-1} > \eta_2$), but linear convergence would only make sense when $\rho < 1$ (see the definition above).

Note 2. The proof of Theorem 4.4 in fact demonstrated that for any $\delta > 0$, there are positive constants M_1, M_2, and integer N that depend only on δ such that, for all $n \geq N$,

$$\mathrm{P}\left[\sup_{m\geq 1}\left\{\frac{|\hat{\beta}^{(m)}-\hat{\beta}^*|}{(p\eta)^{m/2}}\right\}\leq M_1\right]>1-\delta,$$

$$\mathrm{P}\left[\sup_{m\geq 1}\left\{\frac{|\hat{v}^{(m)}-\hat{v}^*|}{(R\eta)^{m/2}}\right\}\leq M_2\right]>1-\delta.$$

2. Asymptotic behavior of IEEE. In Section 4.2.2 we conjectured that the (limiting) IEEE is asymptotically as efficient as the optimal GEE estimator obtained by solving (4.33) with the true V_is. The theorems below show that this conjecture is, indeed, true. The proofs can be found in Jiang et al. (2006a). The first result is about consistency of IEEE.

Theorem 4.5. Under the assumptions of Theorem 4.4, the IEEE $(\hat{\beta}^*, \hat{v}^*)$ is consistent.

To establish the asymptotic efficiency of IEEE, we need to strengthen assumptions *A2* and *A5* a little. Let $L_{2,0}=\max_{1\leq i\leq n}\max_{j\in J_i}\|\partial^2\mu_{ij}/\partial\beta\partial\beta'\|$, $L_3=\max_{1\leq i\leq n}\max_{j\in J_i}d_{ij}$, where

$$d_{ij}=\max_{1\leq a,b,c\leq p}\sup_{|\tilde{\beta}-\beta|\leq\epsilon_1}\left|\frac{\partial^3}{\partial\beta_a\partial\beta_b\partial\beta_c}g_j(X_i,\tilde{\beta})\right|.$$

A2′. Same as *A2* except that $g_j(X_i,\beta)$ are three-times continuously differentiable with respect to β, and that $L_2=O_{\mathrm{P}}(1)$ is replaced by $L_{2,0}\vee L_3=O_{\mathrm{P}}(1)$.

A5′. There is a positive integer γ such that $n/\{n(j,k,l)\}^\gamma\to 0$ for any $1\leq l\leq L_{jk}$, $(j,k)\in D$, as $n\to\infty$.

We also need the following additional assumption.

A6. $n^{-1}\sum_{i=1}^n\dot{\mu}_i'V_i^{-1}\dot{\mu}_i$ is bounded away from zero in probability.

Let $\tilde{\beta}$ be the solution to (4.33) with the true V_is. Note that $\tilde{\beta}$ is efficient, or optimal in the sense discussed in Section 4.2, but not computable, unless the true V_is are known.

Theorem 4.6. Under assumptions *A1*, *A2′*, *A3*, *A4*, *A5′* and *A6*, we have $\sqrt{n}(\hat{\beta}^*-\tilde{\beta})\to 0$ in probability. Thus, asymptotically, $\hat{\beta}^*$ is as efficient as $\tilde{\beta}$.

Note. The proof of Theorem 4.6 also reveals the following asymptotic expansion,

$$\hat{\beta}^*-\beta=\left(\sum_{i=1}^n\dot{\mu}_i'V_i^{-1}\dot{\mu}_i\right)^{-1}\sum_{i=1}^n\dot{\mu}_i'V_i^{-1}(Y_i-\mu_i)+\frac{o_{\mathrm{P}}(1)}{\sqrt{n}},\qquad(4.75)$$

where $o_{\mathrm{P}}(1)$ represents a term that converges to zero (vector) in probability. By Theorem 4.6, (4.72) also holds with $\hat{\beta}^*$ replaced by $\tilde{\beta}$, even though the latter is typically not computable. In the next section, we look at a case where an “exact” version of (4.72), that is, the equation without the term $o_{\mathrm{P}}(1)/\sqrt{n}$, holds for $\tilde{\beta}$.

4.5.3 Incorporating Informative Missing Data in IEE

In Section 4.2.2, we introduced IEE without taking into account the information about the missing data process, which is often available. We now extend the IEE method so that it incorporates such information. We consider a follow-up study, in which the responses are denoted by Y_{it}, $0 \leq t \leq T$ with Y_{i0} being the measurement just prior to the start of the follow-up. Again, let $X_i = (X'_{it})_{0 \leq t \leq T}$ denote a matrix of explanatory variables associated with the ith subject, where $X_{it} = (X_{itl})_{1 \leq l \leq p}$. We assume that X_i is completely observed, $1 \leq i \leq m$. As in Section 4.2.2, we assume a semiparametric regression model:

$$\mathrm{E}(Y_{it}|X_i) = g_t(X_i, \beta). \tag{4.76}$$

The notations μ_{ij} and μ_i are defined here similarly to those in Section 4.2.2, with $J_i = \{0, 1, \ldots, T\}$. Furthermore, we assume that, in addition to Y_{it} and X_i, measures are to be made on a vector of time-dependent covariates V_{it}, $0 \leq t \leq T$. Let $W_{i0} = (X'_{i0}, \ldots, X'_{iT}, Y_{i0}, V'_{i0})'$, and $W_{it} = (Y_{it}, V'_{it})'$, $1 \leq t \leq T$. The notation $\bar{W}_{it}$ denotes $\{W'_{i0}, \ldots, W'_{i(t-1)}\}'$, that is, the vector of all the data up to time $t-1$.

Define $R_{it} = 1$ if subject i is observed at time t, that is, if both Y_{it} and V_{it} are observed, and $R_{it} = 0$ otherwise. We assume that Y_{it} and V_{it} are both observed or both missing, and $R_{i0} = 1$. We also assume that, once a subject leaves the study, the subject will not return. This means that $R_{it} = 1$ implies $R_{i(t-1)} = 1, \ldots, R_{i1} = 1$. The following assumptions are made regarding the missing process R_{it},

$$\mathrm{P}\{R_{it} = 1|R_{i(t-1)} = 1, \bar{W}_{it}, Y_i\} = \mathrm{P}\{R_{it} = 1|R_{i(t-1)} = 1, \bar{W}_{it}\}.$$

Also, denoting the right side of the above by λ_{it}, we have $\lambda_{it} \geq \delta$ and $\lambda_{it} = \lambda_{it}(\bar{W}_{it}, \vartheta)$, where δ is some positive constant, and ϑ an unknown vector of parameters. See Robins et al. (1995) for a discussion of these conditions. The authors also proposed a maximum partial likelihood estimator for ϑ, which does not depend on the estimation of the other parameters. This fact is important to the derivation of IEE below. Write $\pi_{it} = \prod_{s=1}^{t} \lambda_{is}$, and $\Delta_{it} = \pi_{it}^{-1} R_{it}$. Define $\tilde{Y}_{it} = \Delta_{it} Y_{it}$. According to Lemma A.1 of Robins et al. (1995), we have $\mathrm{E}(\tilde{Y}_{it}|X_i) = \mathrm{E}(Y_{it}|X_i) = \mu_{it}$. Also, let $\tilde{V}_i = \mathrm{Var}(\tilde{Y}_i|X_i)$, where $\tilde{Y}_i = (\tilde{Y}_{it})_{0 \leq t \leq T}$. Then, according to Section 4.2.1, when ϑ and $\tilde{V}_i$s are known, the following estimating equation is optimal,

$$\sum_{i=1}^{m} \dot{\mu}'_i \tilde{V}_i^{-1} (\tilde{Y}_i - \mu_i) = 0. \tag{4.77}$$

If ϑ is unknown, because $\tilde{Y}_i$ depends on ϑ, that is, $\tilde{Y}_i = \tilde{Y}_i(\vartheta)$, we replace ϑ by $\hat{\vartheta}$, the maximum partial likelihood estimator, to get $\hat{Y}_i = \tilde{Y}_i(\hat{\vartheta})$. Thus, the estimating equation becomes

$$\sum_{i=1}^{m} \dot{\mu}_i' \tilde{V}_i^{-1} (\hat{Y}_i - \mu_i) = 0. \tag{4.78}$$

Note that there is no need to deal with the ϑ involved in $\tilde{V}_i$, because the latter is unspecified anyway (which is an advantage of this method).

The only real difference between Equations (4.33) and (4.75) is that Y_i is replaced by $\hat{Y}_i$ (V_i and $\tilde{V}_i$ are unspecified anyway). Thus, a very similar iterative procedure can be applied. Namely, given the $\tilde{V}_i$s, an estimator of β is obtained by solving (4.75); given β, the $\tilde{V}_i$s are estimated (by the method of moments) in the same way as (4.36) except that Y_{ij} is replaced by $\hat{Y}_{ij}$, $1 \le i \le n$, $j \in J_i = \{0, 1, \ldots, T\}$; and iterate between the two steps. Once again, we call such a procedure IEE. Note that $\hat{\vartheta}$ is unchanged during the iterations. This is because, as mentioned earlier, the estimation of ϑ does not depend on that of β and V_is. Therefore, there is no need to get ϑ involved in the iterations.

Suppose that $\hat{\vartheta}$ is a $\sqrt{m}$-consistent estimator. Sufficient conditions for the latter property can be found in Robins et al. (1995). Then, under regularity conditions similar to *A1–A5* in the previous Section, linear convergence of the IEE as well as consistency of the limiting estimator, say, $\hat{\beta}^*$, can be established. However, the result for the asymptotic distribution of $\hat{\beta}^*$ is different. More specifically, let $\tilde{\beta}$ be the solution to (4.75), where the $\tilde{V}_i$s are the true conditional covariance matrices. Unlike Theorem 4.6, it is no longer true that $\sqrt{m}(\hat{\beta}^* - \tilde{\beta}) \to 0$ in probability. In fact, the asymptotic covariance matrix of $\hat{\beta}^*$ is different from that of $\tilde{\beta}$. Here is another way to look at it. Suppose that β is the true parameter vector. If the $\tilde{V}_i$s are replaced by consistent estimators, the substitution results in a difference of $o_{\mathrm{P}}(\sqrt{m})$ on the left side of (4.75). However, if $\tilde{Y}_i$ is replaced by $\hat{Y}_i$, $1 \le i \le m$, the difference is $O_{\mathrm{P}}(\sqrt{m})$, if $\hat{\vartheta}$ is $\sqrt{m}$-consistent. A difference of $o_{\mathrm{P}}(\sqrt{m})$ will maintain both the consistency and the asymptotic distribution; a difference of $O_{\mathrm{P}}(\sqrt{m})$ will maintain the consistency but change the asymptotic distribution. For more details, see Jiang and Wang (2005).

A similar result was obtained by Robins et al. (1995; Theorem 1). In fact, the authors showed that the asymptotic covariance matrix of their GEE estimator with estimated missing probabilities (by $\hat{\vartheta}$) is "smaller" than that of the GEE estimator with the true missing probabilities. In other words, the GEE estimator with the estimated missing probabilities is asymptotically at least as efficient as that with the true missing probabilities (see the discussion on page 110 of the above reference). Also see Ying (2003, section 2).

4.5.4 Consistency of MSM Estimator

In this section, we give sufficient conditions for the asymptotic identifiability of the parameters $\varphi = (\beta', \sigma_1^2, \ldots, \sigma_q^2)'$ under the GLMM of Section 4.2.3 as well as consistency of the MSM estimator $\hat{\varphi}$. Here the asymptotic process is such that $n \to \infty$ and $L \to \infty$, where n is the (data) sample size and L the Monte Carlo sample size.

We first give a lemma that establishes convergence of the simulated moments to the corresponding moments after suitable normalizations. Let Q be the set of row vectors v whose components are positive integers ordered decreasingly (i.e., if $v = (v_1, \dots, v_s)$, we have $v_1 \geq \cdots \geq v_s$). Let Q_l be the subset of vectors in Q, whose sum of the components is equal to l. For example, $Q_2 = \{2, (1,1)\}$, $Q_4 = \{4, (3,1), (2,2), (2,1,1), (1,1,1,1)\}$. For $v \in Q$ and $v = (v_1, \dots, v_s)$, define $b^{(v)}(\cdot) = b^{(v_1)}(\cdot) \cdots b^{(v_s)}(\cdot)$, where $b^{(k)}(\cdot)$ represents the kth derivative. For $1 \leq r \leq q$, $1 \leq u \leq n$, let $I_{r,u} = \{1 \leq v \leq n : (u,v) \in I_r\}$, $J_r = \{(u,v,u',v') : (u,v), (u',v') \in I_r, (z_u, z_v)'(z_{u'}, z_{v'}) \neq 0\}$. Let $S = \cup_{r=1}^q I_r = \{(u,v) : 1 \leq u \neq v \leq n, z_u' z_v \neq 0\}$.

Lemma 4.1. Suppose that (i) $b(\cdot)$ is four times differentiable such that

$$\limsup_{n\to\infty} \max_{1\leq i\leq n} \max_{v\in Q_d} \mathrm{E}|b^{(v)}(\xi_i)| < \infty, \qquad d = 2, 4;$$

(ii) the sequences $\{a_{nj}\}$, $1 \leq j \leq p$ and $\{b_{nr}\}$, $1 \leq r \leq q$ are chosen such that the following converge to zero when divided by a_{nj}^2,

$$\sum_{i=1}^n w_i x_{ij}^2, \quad \sum_{(u,v)\in S} w_u w_v |x_{uj} x_{vj}|, \tag{4.79}$$

$1 \leq j \leq p$, and the following converge to zero when divided by b_{nr}^2,

$$\sum_{(u,v)\in I_r} w_u w_v, \quad \sum_{u=1}^n w_u \left(\sum_{v\in I_{r,u}} w_v \right)^2, \quad \sum_{(u,v,u',v')\in J_r} w_u w_v w_{u'} w_{v'}, \tag{4.80}$$

$1 \leq r \leq q$. Then, the following converges to zero in L^2 when divided by a_{nj},

$$\sum_{i=1}^n w_i x_{ij} \{y_i - \mathrm{E}_\theta(y_i)\},$$

$1 \leq j \leq p$, and the following converges to zero in L^2 when divided by b_{rn},

$$\sum_{(u,v)\in I_r} w_u w_v \{y_u y_v - \mathrm{E}_\theta(y_u y_v)\},$$

$1 \leq r \leq q$.

The proof is given in Jiang (1998a). We now define the normalized moments, simulated moments, and sample moments. Let

$$M_{\mathrm{N},j}(\theta) = a_{nj}^{-1} \sum_{i=1}^n w_i x_{ij} \mathrm{E}\{b'(\xi_i)\},$$

$$\tilde{M}_{\mathrm{N},j} = (a_{nj}L)^{-1}\sum_{i=1}^{n} w_i x_{ij} \sum_{l=1}^{L} b'(\xi_{il}),$$

$$\hat{M}_{\mathrm{N},j} = a_{nj}^{-1}\sum_{i=1}^{n} w_i x_{ij} y_i,$$

$1 \le j \le p$, where $\xi_{il} = x_i'\beta + z_i'Du^{(l)}$, and $u^{(1)}, \ldots, u^{(L)}$ are generated independently from the m-dimensional standard normal distribution. Here the subscript N refers to normalization. Similarly, we define

$$M_{\mathrm{N},p+r} = b_{nr}^{-1} \sum_{(u,v)\in I_r} w_u w_v \mathrm{E}\{b'(\xi_u)b'(\xi_v)\},$$

$$\tilde{M}_{\mathrm{N},p+r} = (b_{nr}L)^{-1} \sum_{(u,v)\in I_r} w_u w_v \sum_{l=1}^{L} b'(\xi_{ul})b'(\xi_{vl}),$$

$$\hat{M}_{\mathrm{N},p+r} = b_{nr}^{-1} \sum_{(u,v)\in I_r} w_u w_v y_u y_v,$$

$1 \le r \le q$. Let A_{nj}, $1 \le j \le p$ and B_{nr}, $1 \le r \le q$ be sequences of positive numbers such that $A_{nj} \to \infty$ and $B_{nr} \to \infty$ as $n \to \infty$. Let $\hat{\theta}$ be any $\theta \in \Theta_n = \{\theta : |\beta_j| \le A_{nj}, 1 \le j \le p; |\sigma_r| \le B_{nr}, 1 \le r \le q\}$ satisfying

$$|\tilde{M}_{\mathrm{N}}(\theta) - \hat{M}_{\mathrm{N}}| \le \delta_n, \tag{4.81}$$

where $\tilde{M}_{\mathrm{N}}(\theta)$ is the $(p+q)$-dimensional vector whose jth component is $\tilde{M}_{\mathrm{N},j}(\theta)$, $1 \le j \le p+q$, $\hat{M}_{\mathrm{N}}$ is defined similarly, and $\delta_n \to 0$ as $n \to \infty$. For any vector $v = (v_r)_{1\le r\le s}$, define $\|v\| = \max_{1\le r\le s}|v_r|$.

Theorem 4.7. Suppose that the conditions of Lemma 4.1 are satisfied.

a. Let ϵ_n be the maximum of the terms in (4.76) divided by a_{nj}^2 and the terms in (4.77) divided by b_{nr}^2 over $1 \le j \le p$ and $1 \le r \le q$. If $\epsilon_n/\delta_n^2 \to 0$, $\hat{\theta}$ exists with probability tending to one as $n \to \infty$.

b. If, furthermore, the first derivatives of $\mathrm{E}_\theta(y_i)$ and $\mathrm{E}(y_u y_v)$ $(u \ne v)$ with respect to components of θ can be taken under the expectation sign; the quantities

$$\sup_{\|\theta\|\le B} \mathrm{E}\{b'(\xi_i)\}^4, \ \mathrm{E}\left\{\sup_{\|\theta\|\le B}|b''(\xi_i)|\right\}, \ \mathrm{E}\left\{\sup_{\|\theta\|\le B}|b''(\xi_u)b''(\xi_v)|\right\},$$

$1 \le i \le n$, $(u,v) \in S$ are bounded for any $B > 0$; and

$$\liminf_{n\to\infty} \inf_{\|\tilde{\varphi}-\varphi\|>\epsilon} |M_{\mathrm{N}}(\tilde{\theta}) - M_{\mathrm{N}}(\theta)| > 0 \tag{4.82}$$

for any $\epsilon > 0$, there exists a sequence $\{d_n\}$ such that, as $n, L \to \infty$ with $L \ge d_n$, $\hat{\varphi}$ is a consistent estimator of φ.

Note that condition (4.79) ensures the identifiability of the true θ. Similar conditions can be found in, for example, McFadden (1989) and Lee (1992). Suppose that the function $M_{\mathrm{N}}(\cdot)$ is continuous and injective. Then, $\inf_{\|\tilde{\varphi}-\varphi\|>\epsilon}|M_{\mathrm{N}}(\tilde{\theta})-M_{\mathrm{N}}(\theta)|>0$. If the lower bound stays away from zero as $n\to\infty$, then (4.79) is satisfied. We consider an example.

Example 4.8 (Continued). Suppose that $\sigma^2>0$, $m\to\infty$ and k remains fixed and $k>1$. Then, it can be shown (Exercise 4.16) that all the conditions of Theorem 4.7 are satisfied in this case. In particular, we verify condition (4.79) here. Note that in this case $M_{\mathrm{N}}(\cdot)$ does not depend on n. Write $M_1(\theta)=\mathrm{E}\{h_\theta(\zeta)\}$, $M_2(\theta)=\mathrm{E}\{h_\theta^2(\zeta)\}$. It is easy to show that $\sup_\mu|M_1(\theta)-M_2(\theta)|\to 0$ as $\sigma\to\infty$ and $\sup_\mu|M_1^2(\theta)-M_2(\theta)|\to 0$ as $\sigma\to 0$. Therefore, there exist $0<a<b$ and $A>0$ such that $\inf_{\tilde{\theta}\notin[-A,A]\times[a,b]}|M(\tilde{\theta})-M(\theta)|>0$, where $M(\theta)=(M_1(\theta),M_2(\theta))'$. By continuity, it suffices to show that $M(\cdot)$ in injective. Let $0<c<1$ and consider the equation

$$M_1(\theta)=c. \tag{4.83}$$

For any $\sigma>0$, there is a unique $\mu=\mu_c(\sigma)$ that satisfies (4.80). The function $\mu_c(\cdot)$ is continuously differentiable. Write $\mu_c=\mu_c(\sigma)$, $\mu_c'=\mu_c'(\sigma)$. By (4.80), one has

$$\mathrm{E}\left[\frac{\exp(\mu_c+\sigma\zeta)}{\{1+\exp(\mu_c+\sigma\zeta)\}^2}(\mu_c'+\zeta)\right]=0. \tag{4.84}$$

Now consider $M_2(\theta)$ along the curve determined by (4.80); that is, $M_c(\sigma)=M_2(\mu_c,\sigma)$. We use the following covariance inequality. For continuous functions f. g and h with f and g strictly increasing and $h>0$, we have

$$\int f(x)g(x)h(x)dx\int h(x)dx>\int f(x)h(x)dx\int g(x)h(x)dx,$$

provided that the integrals are finite. By (4.81) and the covariance inequality, we have

$$\begin{aligned}M_c'(\sigma)&=2\mathrm{E}\left[\frac{\{\exp(\mu_c+\sigma\zeta)\}^2}{\{1+\exp(\mu_c+\sigma\zeta)\}^3}(\mu_c'+\zeta)\right]\\&>0.\end{aligned}$$

The injectivity of $M(\cdot)$ then follows.

The constant d_n in Theorem 4.7 can be determined by the proof of the theorem, given in Jiang (1998a).

4.5.5 Asymptotic Properties of First and Second-Step Estimators

In this section, we specify the conditions in (iii) of Section 4.2.4, which are sufficient for the existence, consistency, and asymptotic normality of the first-

and second-step estimators. Note that the results proved here do not require the assumptions of GLMM in Section 3.2 or its extended version in Section 4.2.4.

Let the responses be $y_1, \ldots, y_n$, and let Θ be the parameter space. First, note that B, S, and $u(\theta)$ in (4.52) may depend on n, and hence in the sequel we use the notation B_n, S_n, and $u_n(\theta)$. Also, the solution to (4.52) is unchanged if B_n is replaced by $C_n^{-1}B_n$, where $C_n = \text{diag}(c_{n,1}, \ldots, c_{n,r})$, $c_{n,j}$ is a sequence of positive constants, $1 \leq j \leq r$, and r is the dimension of θ. Write $M_n = C_n^{-1}B_nS_n$, and $M_n(\theta) = C_n^{-1}B_nu_n(\theta)$. Then the first-step estimator $\tilde{\theta} = \tilde{\theta}_n$ is the solution to the equation

$$M_n(\theta) = M_n. \tag{4.85}$$

Consider $M_n(\cdot)$ as a map from Θ, the parameter space, to a subset of R^r. Let θ denote the true θ everywhere except when defining a function of θ, such as in (4.82), and $M_n(\Theta)$ be the image of Θ under $M_n(\cdot)$. For $x \in R^r$ and $A \subset R^r$, define $d(x, A) = \inf_{y\in A} |x - y|$. Obviously, $M_n(\theta) \in M_n(\Theta)$. Furthermore, if $M_n(\theta)$ is in the interior of $M_n(\Theta)$, we have $d(M_n(\theta), M_n^c(\Theta)) > 0$. In fact, the latter essentially ensures the existence of the solution to (4.82).

Theorem 4.8. Suppose that, as $n \to \infty$,

$$M_n - M_n(\theta) \longrightarrow 0 \tag{4.86}$$

in probability, and

$$\liminf d\{M_n(\theta), M_n^c(\Theta)\} > 0. \tag{4.87}$$

Then, with probability tending to one, the solution to (4.82) exists and is in Θ. If, in addition, there is a sequence $\Theta_n \subset \Theta$ such that

$$\liminf \inf_{\theta_* \notin \Theta_n} |M_n(\theta_*) - M_n(\theta)| > 0, \tag{4.88}$$

$$\liminf \inf_{\theta_* \in \Theta_n, \theta_* \neq \theta} \frac{|M_n(\theta_*) - M_n(\theta)|}{|\theta_* - \theta|} > 0, \tag{4.89}$$

then, any solution $\tilde{\theta}_n$ to (4.82) is consistent.

Proof. The solution to (4.82) exists and is in Θ if and only if $M_n \in M_n(\Theta)$. Inequality (4.84) implies that there is $\epsilon > 0$ such that $d\{M_n(\theta), M_n^c(\Theta)\} \geq \epsilon$ for large n. Thus, $\text{P}\{M_n \notin M_n(\Theta)\} \leq \text{P}\{|M_n - M_n(\theta)| \geq \epsilon\}$. Therefore, $\tilde{\theta}_n$ exists with probability tending to one.

We now show that $\tilde{\theta}_n$ is consistent. By (4.85), there is $\epsilon_1 > 0$ such that, for large n, $\text{P}(\tilde{\theta}_n \notin \Theta_n) \leq \text{P}\{|M_n(\tilde{\theta}_n) - M_n(\theta)| \geq \epsilon_1\}$. On the other hand, by (4.86), there is $\epsilon_2 > 0$ such that, for large n and any $\epsilon > 0$, $\text{P}(|\tilde{\theta}_n - \theta| \geq \epsilon) \leq \text{P}(\tilde{\theta}_n \notin \Theta_n) + \text{P}\{|M_n(\tilde{\theta}_n) - M_n(\theta)| \geq \epsilon_2\epsilon\}$. The result follows by the fact that $M_n(\tilde{\theta}_n) = M_n$ with probability tending to one and the above argument. ∎

The following lemmas give sufficient conditions for (4.83)–(4.86). Let V_n be the covariance matrix of S_n.

Lemma 4.2. (4.83) holds provided that, as $n \to \infty$,

$$\mathrm{tr}(C_n^{-1} B_n V_n B_n' C_n^{-1}) \longrightarrow 0.$$

Lemma 4.3. Suppose that there is a vector-valued function $M_0(\theta)$ such that $M_n(\theta) \to M_0(\theta)$ as $n \to \infty$. Furthermore, suppose that there exist $\epsilon > 0$ and $N_\epsilon \geq 1$ such that $y \in M_n(\Theta)$ whenever $|y - M_0(\theta)| < \epsilon$ and $n \geq N_\epsilon$. Then (4.84) holds. In particular, if $M_n(\theta)$ does not depend on n, $M_n(\theta) = M(\theta)$, say, then (4.84) holds provided that $M(\theta)$ is in the interior of $M(\Theta)$, the image of $M(\cdot)$.

Lemma 4.4. Suppose that there are continuous functions $f_j(\cdot)$, $g_j(\cdot)$, $1 \leq j \leq r$, such that $f_j\{M_n(\theta)\} \to 0$ if $\theta \in \Theta$ and $\theta_j \to -\infty$, $g_j\{M_n(\theta)\} \to 0$ if $\theta \in \Theta$ and $\theta_j \to \infty$, $1 \leq j \leq r$, uniformly in n. If, as $n \to \infty$,

$$\limsup |M_n(\theta)| < \infty,$$
$$\liminf \min[|f_j\{M_n(\theta)\}|, |g_j\{M_n(\theta)\}|] > 0, \qquad 1 \leq j \leq r,$$

then there is a compact subset $\Theta_0 \subset \Theta$ such that (4.85) holds with $\Theta_n = \Theta_0$.

Write $U_n = \partial u_n / \partial \theta'$. Let $H_{n,j}(\theta) = \partial^2 u_{n,j} / \partial\theta\partial\theta'$, where $u_{n,j}$ is the jth component of $u_n(\theta)$, and $H_{n,j,\epsilon} = \sup_{|\theta_* - \theta| \leq \epsilon} \|H_{n,j}(\theta_*)\|$, $1 \leq j \leq L_n$, where L_n is the dimension of u_n.

Lemma 4.5. Suppose that $M_n(\cdot)$ is twice continuously differentiable, and that, as $n \to \infty$,

$$\liminf \lambda_{\min}(U_n' B_n' C_n^{-2} B_n U_n) > 0,$$

and there is $\epsilon > 0$ such that

$$\limsup \frac{\max_{1 \leq i \leq r} c_{n,i}^{-2} (\sum_{j=1}^{L_n} |b_{n,ij}| H_{n,j,\epsilon})^2}{\lambda_{\min}(U_n' B_n' C_n^{-2} B_n U_n)} < \infty,$$

where $b_{n,ij}$ is the (i,j) element of B_n. Furthermore suppose, for any compact subset $\Theta_1 \subset \Theta$ such that $d(\theta, \Theta_1) > 0$, that

$$\liminf \inf_{\theta_* \in \Theta_1} |M_n(\theta_*) - M_n(\theta)| > 0,$$

as $n \to \infty$. Then (4.86) holds for $\Theta_n = \Theta_0$, where Θ_0 is any compact subset of Θ that includes θ as an interior point.

The proofs of these lemmas are fairly straightforward.

Example 4.8 (Continued). As noted in Example 4.10, both the first and second-step estimators of $\theta = (\mu, \sigma)'$ correspond to $B_n = \mathrm{diag}(1, 1_m')$. Then it can be shown that, by choosing $C_n = \text{diag}\{mk, mk(k-1)\}$, the conditions of Lemmas 4.2–4.5 are satisfied.

We now consider the asymptotic normality of the first-step estimator. We say that an estimator $\tilde{\theta}_n$ is asymptotically normal with mean θ and asymptotic covariance matrix $(\Gamma_n'\Gamma_n)^{-1}$ if $\Gamma_n(\tilde{\theta}_n - \theta) \longrightarrow N(0, I_r)$ in distribution. Let $\lambda_{n,1} = \lambda_{\min}(C_n^{-1}B_nV_nB_n'C_n^{-1})$ and $\lambda_{n,2} = \lambda_{\min}\{U_n'B_n'(B_nV_nB_n')^{-1}B_nU_n\}$.

Theorem 4.9. Suppose that (i) the components of $u_n(\theta)$ are twice continuously differentiable; (ii) $\tilde{\theta}_n$ satisfies (4.82) with probability tending to one and is consistent; (iii) there exists $\epsilon > 0$ such that

$$\frac{|\tilde{\theta}_n - \theta|}{(\lambda_{n,1}\lambda_{n,2})^{1/2}} \max_{1\leq i\leq r} c_{n,i}^{-1}\left(\sum_{j=1}^{L_n}|b_{n,ij}|H_{n,j,\epsilon}\right) \longrightarrow 0$$

in probability; and (iv)

$$\{C_n^{-1}B_nV_nB_n'C_n^{-1}\}^{-1/2}[M_n - M_n(\theta)] \longrightarrow N(0, I_r) \tag{4.90}$$

in distribution. Then $\tilde{\theta}$ is asymptotically normal with mean θ and asymptotic covariance matrix

$$(B_nU_n)^{-1}B_nV_nB_n'(U_n'B_n')^{-1}. \tag{4.91}$$

Proof. Write $s_n(\theta) = S_n - u_n(\theta)$. By Taylor expansion, it is easy to show that, with probability tending to one,

$$0 = C_n^{-1}B_ns_n(\theta) - C_n^{-1}B_n(U_n + R_n)(\tilde{\theta}_n - \theta), \tag{4.92}$$

where the jth component of R_n is $(1/2)(\tilde{\theta}_n - \theta)'H_{n,j}(\theta^{(n,j)})$, and $\theta^{(n,j)}$ lies between θ and $\tilde{\theta}_n$, $1 \leq j \leq L_n$. Write $W_n = C_n^{-1}B_nV_nB_n'C_n^{-1}$. Then, by (4.87) and (4.89), we have

$$W_n^{-1/2}C_n^{-1}B_n(U_n + R_n)(\tilde{\theta}_n - \theta) \longrightarrow N(0, I_r) \text{ in distribution.}$$

Also, we have $W_n^{-1/2}C_n^{-1}B_n(U_n + R_n) = (I_r + K_n)W_n^{-1/2}C_n^{-1}B_nU_n$, where

$$K_n = W_n^{-1/2}C_n^{-1}B_nR_n(W_n^{-1/2}C_n^{-1}B_nU_n)^{-1}.$$

On the other hand, it is easy to show that $\|K_n\| \leq (\lambda_{n,1}\lambda_{n,2})^{-1/2}\|C_n^{-1}B_nR_n\|$, and

$$\|C_n^{-1}B_nR_n\|^2 \leq \frac{r}{4}|\tilde{\theta}_n - \theta|^2 \max_{1\leq i\leq r} c_{n,i}^{-2}\left(\sum_{j=1}^{L_n} b_{n,ij}|H_{n,j,\epsilon}\right)^2.$$

The result then follows. ∎

Sufficient conditions for existence, consistency, and asymptotic normality of the second-step estimators can be obtained by replacing the conditions of Theorems 4.8 and 4.9 by corresponding conditions with a probability statement. Let ξ_n be a sequence of nonnegative random variables. We say that $\liminf \xi_n > 0$ with probability tending to one if for any $\epsilon > 0$ there is $\delta > 0$ such that $\mathrm{P}(\xi_n > \delta) \geq 1 - \epsilon$ for all sufficiently large n. Note that this is equivalent to $\xi_n^{-1} = O_P(1)$. Then, for example, (4.85) is replaced by (4.85) with probability tending to one.

Note that the asymptotic covariance matrix of the second-step estimator is given by (4.88) with $B_n = U_n' V_n^{-1}$, which is $(U_n' V_n U_n)^{-1}$. This is the same as the asymptotic covariance matrix of the solution to (4.52) [(4.82)] with the optimal B (B_n). In other words, the second-step estimator is asymptotically optimal.

4.5.6 Further Results of the Fence Method

1. Estimation of σ_{M,M^*} in the case of clustered observations. Clustered data arise naturally in many fields, including analysis of longitudinal data (e.g., Diggle et al. 1996) and small area estimation (e.g., Rao 2003). Let $y_i = (y_{ij})_{1 \leq j \leq k_i}$ represent the vector of observations in the ith cluster, and $y = (y_i)_{1 \leq i \leq m}$. We assume that $y_1, \ldots, y_m$ are independent.

Furthermore, we assume that Q_M is additive in the sense that

$$Q_M = \sum_{i=1}^{m} Q_{M,i}, \tag{4.93}$$

where $Q_{M,i} = Q_{M,i}(y_i, \theta_M)$. We consider some examples.

Example 4.12. For ML model selection (Section 4.3.1.1), because, for clustered data, $f_M(y|\theta_M) = \prod_{i=1}^{m} f_{M,i}(y_i|\theta_M)$, where $f_{M,i}(\cdot|\theta_M)$ is the joint pdf of y_i under M and θ_M, we have

$$Q_M = -\sum_{i=1}^{m} \log\{f_{M,i}(y_i|\theta_M)\}.$$

Thus, (4.90) holds with $Q_{M,i} = -\log\{f_{M,i}(y_i|\theta_M)\}$.

Example 4.13. Now consider the case of MVC model selection (Section 4.3.1.2). If we choose $T = \mathrm{diag}(T_1, \ldots .T_m)$, where T_i is $k_i \times s_i$ and $1 \leq s_i \leq k_i$, we have

$$Q_M = \sum_{i=1}^{m} |(T_i' V_{M,i}^{-1} T_i)^{-1} T_i' V_{M,i}^{-1} (y_i - \mu_{M,i})|^2,$$

where $\mu_{M,i}$ and $V_{M,i}$ are the mean vector and covariance matrix of y_i under M and θ_M. Thus, (4.90) holds with $Q_{M,i} = |(T_i' V_{M,i}^{-1} T_i)^{-1} T_i' V_{M,i}^{-1} (y_i - \mu_{M,i})|^2$.

Example 4.14. Note that the Q_M defined for extended GLMM selection (Section 4.3.1.3) always satisfies (4.90), even if the data are not clustered.

Denote, with a little abuse of the notation, the minimizer of $\mathrm{E}(Q_M)$ over $\theta_M \in \Theta_M$ by θ_M. Let M^* denote a correct model. The following lemma gives approximations to $\mathrm{E}(\hat{Q}_M - \hat{Q}_{M^*})^2$ in two different situations.

Lemma 4.6. Suppose that the following regularity conditions are satisfied: i) $\mathrm{E}(\partial Q_M/\partial\theta_M) = 0$, and $\mathrm{tr}\{\mathrm{Var}(\partial Q_{M,i}/\partial\theta_M)\} \le c$ for some constant c; ii) there is a constant B_M such that $Q_M(\tilde{\theta}_M) > Q_M(\theta_M)$, if $|\tilde{\theta}_M| > B_M$; iii) there are constants $c_j > 0$, $j = 1, 2, 3$ such that $\mathrm{E}(|\hat{\theta}_M - \theta_M|^8) \le c_1 m^{-4}$, $\mathrm{E}(|\partial Q_M/\partial\theta_M|^4) \le c_2 m^2$, and

$$\mathrm{E}\left(\sup_{|\tilde{\theta}_M| \le B_M} \left\| \frac{\partial^2 \tilde{Q}_M}{\partial\theta_M \partial\theta_M'} \right\|^4\right) \le c_3 m^4;$$

and iv) there are constants $a, b > 0$ such that $am \le \mathrm{var}(Q_M - Q_{M^*}) \le bm$, if $M \ne M^*$; v) for any incorrect model M, we have $\mathrm{E}(Q_M - Q_{M^*}) = O(m)$. Then, we have $\mathrm{E}(\hat{Q}_M - \hat{Q}_{M^*})^2 = \mathrm{var}(Q_M - Q_{M^*})\{1 + o(1)\} = O(m)$, if M is correct; and $\mathrm{E}(\hat{Q}_M - \hat{Q}_{M^*})^2 = \mathrm{var}(Q_M - Q_{M^*}) + O(m^2) = O(m^2)$, if M is incorrect.

The proof is omitted (see Jiang et al. 2006b). Note that (i) is satisfied if $\mathrm{E}(Q_M)$ can be differentiated inside the expectation; that is, $\partial\mathrm{E}(Q_M)/\partial\theta_M = \mathrm{E}(\partial Q_M/\partial\theta_M)$. Also note that (ii) implies that $|\hat{\theta}_M| \le B_M$.

Because a measure of the difference $\hat{Q}_M - \hat{Q}_{M^*}$ is its L^2-norm, $\|\hat{Q}_M - \hat{Q}_{M^*}\|_2 = \sqrt{\mathrm{E}(\hat{Q}_M - \hat{Q}_{M^*})^2}$, Lemma 4.6 suggests a difference between a true model and an incorrect one: If M is a true model, $\hat{Q}_M - \hat{Q}_{M^*}$ may be measured by $\sigma_{M,M^*} = \sqrt{\mathrm{var}(Q_M - Q_{M^*})} = \mathrm{sd}(Q_M - Q_{M^*})$; otherwise, $\hat{Q}_M - \hat{Q}_{M^*}$ is expected to be much larger because $\mathrm{sd}(Q_M - Q_{M^*}) = O(\sqrt{m})$.

Furthermore, it is not difficult to obtain an estimator of σ_{M,M^*}. By (4.90) and independence, we have

$$\begin{aligned}
\sigma^2_{M,M^*} &= \sum_{i=1}^m \mathrm{var}(Q_{M,i} - Q_{M^*,i}) \\
&= \sum_{i=1}^m [\mathrm{E}(Q_{M,i} - Q_{M^*,i})^2 - \{\mathrm{E}(Q_{M,i}) - \mathrm{E}(Q_{M^*,i})\}^2] \\
&= \mathrm{E}\left[\sum_{i=1}^m (Q_{M,i} - Q_{M^*,i})^2 - \sum_{i=1}^m \{\mathrm{E}(Q_{M,i}) - \mathrm{E}(Q_{M^*,i})\}^2\right].
\end{aligned}$$

Thus, an estimator of σ^2_{M,M^*} is the observed variance given by

$$\hat{\sigma}^2_{M,M^*} = \sum_{i=1}^m (\hat{Q}_{M,i} - \hat{Q}_{M^*,i})^2 - \sum_{i=1}^m \{\hat{\mathrm{E}}(Q_{M,i}) - \hat{\mathrm{E}}(Q_{M^*,i})\}^2, \quad (4.94)$$

where $\hat{Q}_{M,i} = Q_{M,i}(y_i, \hat{\theta}_M)$, $\hat{Q}_{M^*,i} = Q_{M^*,i}(y_i, \hat{\theta}_{M^*})$, and

$$\hat{\mathrm{E}}(Q_{M,i}) = \mathrm{E}_{M^*,\hat{\theta}_{M^*}}\{Q_{M,i}(y_i, \hat{\theta}_M)\},$$
$$\hat{\mathrm{E}}(Q_{M^*,i}) = \mathrm{E}_{M^*,\hat{\theta}_{M^*}}\{Q_{M^*,i}(y_i, \hat{\theta}_{M^*})\},$$

where the expectations are with respect to y_i under model M^* and evaluated at $\hat{\theta}_{M^*}$. Again, we consider some examples.

Example 4.12 (Continued). In the case of ML model selection, we have

$$\mathrm{E}(Q_{M,i}) = -\int \log\{f_{M,i}(y_i|\theta_M)\} f_i(y_i)\nu(dy_i),$$
$$\mathrm{E}(Q_{M^*,i}) = -\int \log\{f_i(y_i)\} f_i(y_i)\nu(dy_i),$$
$$\hat{\mathrm{E}}(Q_{M,i}) = -\int \log\{f_{M,i}(y_i|\hat{\theta}_M)\} f_{M^*,i}(y_i|\hat{\theta}_{M^*})\nu(dy_i),$$
$$\hat{\mathrm{E}}(Q_{M^*,i}) = -\int \log\{f_{M^*,i}(y_i|\hat{\theta}_{M^*})\} f_{M^*,i}(y_i|\hat{\theta}_{M^*})\nu(dy_i).$$

Therefore,

$$\hat{\mathrm{E}}(Q_{M,i}) - \hat{\mathrm{E}}(Q_{M^*,i}) = \int \log\left\{\frac{f_{M^*,i}(y_i|\hat{\theta}_{M^*})}{f_{M,i}(y_i|\hat{\theta}_M)}\right\} f_{M^*,i}(y_i|\hat{\theta}_{M^*})\nu(dy_i).$$

Example 4.13 (Continued). In the case of MVC model selection, we have

$$\begin{aligned}\mathrm{E}(Q_{M,i}) = {} & \mathrm{tr}\{(T_i' V_{M,i}^{-1} T_i)^{-1} T_i' V_{M,i}^{-1} V_i V_{M,i}^{-1} T_i (T_i' V_{M,i}^{-1} T_i)^{-1}\} \\ & + |(T_i' V_{M,i}^{-1} T_i)^{-1} T_i' V_{M,i}^{-1}(\mu_{M,i} - \mu_i)|^2, \end{aligned} \tag{4.95}$$

and $\mathrm{E}(Q_{M^*,i}) = \mathrm{tr}\{(T_i' V_{M^*,i}^{-1} T_i)^{-1}\}$. Thus, $\hat{E}(Q_{M,i})$ is given by (4.92) with $\mu_{M,i}$ replaced by $\hat{\mu}_{M,i} = \mu_{M,i}(\hat{\theta}_M)$, $V_{M,i}$ by $\hat{V}_{M,i} = V_{M,i}(\hat{\theta}_M)$, μ_i by $\hat{\mu}_{M^*,i} = \mu_{M^*,i}(\hat{\theta}_{M^*})$, and V_i by $\hat{V}_{M^*,i} = V_{M^*,i}(\hat{\theta}_{M^*})$; and $\hat{\mathrm{E}}(Q_{M^*,i}) = \mathrm{tr}\{(T_i' \hat{V}_{M^*,i}^{-1} T_i)^{-1}\}$.

Example 4.14 (Continued). In the case of clustered data, this is a special case of Example 4.13 (continued) with $T = I$, the identity matrix.

2. Consistency of the fence. We now give conditions for the consistency of the fence procedure. The results given below do not require that the data be clustered.

We assume that the following A1–A4 hold for each $M \in \mathcal{M}$, where, as before, θ_M represents a parameter vector at which $\mathrm{E}(Q_M)$ attains its minimum, and $\partial Q_M/\partial\theta_M$, and so on. represent derivatives evaluated at θ_M. Similarly, $\partial\tilde{Q}_M/\partial\theta_M$, and so on. represent derivatives evaluated at $\tilde{\theta}_M$.

A1. Q_M is three-times continuously differentiable with respect to θ_M; and the following holds,

$$\mathrm{E}\left(\frac{\partial Q_M}{\partial \theta_M}\right) = 0. \tag{4.96}$$

A2. There is a constant B_M such that $Q_M(\tilde{\theta}_M) > Q_M(\theta_M)$, if $|\tilde{\theta}_M| > B_M$.

A3. The equation $\partial Q_M/\partial \theta_M = 0$ has an unique solution.

A4. There is a sequence of positive numbers $a_n \to \infty$ and $0 \le \gamma < 1$ such that

$$\begin{aligned}
\frac{\partial Q_M}{\partial \theta_M} - \mathrm{E}\left(\frac{\partial Q_M}{\partial \theta_M}\right) &= O_{\mathrm{P}}(a_n^{\gamma}),\\
\frac{\partial^2 Q_M}{\partial \theta_M \partial \theta_M'} - \mathrm{E}\left(\frac{\partial^2 Q_M}{\partial \theta_M \partial \theta_M'}\right) &= O_{\mathrm{P}}(a_n^{\gamma}),\\
\liminf \frac{1}{a_n} \lambda_{\min}\left\{\mathrm{E}\left(\frac{\partial^2 Q_M}{\partial \theta_M \partial \theta_M'}\right)\right\} &> 0,\\
\limsup \frac{1}{a_n} \lambda_{\max}\left\{\mathrm{E}\left(\frac{\partial^2 Q_M}{\partial \theta_M \partial \theta_M'}\right)\right\} &< \infty,
\end{aligned}$$

and there is $\delta_M > 0$ such that

$$\sup_{|\tilde{\theta}_M - \theta_M| \le \delta_M} \left|\frac{\partial^3 \tilde{Q}_M}{\partial \theta_{M,j} \partial \theta_{M,k} \partial \theta_{M,l}}\right| = O_{\mathrm{P}}(a_n), \qquad 1 \le j, k, l \le p_M,$$

where p_M is the dimension of θ_M.

In addition, we assume the following. Recall that c_n is the constant used in (4.69).

A5. $c_n \to \infty$; for any true model M^* and incorrect model M, we have $\mathrm{E}(Q_M) > \mathrm{E}(Q_{M^*})$,

$$\begin{aligned}
\liminf \left(\frac{\sigma_{M,M^*}}{a_n^{2\gamma-1}}\right) &> 0,\\
\frac{c_n \sigma_{M,M^*}}{\mathrm{E}(Q_M) - \mathrm{E}(Q_{M^*})} &\longrightarrow 0.
\end{aligned}$$

A6. $\hat{\sigma}_{M,M^*} > 0$ and $\hat{\sigma}_{M,M^*} = \sigma_{M,M^*} O_{\mathrm{P}}(1)$ if M^* is true and M incorrect; and $\sigma_{M,M^*} \vee a_n^{2\gamma-1} = \hat{\sigma}_{M,M^*} O_{\mathrm{P}}(1)$ if both M and M^* are true.

Note. Equation (4.93) is satisfied if $\mathrm{E}(Q_M)$ can be differentiated inside the expectation; that is, $\partial \mathrm{E}(Q_M)/\partial \theta_M = \mathrm{E}(\partial Q_M/\partial \theta_M)$. Also note that A2 implies that $|\hat{\theta}_M| \le B_M$. To illustrate A4 and A5, consider the case of clustered responses (see earlier discussions). Then, under regularity conditions, A4 holds with $a_n = m$ and $\gamma = 1/2$. Furthermore, we have $\sigma_{M,M^*} = O(\sqrt{m})$ and $\mathrm{E}(Q_M) - \mathrm{E}(Q_{M^*}) = O(m)$, provided that M^* is true, M is incorrect, and some regularity conditions hold. Thus, A5 holds with $\gamma = 1/2$ and c_n being any sequence satisfying $c_n \to \infty$ and $c_n/\sqrt{m} \to 0$. Finally, A6 does not require that $\hat{\sigma}_{M,M^*}$ be a consistent estimator of σ_{M,M^*}, only that it has the same order as σ_{M,M^*}. However, see our earlier discussion.

Lemma 4.8. Under A1–A4, we have $\hat{\theta}_M - \theta_M = O_{\rm P}(a_n^{\gamma-1})$ and $\hat{Q}_M - Q_M = O_{\rm P}(a_n^{2\gamma-1})$.

Recall that M_0 is the model selected by the fence (see Section 4.3.1). The following theorem establishes the consistency of the fence.

Theorem 4.10. Under assumptions A1–A6, we have with probability tending to one that M_0 is a true model with minimum dimension.

The proofs of Lemma 4.8 and Theorem 4.10 are omitted (see Jiang et al. 2006b).

4.6 Exercises

4.1. Show that in Example 4.1, the log-likelihood function under the assumed model is given by (4.1).

4.2. Write a simple routine based on the simple algorithm to numerically evaluate the likelihood function in Example 4.1. Use simulated data for the evaluation.

4.3. Show that the threshold model introduced at the beginning of Section 4.1.1 is a special case of GLMM with binary responses.

4.4. Using the results of Appendix C, verify expressions (4.3) and (4.4) in the E-step of the maximum likelihood estimation under a Gaussian mixed model. Also verify the M-step (4.5) and (4.6).

4.5. Verify the expressions (4.7)–(4.9).

4.6. Consider Example 4.3 on the Gaussian copula distribution.

a. Verify that the joint pdf of the Gaussian copula is given by (4.14).

b. Show that the marginal cdf and pdf of y_{ij} are $F_j(\cdot|\theta_j)$ and $f_j(\cdot|\theta_j)$, respectively.

c. Verify that the joint pdf of y_i is given by (4.15).

4.7. Verify that, in Example 4.3 (Continued), the likelihood equation under the working independence model is unbiased; that is, $\mathrm{E}_\theta\{\dot{l}_w(\theta)\} = 0$.

4.8. Verify the expressions of partial derivatives in Example 4.4.

4.9. Verify that in Section 4.1.6, the (joint) posterior of β and G under the assumed model is given by (4.25).

4.10. Consider the following linear model for longitudinal data: $y_1, \dots, y_m$ are independent with $\mathrm{E}(y_i) = X_i\beta$ and $\mathrm{Var}(y_i) = V_i$, where $y_i = (y_{ij})_{j\in J_i}$ and X_i is a matrix of fixed covariates (see Section 1.4.3). Show that in this case, the GEE estimator is the same as the WLS estimator of Section 1.4.3, that is, (1.43), with $W = V^{-1}$, where $V = \mathrm{diag}(V_1, \dots, V_m)$, provided that the V_is are nonsingular and $\sum_{i=1}^n X_i'V_i^{-1}X_i$ is nonsingular.

4.11. Show that in Example 4.8, a set of sufficient statistics for μ and σ are $y_{1\cdot}, \dots, y_{m\cdot}$. Also verify the following: $\mathrm{E}(y_{1\cdot}) = n\mathrm{E}\{h_\theta(\xi)\}$, $\mathrm{E}(y_{1\cdot}^2) =$

$n\mathrm{E}\{h_\theta(\xi)\}+n(n-1)\mathrm{E}\{h_\theta^2(\xi)\}$, where $h_\theta(x)=\exp(\mu+\sigma x)/\{1+\exp(\mu+\sigma x)\}$ and $\xi \sim N(0,1)$.

4.12. Show that under the GLMM of Section 4.2.3, the marginal density of y can be expressed as (4.43). Therefore, a set of sufficient statistics for θ is given by the S_j, $1 \le j \le p+m$ below (4.43), where $m = m_1 + \cdots + m_q$.

4.13. Verify (4.46). Also show that the first term on the right side of (4.46) depends on ϕ, and the second term does not depend on ϕ.

4.14 This exercise has two parts.

(i) Show that the right side of (4.44) can be expressed as

$$X_j' W \mathrm{E}\{e(\theta,u)\},$$

where X_j is the jth column of X, $W = \mathrm{diag}(w_i, 1 \le i \le n)$, and $e(\theta,u) = \{b'(\xi_i)\}_{1 \le i \le n}$ with ξ_i given by (3.2) with $\eta_i = \xi_i$.

(ii) Show that the right side of (4.47) can be expressed as

$$\mathrm{E}\{e(\theta,u)' W H_r W e(\theta,u)\},$$

where H_r is the $n \times n$ symmetric matrix whose (s,t) entry is $1_{\{(s,t)\in S_r\}}$.

4.15. Verify that in Example 4.9, the MSM equations are given by (4.50).

4.16. Show that all the conditions of Theorem 4.7 are satisfied in the case of Example 4.8, provided that $\sigma^2 > 0$, $m \to \infty$ and k remains fixed and $k > 1$. You may skip condition (4.79) inasmuch as it has been verified in Section 4.5.4.

4.17. Consider the base statistics given in Example 4.10. Show that in the special case of Example 4.8 (i.e., $k_i = k$, $1 \le i \le m$), (4.39) and (4.40) correspond to the first-step estimating equation (4.52) with $B = \mathrm{diag}(1, 1_m')$. Show that this B is, in fact, optimal in the sense of Section 4.2.4. Note that in this case the optimal B does not depend on θ. In other words, the first-step estimators are the same as the second-step ones in this case.

4.18. Verify the marginal distribution of Y in Example 4.11, that is, (4.59). Also verify the mean and variance expressions of Y.

A

List of Notations

(The list is in alphabetical order.)

$a \wedge b$: $= \min(a, b)$.

$a \vee b$: $= \max(a, b)$.

a': transpose of vector a.

$\dim(a)$: the dimension of vector a.

$|A|$: the determinant of matrix A.

$\lambda_{\min}(A)$: the smallest eigenvalue of matrix A.

$\lambda_{\max}(A)$: the largest eigenvalue of matrix A.

$\text{tr}(A)$: the trace of matrix A.

$\|A\|$: the spectral norm of matrix A defined as $\|A\| = \{\lambda_{\max}(A'A)\}^{1/2}$.

$\|A\|_2$: the 2-norm of matrix A defined as $\|A\|_2 = \{\text{tr}(A'A)\}^{1/2}$.

$\text{rank}(A)$: the (column) rank of matrix A.

$A^{1/2}$: the square root of a nonnegative definite matrix A defined in Appendix B.

$\mathcal{L}(A)$: the linear space spanned by the columns of matrix A.

P_A: the projection matrix to $\mathcal{L}(A)$ defined as $P_A = A(A'A)^- A'$, where A^- is the generalized inverse of A (see Appendix B).

$P_{A^\perp}$: the projection matrix with respect to the linear space orthogonal to $\mathcal{L}(A)$, defined as $P_A = I - P_A$, where I is the identity matrix.

If A is a set, $|A|$ represents the cardinality of A.

$\text{Cov}(\xi, \eta)$: the covariance matrix between random vectors ξ and η, defined as $\text{Cov}(\xi, \eta) = (\text{cov}(\xi_i, \eta_j))_{1 \le i \le k, 1 \le j \le l}$, where ξ_i is the ith component of ξ, η_j is the jth component of η, $k = \dim(\xi)$, and $l = \dim(\eta)$.

$\xrightarrow{\mathcal{D}}$: convergence in distribution.

$\text{diag}(A_1, \dots, A_k)$: the block-diagonal matrix with $A_1, \dots, A_k$ on its diagonal; note that this also includes the diagonal matrix, when $A_1, \dots, A_k$ are numbers.

I_n: the n-dimensional identity matrix.

J_n: the $n \times n$ matrix of 1s, or $J_n = 1_n 1_n'$.

$\bar{J}_n$: $= n^{-1} J_n$.

$N(\mu, \Sigma)$: the multivariate normal distribution with mean vector μ and covariance matrix Σ.

MSA: for balanced data y_{ij}, $1 \le i \le m$, $1 \le j \le k$, MSA $=$ SSA$/(k-1)$.

MSE: for balanced data y_{ij}, $1 \le i \le m$, $1 \le j \le k$, MSE $=$ SSE$/m(k-1)$.

1_n: the n-dimensional vector of 1s.

1_n^0: $= I_n$.

1_n^1: $= 1_n$.

$\partial\xi/\partial\eta'$: when $\xi = (\xi_i)_{1\le i\le a}$, $\eta = (\eta_j)_{1\le j\le b}$, this notation means the matrix $(\partial\xi_i/\partial\eta_j)_{1\le i\le a, 1\le j\le b}$.

$\partial^2\xi/\partial\eta\partial\eta'$: when ξ is a scalar, $\eta = (\eta_j)_{1\le j\le b}$, this notation means the matrix $(\partial^2\xi/\partial\eta_j\partial\eta_k)_{1\le j,k\le b}$.

SSA: for balanced data y_{ij}, $1 \le i \le m$, $1 \le j \le k$, SSA $= k\sum_{i=1}^m (\bar{y}_{i\cdot} - \bar{y}_{\cdot\cdot})^2$.

SSE: for balanced data y_{ij}, $1 \le i \le m$, $1 \le j \le k$, SSE $= \sum_{i=1}^m \sum_{j=1}^k (y_{ij} - y_{i\cdot})^2$.

Var(ξ): The covariance matrix of random vector ξ defined as $\mathrm{Var}(\xi) = (\mathrm{cov}(\xi_i, \xi_j))_{1\le i,j\le k}$, where ξ_i is the ith component of ξ and $k = \dim(\xi)$.

$(X_i)_{1\le i\le m}$: when $X_1, \ldots, X_m$ are matrices with the same number of columns, means the matrix that combines the rows of $X_1, \ldots, X_m$, one after the other.

$(y_i)_{1\le i\le m}$: when $y_1, \ldots, y_m$ are column vectors, this notation means the column vector $(y_1', \ldots, y_m')'$.

$(y_{ij})_{1\le i\le m, 1\le j\le n_i}$: in the case of clustered data, where y_{ij}, $j = 1, \ldots, n_i$ denote the observations from the ith cluster, this notation represents the vector $(y_{11}, \ldots, y_{1n_1}, y_{21}, \ldots, y_{2n_2}, \ldots, y_{m1}, \ldots, y_{mn_m})'$.

$y_{i\cdot}$, $\bar{y}_{i\cdot}$, $y_{\cdot j}$, $\bar{y}_{\cdot j}$, $y_{\cdot\cdot}$ and $\bar{y}_{\cdot\cdot}$: in the case of clustered data y_{ij}, $i = 1, \ldots, m$, $j = 1, \ldots, n_i$, $y_{i\cdot} = \sum_{j=1}^{n_i} y_{ij}$, $\bar{y}_{i\cdot} = n_i^{-1} y_{i\cdot}$, $y_{\cdot\cdot} = \sum_{i=1}^m \sum_{j=1}^{n_i} y_{ij}$, $\bar{y}_{\cdot\cdot} = (\sum_{i=1}^m n_i)^{-1} y_{\cdot\cdot}$; in the case of balanced data y_{ij}, $1 \le i \le a$, $j = 1, \ldots, b$, $y_{i\cdot} = \sum_{j=1}^b y_{ij}$, $\bar{y}_{i\cdot} = b^{-1} y_{i\cdot}$, $y_{\cdot j} = \sum_{i=1}^a y_{ij}$, $\bar{y}_{\cdot j} = a^{-1} y_{\cdot j}$, $y_{\cdot\cdot} = \sum_{i=1}^a \sum_{j=1}^b y_{ij}$, $\bar{y}_{\cdot\cdot} = (ab)^{-1} y_{\cdot\cdot}$.

$y|\eta \sim$: the distribution of y given η is ...; note that here η may represent a vector of parameters or random variables, or a combination of both.

B

Matrix Algebra

B.1 Kronecker Products

Let $A = (a_{ij})_{1 \le i \le m, 1 \le j \le n}$ be a matrix. Then, for any matrix B, the Kronecker product, $A \otimes B$ is defined as the partitioned matrix $(a_{ij}B)_{1 \le i \le m, 1 \le j \le n}$. For example, if $A = I_m$ and $B = 1_n$, then $A \otimes B = \text{diag}(1_n, \ldots, 1_n)$. Below are some well-known and useful properties of Kronecker products:

2mm

(i) $(A_1 + A_2) \otimes B = A_1 \otimes B + A_2 \otimes B$.

(ii) $A \otimes (B_1 + B_2) = A \otimes B_1 + A \otimes B_2$.

(iii) $c \otimes A = A \otimes c = cA$, where c is a real number.

(iv) $A \otimes (B \otimes C) = (A \otimes B) \otimes C$.

(v) $(A \otimes B)' = A' \otimes B'$.

(vi) If A is partitioned as $A = [A_1 \ A_2]$, then $[A_1 \ A_2] \otimes B = [A_1 \otimes B \ A_2 \otimes B]$. However, if B is partitioned as $[B_1 \ B_2]$, then $A \otimes [B_1 \ B_2] \neq [A \otimes B_1 \ A \otimes B_2]$.

(vii) $(A_1 \otimes B_1)(A_2 \otimes B_2) = (A_1 A_2) \otimes (B_2 B_2)$.

(viii) If A, B are nonsingular, so is $A \otimes B$, and $(A \otimes B)^{-1} = A^{-1} \otimes B^{-1}$.

(ix) $\text{rank}(A \otimes B) = \text{rank}(A)\text{rank}(B)$.

(x) $\text{tr}(A \otimes B) = \text{tr}(A)\text{tr}(B)$.

(xi) If A is $m \times m$ and B is $k \times k$, then $|A \otimes B| = |A|^m |B|^k$.

(xii) The eigenvalues of $A \otimes B$ are all possible products of an eigenvalue of A and an eigenvalue of B.

B.2 Matrix Differentiation

If A is a matrix whose elements are functions of θ, a real-valued variable, then $\partial A/\partial\theta$ represents the matrix whose elements are the derivatives of the corresponding elements of A with respect to θ. For example, if

$$A = \begin{pmatrix} a_{11} \ a_{12} \\ a_{21} \ a_{22} \end{pmatrix}, \quad \text{then} \quad \frac{\partial A}{\partial \theta} = \begin{pmatrix} \partial a_{11}/\partial\theta \ \partial a_{12}/\partial\theta \\ \partial a_{21}/\partial\theta \ \partial a_{22}/\partial\theta \end{pmatrix}.$$

If $a = (a_i)_{1 \le i \le k}$ is a vector whose components are functions of $\theta = (\theta_j)_{1 \le j \le l}$, a vector-valued variable, then $\partial a / \partial \theta'$ is defined as the matrix $(\partial a_i / \partial \theta_j)_{1 \le i \le k, 1 \le j \le l}$. Similarly, $\partial a' / \partial \theta$ is defined as the matrix $(\partial a / \partial \theta')'$.

The following are some useful results.

(i) (Inner-product) If a, b, and θ are vectors, then

$$\frac{\partial (a'b)}{\partial \theta} = \left(\frac{\partial a'}{\partial \theta} \right) b + \left(\frac{\partial b'}{\partial \theta} \right) a.$$

(ii) (Quadratic form) If x is a vector and A is a symmetric matrix, then

$$\frac{\partial}{\partial x} x' A x = 2Ax.$$

(iii) (Inverse) If the matrix A depends on a vector θ and is nonsingular, then, for any component θ_i of θ,

$$\frac{\partial A^{-1}}{\partial \theta_i} = -A^{-1} \left(\frac{\partial A}{\partial \theta_i} \right) A^{-1}.$$

(iv) (Log-determinant) If the matrix A above is also positive definite, then, for any component θ_i of θ,

$$\frac{\partial}{\partial \theta_i} \log(|A|) = \mathrm{tr} \left(A^{-1} \frac{\partial A}{\partial \theta_i} \right).$$

B.3 Projection

For any matrix X, the matrix $P_X = X(X'X)^{-1}X'$ is called the projection matrix to $\mathcal{L}(X)$ (see Appendix A). Here it is assumed that $X'X$ is nonsingular; otherwise, $(X'X)^{-1}$ will be replaced by $(X'X)^{-}$, the generalized inverse (see the next section).

To see why P_X is given such a name, note that any vector in $\mathcal{L}(X)$ can be expressed as $v = Xb$, where b is a vector of the same dimension as the number of columns of X. Then, we have $P_X v = X(X'X)^{-1}X'Xb = Xb = v$, that is, P_X keeps v unchanged.

The orthogonal projection to $\mathcal{L}(X)$ is defined as $P_{X^\perp} = I - P_X$, where I is the identity matrix. Then, for any $v \in \mathcal{L}(X)$, we have $P_{X^\perp} v = v - P_X v = v - v = 0$. In fact, $P_{X^\perp}$ is the projection matrix to the orthogonal space of X, denoted by $\mathcal{L}(X)^\perp$.

If we define the projection of any vector v to $\mathcal{L}(X)$ as $P_X v$, then, if $v \in \mathcal{L}$, the projection of v is itself; if $v \in \mathcal{L}(X)^\perp$, the projection of v is zero (vector). In general, we have the orthogonal decomposition $v = v_1 + v_2$, where $v_1 = P_X v \in \mathcal{L}(X)$, $v_2 = P_{X^\perp} v \in \mathcal{L}(X)^\perp$ such that $v_1' v_2 = v' P_X P_{X^\perp} v = 0$, because $P_X P_{X^\perp} = P_X(1 - P_X) = P_X - P_X^2 = 0$.

The last equation recalls an important property of a projection matrix; that is, any projection matrix is idempotent; that is, $P_X^2 = P_X$.

Example B.1. If $X = 1_n$ (see Appendix A), then $P_X = 1_n(1_n'1_n)^{-1}1_n' = n^{-1}J_n = \bar{J}_n$. The orthogonal projection is thus $I_n - \bar{J}_n$. It is easy to verify that $\bar{J}_n^2 = \bar{J}_n$ and $(I_n - \bar{J}_n)^2 = I_n - \bar{J}_n$.

Another useful result involving projections is the following. Suppose that X is $n \times p$ such that $\text{rank}(X) = p$, and V is $n \times n$ and positive definite. For any $n \times (n-p)$ matrix A such that $\text{rank}(A) = n - p$ and $A'X = 0$, we have

$$A(A'VA)^{-1}A' = V^{-1} - V^{-1}X(X'V^{-1}X)^{-1}X'V^{-1}. \tag{B.1}$$

Equation (B.1) may be expressed in a different way: $P_{V^{1/2}A} = I - P_{V^{-1/2}X}$, where $V^{1/2}$ and $V^{-1/2}$ are the square root matrices of V and V^{-1}, respectively (see Section B.5). In particular, if $V = I$, we have $P_A = I - P_X = P_{X^\perp}$. If X is not of full rank, (B.1) holds with $(X'V^{-1}X)^{-1}$ replaced by $(X'V^{-1}X)^-$ (see below).

B.4 Generalized Inverse

For any matrix A, whether it is nonsingular or not, there always exists a matrix A^- satisfying $AA^-A = A$. Such an A^- is called a generalized inverse of A. Note that here we use the term "a generalized inverse" instead of "the generalized inverse", because such an A^- may not be unique. Two special kinds of generalized inverse are often of interest.

Any matrix A^- satisfying

$$AA^-A = A \quad \text{and} \quad A^-AA^- = A^-$$

is called a reflexible generalized inverse of A. Given a generalized inverse A^- of A, one can produce a generalized inverse that is reflexible by $A_\mathrm{r}^- = A^-AA^-$.

If the generalized inverse is required to satisfy the following conditions, known as the Penrose conditions, (i) $AA^-A = A$, (ii) $A^-AA^- = A^-$, (iii) AA^- is symmetric, and (iv) A^-A is symmetric, it is called the Moore–Penrose inverse. In other words, a reflexible generalized inverse that satisfies the symmetry conditions (iii) and (iv) is the Moore–Penrose inverse. It can be shown that for any matrix A, its Moore-Penrose inverse exists and is unique. See Searle (1971; Section 1.3) for more details.

B.5 Decompositions of Matrices

There are various decompositions of a matrix satisfying certain conditions. Two of them are most relevant to this book.

The first is Choleski's decomposition. Let A be a nonnegative definite matrix. Then, there exists an upper-triangular matrix U such that $A = U'U$.

An application of Choleski decomposition is the following. For any $k \times 1$ vector μ and $k \times k$ covariance matrix V, one can generate a k-variate normal random vector with mean μ and covariance matrix V. Simply let $\xi = \mu + U'\eta$, where η is a $k \times 1$ vector whose components are independent $N(0,1)$ random variables, and U is the upper-triangular matrix in the Choleski's decomposition of V.

Another decomposition is the eigenvalue decomposition. For any $k \times k$ symmetric matrix A, there exists an orthogonal matrix T such that $A = TDT'$, where $D = \mathrm{diag}(\lambda_1, \dots, \lambda_k)$, and $\lambda_1, \dots, \lambda_k$ are the eigenvalues of A. In particular, if A is nonnegative definite, in which case the eigenvalues are nonnegative, we define $D^{1/2} = \mathrm{diag}(\sqrt{\lambda_1}, \dots, \sqrt{\lambda_k})$, and $A^{1/2} = TD^{1/2}T'$, called the square root matrix of A. It follows that $(A^{1/2})^2 = A$. If A is positive definite, then we write $A^{-1/2} = (A^{1/2})^{-1}$, which is identical to $(A^{-1})^{1/2}$. Thus, for example, an alternative way of generating the k-variate normal random vector is to let $\xi = \mu + V^{1/2}\eta$, where η is the same as above.

B.6 The Eigenvalue Perturbation Theory

If A and B are symmetric matrices, whose eigenvalues, arranged in decreasing orders, are $\alpha_1 \geq \cdots \geq \alpha_k$ and $\beta_1 \geq \cdots \geq \beta_k$, respectively, then Weyl's perturbation theorem states that

$$\max_{1 \leq i \leq k} |\alpha_i - \beta_i| \leq \|A - B\|.$$

An application of Weyl's theorem is the following. If A_n is a sequence of symmetric matrices such that $\|A_n - A\| \to 0$ as $n \to \infty$, where A is a symmetric matrix, then the eigenvalues of A_n converge to those of A as $n \to \infty$.

C

Some Results in Statistics

C.1 Multivariate Normal Distribution

A random vector ξ is said to have a multivariate normal distribution with mean vector μ and covariance matrix Σ, or $\xi \sim N(\mu, \Sigma)$, if the (joint) pdf of ξ is given by

$$f(x) = \frac{1}{(2\pi)^{k/2}|\Sigma|^{1/2}} \exp\left\{-\frac{1}{2}(x-\mu)'\Sigma^{-1}(x-\mu)\right\}, \qquad x \in R^k,$$

where k is the dimension of ξ. Below are some useful results.

1. For any matrices $r \times k$ matrix A and $s \times k$ matrix B, $A\xi$ and $B\xi$ are independent if and only if $A\Sigma B' = 0$.
2. If

$$\xi = \begin{pmatrix} \xi_1 \\ \xi_2 \end{pmatrix}, \qquad \mu = \begin{pmatrix} \mu_1 \\ \mu_2 \end{pmatrix}, \qquad \Sigma = \begin{pmatrix} \Sigma_{11} & \Sigma_{12} \\ \Sigma_{21} & \Sigma_{22} \end{pmatrix},$$

then the conditional distribution of ξ_1 given ξ_2 is

$$N\left(\mu_1 + \Sigma_{12}\Sigma_{22}^{-1}(\xi_2 - \mu_2),\ \Sigma_{11} - \Sigma_{12}\Sigma_{22}^{-1}\Sigma_{21}\right).$$

Note that $\Sigma_{21} = \Sigma_{12}'$.

C.2 Quadratic Forms

Let ξ be a random vector such that $\mathrm{E}(\xi) = \mu$ and $\mathrm{Var}(\xi) = \Sigma$. Then, for any nonrandom symmetric matrix A, we have

$$\mathrm{E}(\xi' A \xi) = \mu' A \mu + \mathrm{tr}(A\Sigma).$$

If $\xi \sim N(0, \Sigma)$, $\xi' A \xi$ is distributed as χ_r^2 if and only if $A\Sigma$ is idempotent and $r = \mathrm{rank}(A)$.

If $\xi \sim N(\mu, \Sigma)$, then $\xi' A \xi$ and $b'\xi$ are independent if and only if $b'\Sigma A = 0$; $\xi' A \xi$ and $\xi' B \xi$ are independent if and only if $A\Sigma B = 0$, where B is another nonrandom symmetric matrix.

Furthermore, if $\xi \sim N(\mu, \Sigma)$, then

$$\mathrm{cov}(\xi' A \xi, b'\xi) = 2b'\Sigma A\mu,$$
$$\mathrm{cov}(\xi' A \xi, \xi' B \xi) = 4\mu' A\Sigma B\mu + 2\mathrm{tr}(A\Sigma B\Sigma).$$

Finally, if $\xi \sim N(0, \Sigma)$, then the distribution of $\xi' A \xi$ is χ^2_r, where $r = \mathrm{rank}(A)$, if and only if $A\Sigma$ is idempotent.

For more details, see Searle (1971, Section 2.5).

C.3 $\mathrm{O_P}$ and $\mathrm{o_P}$

A sequence of random vectors (including random variables), ξ_n, is said to be bounded in probability, denoted by $\mathrm{O_P}(1)$, if for any $\epsilon > 0$, there is $M > 0$ such that $\mathrm{P}(|\xi_n| > M) < \epsilon$, $n = 1, 2, \ldots$. If a_n is a sequence of positive numbers, the notation $\xi_n = \mathrm{O_P}(a_n)$ means that $\xi_n / a_n = \mathrm{O_P}(1)$.

A sequence of random vectors (including random variables), ξ_n, is $\mathrm{o_P}(1)$ if $|\xi_n|$ converges to zero in probabilirty. If a_n is a sequence of positive numbers, the notation $\xi_n = \mathrm{o_P}(a_n)$ means that $\xi_n / a_n = \mathrm{o_P}(1)$.

Some important results regarding $\mathrm{O_P}$ and $\mathrm{o_P}$ are the following.

1. If there is a number $k > 0$ such that $\mathrm{E}(|\xi_n|^k)$ is bounded, then $\xi_n = \mathrm{O_P}(1)$; similarly, if $\mathrm{E}(|\xi_n|^k) \leq c a_n$, where c is a constant and a_n a sequence of positive numbers, then $\xi_n = \mathrm{O_P}(a_n^{1/k})$.

2. If there is a number $k > 0$ such that $\mathrm{E}(|\xi_n|^k) \to 0$, then $\xi_n = \mathrm{o_P}(1)$; similarly, if $\mathrm{E}(|\xi_n|^k) \leq c a_n$, where c is a constant and a_n a sequence of positive numbers, then $\xi_n = \mathrm{o_P}(b_n)$ for any sequence $b_n > 0$ such that $b_n^{-1} a_n^{1/k} \to 0$.

3. If there are sequences of vectors $\{\mu_n\}$ and nonsingular matrices $\{A_n\}$ such that $A_n(\xi_n - \mu_n)$ converges in distribution, then $\xi_n = \mu_n + \mathrm{O_P}(\|A_n^{-1}\|)$.

C.4 Convolution

If X and Y are random variables with cdfs F and G, respectively, the cdf of $X + Y$ is given by

$$F * G(z) = \int F(z - y) dG(y),$$

which is called the convolution of F and G. In particular, if F and G have pdfs f and g, respectively, the pdf of $X + Y$ is given by

$$f * g(z) = \int f(z - y) g(y) dy,$$

which is called the convolution of f and g.

The definition can be extended to the sum of more than two random variables. Let F_j (f_j) denote the cdf (pdf) of X_j, $1 \le j \le k$. Then, the cdf of $X_1 + X_2 + X_3$ is $F_1 * F_2 * F_3 = F_1 * (F_2 * F_3)$; the pdf of $X_1 + X_2 + X_3$ is $f_1 * f_2 * f_3 = f_1 * (f_2 * f_3)$, and so on.

C.5 Exponential Family and Generalized Linear Models

The concept of generalized linear models, or GLM, is closely related to that of the exponential family. The distribution of a random variable Y is a member of the exponential family, if its pdf or pmf can be expressed as

$$f(y;\theta) = \exp\left\{\frac{y\theta - b(\theta)}{a(\phi)} + c(y,\phi)\right\}, \tag{C.1}$$

where $a(\cdot)$, $b(\cdot)$ and $c(\cdot,\cdot)$ are known functions, θ is an unknown parameter, and ϕ is an additional dispersion parameter, which may or may not be known. Many of the well-known distributions are members of the exponential family. These include normal, Gamma, binomial and Poisson distributions.

An important fact regarding the exponential family is the following relationship between the mean of Y and θ,

$$\mu = \mathrm{E}(Y) = b'(\theta).$$

In many cases, this establishes an 1–1 correspondence between μ and θ. Another relationship between θ, ϕ, and the variance of Y is

$$\mathrm{var}(Y) = b''(\theta)a(\phi).$$

The following is an example.

Example C.1. Suppose that $Y \sim$ binomial(n,p). Then, the pmf of Y can be expressed as (C.1) with

$$\theta = \log\left(\frac{p}{1-p}\right), \qquad b(\theta) = n\log(1+e^\theta), \qquad a(\phi) = \log\binom{n}{y}.$$

Note that in this case $\phi = 1$. It follows that $b'(\theta) = ne^\theta/(1+e^\theta) = np = \mathrm{E}(Y)$, $b''(\theta) = ne^\theta/(1+e^\theta)^2 = np(1-p) = \mathrm{var}(Y)$.

McCullagh and Nelder (1989) introduced GLM as an extension of the classical linear models. Suppose that

(i) The observations $y_1, \dots, y_n$ are independent;
(ii) The distribution of y_i is a member of the exponential family, which can be expressed as

$$f_i(y) = \exp\left\{\frac{y\theta_i - b(\theta_i)}{a_i(\phi)} + c_i(y, \phi)\right\},$$

(iii) the mean of y_i, μ_i, is associated with a linear predictor $\eta_i = x_i'\beta$ through a link function, that is,

$$\eta_i = g(\mu_i),$$

where x_i is a vector of known covariates, β is a vector of unknown regression coefficients, and $g(\cdot)$ is a link function.

Assumptions (i)–(iii) define a GLM. By the properties of the exponential family mentioned above, θ_i is associated with η_i. In particular, if

$$\theta_i = \eta_i,$$

the link function $g(\cdot)$ is called canonical.

The function $a_i(\phi)$ typically takes the form $a_i(\phi) = \phi/w_i$, where w_i is a weight. For example, if the observation y_i is the average of k_i observations (e.g., a binomial proportion, where k_i is the number of Bernoulli trials), then $w_i = k_i$.

References

Akaike, H. (1972), Use of an information theoretic quantity for statistical model identification, *Proc. 5th Hawaii Inter. Conf. Syst. Sci.*, 249–250.

Anderson, R. D. (1979), Estimating variance components from balanced data: Optimum properties of REML solutions and MIVQUE estimators, in *Variance Components and Animal Breeding* (L. D. VanVleck and S. R. Searle, eds.), 205–216, Dept. of Animal Sci., Cornell Univ.

Anderson, T. W. (1969), Statistical inference for covariance matrices with linear structure, *Proc. 2nd Internat. Symp. Multivariate Anal.* (P. R. Krishnaiah, ed.), 55–66, Academic Press, New York.

Anderson, T. W. (1971a), Estimation of covariance matrices with linear structure and moving average process of finite order, Tech. Report No. 6, Dept. of Statist., Stanford Univ.

Anderson, T. W. (1971b), *The Statistical Analysis of Time Series*, Wiley, New York.

Arora, V., Lahiri, P., and Mukherjee, K. (1997), Empirical Bayes estimation of finite population means from complex surveys, *J. Amer. Statist. Assoc.* 92, 1555–1562.

Baker, G. A. (1935), The probability that the mean of a second sample will differ from the mean of a first sample by less than a certain multiple of the standard deviation of the first sample. *Ann. Math. Statist.* 6, 197–201.

Barndorff-Nielsen, O. (1983), On a formula for the distribution of the maximum likelihood estimator, *Biometrika* 70, 343–365.

Battese, G. E., Harter, R. M., and Fuller, W. A. (1988), An error-components model for prediction of county crop areas using survey and satellite data, *J. Amer. Statist. Assoc.* 80, 28–36.

Bernardo, R. (1996), Best linear unbiased prediction of maize single-cross performance given erroneous inbred relationships, *Crop Sci.* 36, 862–866.

Bhat, B. R. and Nagnur, B. N. (1965), Locally asymptotically most stringent tests and Lagrangian multiplier tests of linear hypothesis, *Biometrika* 53, 459–468.

Bickel, P. J., and Zhang, P. (1992), Variable selection in nonparametric regression with categorical covariates, *J. Amer. Statist. Assoc.* 87, 90–97.

Booth, J. G. and Hobert, J. P. (1999), Maximum generalized linear mixed model likelihood with an automated Monte Carlo EM algorithm, *J. Roy. Statist. Soc. B* 61, 265–285.

Box, G. E. and Tiao, G. C. (1973), *Bayesian Inference in Statistical Analysis*, Addison-Wesley, Reading, MA.

Breslow, N. E. and Clayton, D. G. (1993), Approximate inference in generalized linear mixed models, *J. Amer. Statist. Assoc.* 88, 9–25.

Breslow, N. E. and Lin, X. (1995), Bias correction in generalized linear mixed models with a single component of dispersion, *Biometrika* 82, 81–91.

Brewer, K.R. (1963), Ratio estimation and finite populations: Some results deductible from the assumption of an underlying stochastic process, *Australian J. Statist.* 5, 93–105.

Brooks, S. P., Morgan, B. J. T., Ridout, M. S., and Pack, S. E. (1997), Finite mixture models for proportions, *Biometrics* 53, 1097–1115.

Brown, H. and Prescott, R. (1999), *Applied Mixed Models in Medicine*, Wiley, New York.

Brown, K. G. (1976), Asymptotic behavior of MINQUE-type estimators of variance components, *Ann. Statist.* 4, 746–754.

Burdick, R. K. and Graybill, F. A. (1992), *Confidence Intervals on Variance Components*, Marcel Dekker, New York.

Burdick, R. K. and Sielken, Jr. R. L. (1978), Exact condifence intervals for linear combinations of variance components in nested classifications, *J. Amer. Statist. Assoc.* 73, 632–635.

Calvin, J. A., and Sedransk, J. (1991), Bayesian and frequentist predictive inference for the patterns of care studies, *J. Amer. Statist. Assoc.* 86, 36–48.

Casella, G. and Berger, R. L. (2002), *Statistical Inference*, 2nd ed., Duxbury.

Chen, C. F. (1985), Robustness aspects of score tests for generalized linear and partially linear regression models, *Technometrics* 27, 277–283.

Chernoff, H. and Lehmann, E. L. (1954), The use of maximum-likelihood estimates in χ^2 tests for goodness of fit, *Ann. Math. Statist.* 25, 579–586.

Choi, B. S. (1992), *ARMA Model Identification*, Springer, New York.

Clayton, D. (1996), Comment on Lee and Nelder: Hierarchical generalized linear models, *J. Roy. Statist. Soc. B*, 657-659.

Cochran, W.G. (1977), *Sampling Techniques*, 3rd ed., Wiley, New York.

Cox, D. R. and Hinkley, D. V. (1974), *Theoretical Statistics*, Chapman & Hall,

London.

Cramér, H. (1946), *Mathematical methods of statistics*, Princeton Univ. Press, Princeton, NJ.

Cressie, N. and Lahiri, S. N. (1993), The asymptotic distribution of REML estimators, *J. Multivariate Anal.* 45, 217–233.

Crouch, E. A. C. and Spiegelman, D. (1990), The evaluation of integrals of the form $\int f(t)\exp(-t^2)dt$: Application to logistic normal models, *J. Amer. Statist. Assoc.* 85, 464–469.

Daiger, S. P., Miller M., and Chakraborty (1984), Heritability of quantitative variation at the group-specific component (Gc) Locus, *Amer. J. Hum. Genet.* 36, 663-676.

Dalal, S. R., Fowlkes, E. B., and Hoadley, B. (1989), Risk analysis of the space shuttle: Pre-Challenger prediction of failure, *J. Amer. Statist. Assoc.* 84, 945–957.

Das, K. (1979), Asymptotic optimality of restricted maximum likelihood estimates for the mixed model, *Calcutta Statist. Assoc. Bull.* 28, 125–142.

Das, K., Jiang, J., and Rao, J. N. K. (2004), Mean squared error of empirical predictor, *Ann. Statist.* 32, 818–840.

Datta, G. S. and Lahiri, P. (2000), A unified measure of uncertainty of estimated best linear unbiased predictors in small area estimation problems, *Statist. Sinica* 10, 613–627.

de Bruijn, N. G. (1981), *Asymptotic Methods in Analysis*, Dover, New York.

Demidenko, E. (2004), *Mixed Models—Theory and Applications*, Wiley Series in Probability and Statistics, Wiley, New York.

Dempster, A. P., Laird, N. M., and Rubin, D. B. (1977), Maximum likelihood from incomplete data via de EM algorithm (with discussion), *J. Roy. Statist. Soc. B* 39, 1–38.

Dempster, A. P. and Ryan, L. M. (1985), Weighted normal plots, *J. Amer. Ststist. Assoc.* 80, 845–850.

Diggle, P. J., Liang, K. Y., and Zeger, S. L. (1996). *Analysis of Longitudinal Data.* Oxford Univ. Press.

Drum, M. L. and McCullagh, P. (1993), REML estimation with exact covariance in the logistic mixed model, *Biometrics* 49, 677–689.

Efron, B. (1975), Biased versus unbiased estimation, *Advances in Mathematics*, 16, 259–277.

Efron, B. (1979), Bootstrap methods: Another look at the jackknife, *Ann. Statist.* 7, 1–26.

Efron, B. and Hinkley, D. V. (1978), Assessing the accuracy of the maximum likelihood estimator: observed versus expected Fisher information, *Biometrika* 65, 457–487.

Efron, B., and Morris, C. (1973), Stein's estimation rule and its competitors-an empirical Bayes approach, *J. Amer. Statist. Assoc.* 68, 117–130.

Efron, B., and Morris, C. (1975), Data analysis using Stein's estimator and its generalizations, *J. Amer. Statist. Asso.*, 70, 311–319.

Fay, R. E. and Herriot, R. A. (1979), Estimates of income for small places: An application of James-Stein procedures to census data, *J. Amer. Statist. Assoc.* 74, 269–277.

Fisher, R. A. (1922), On the mathematical foundations of theoretical statistics, *Phil. Trans. R. Soc. Lond., A* 222, 309–368.

Foutz, R. V. and Srivastava, R. C. (1977), The performance of the likelihood ratio test when the model is incorrect, *Ann. Statist.* 5, 1183–1194.

Gan, L. and Jiang, J. (1999), A test for global maximum, *J. Amer. Statist. Assoc.* 94, 847–854.

Gelman, A., Bois, F., and Jiang, J. (1996), Physiological pharmacokinetic analysis using population modeling and informative prior distribution, *J. Amer. Statist. Assoc.* 91, 1400–1412.

Gelman, A., Carlin, J. B., Stern, H. S., and Rubin, D. B. (1995), *Bayesian Data Analysis*, Chapman & Hall.

Gelman, A., Carlin, J. B., Stern, H. S., and Rubin, D. B. (2003), *Bayesian Data Analysis*, 2nd ed., Chapman & Hall/CRC.

Geweke, J. (1996), *Handbook of Computational Economics*, North-Holland, Amsterdam.

Ghosh, M. and Meeden, G. (1986), Empirical Bayes Estimation in Finite Population Sampling, *J. Amer. Statist. Assoc.*, 81, 1058–1062.

Ghosh, M. and Rao, J.N.K. (1994), Small area estimation: An appraisal (with discussion), *Statist. Sci.* 9, 55–93.

Ghosh, M., Natarajan, K., Stroud, T. W. F. and Carlin, B. P. (1998), Generalized linear models for small-area estimation, *J. Amer. Statist. Assoc.* 93, 273–282.

Godambe, V. P. (1960), An optimum property of regular maximum-likelihood estimation, *Ann. Math. Statist.* 31, 1208–1211.

Godambe, V. P. (1991), *Estimating Functions*, Oxford Science, Oxford.

Goldstein, H. (1986), Multilevel mixed linear model analysis using iterative generalized least squares, *Biometrika* 73, 43–56.

Goodwin, E. T. (1949), The evaluation of integrals of the form $\int_{-\infty}^{\infty} f(x)e^{-x^2}dx$, *Proc. Cambridge Philosoph. Soc.* 45, 241–245.

Graybill, F. A. and Wang, C. M. (1980), Confidence intervals for nonnegative linear combinations of variances, *J. Amer. Statist. Assoc.* 75, 869–873.

Green, P. J. (1987), Penalized likelihood for general semi-parametric regression models, *International Statist. Rew.* 55, 245–259.

Hahn, G. J., and Meeker, W. Q. (1991). *Statistical Intervals - A Guide for Practitioners.* John Wiley, New York.

Hajek, J. (1971), Comment, in *Foundations of Statistical Inference* (V. P. Godambe and D. A. Sprott, eds.), Holt, Rinchart and Winston, Toronto.

Hampel, F. R., Ronchetti, E. M., Rousseeuw, P. J., and Stahe, W. A. (1986), *Robust Statistics: The Approach Based on Influence Functions*, Wiley, New York.

Hand, D. and Crowder, M. (1996), *Practical Longitudinal Data Analysis*, Chapman and Hall, London.

Hartley, H. O. and Rao, J. N. K. (1967), Maximum likelihood estimation for the mixed analysis of variance model, *Biometrika* 54, 93–108.

Harville, D. A. (1974), Bayesian inference for variance components using only error contrasts, *Biometrika* 61, 383–385.

Harville, D. A. (1977), Maximum likelihood approaches to variance components estimation and related problems, *J. Amer. Statist. Assoc.* 72, 320–340.

Harville, D. A. (1990), BLUP (best linear unbiased prediction) and beyond, in *Advances in Statistical Methods for Genetic Improvement of Livestock* (D. Gianola and K. Hammond, eds.) 239–276, Springer, New York.

Harville, D. A. (1991), Comment on Robinson: Estimation of random effects, *Statist. Sci.* 6, 35–39.

Harville, D. A. and Fenech, A. P. (1985), Confidence intervals for a variance ratio, or for heritability, in an unbalanced mixed linear model, *Biometrics* 41, 137–152.

Healy, Jr. W. C. (1961), Limit for a variance component with an exact confidence coefficient, *Ann. Math. Statist.* 32. 466–476.

Henderson, C. R. (1948), Estimation of general, specific and maternal combining abilities in crosses among inbred lines of swine, Ph. D. Thesis, Iowa State Univ., Ames, Iowa.

Henderson, C. R. (1950), Estimation of genetic parameters (abstract), *Ann. Math. Statist.* 21, 309–310.

Henderson, C. R. (1953), Estimation of variance and covariance components, *Biometrics* 9, 226–252.

Henderson, C. R. (1963), Selection index and expected genetic advance, in *Statistical Genetics and Plant Breeding* 141–163, Nat. Acad. Sci., Nat. Res. Council, Publication 982, Washington, D. C.

Henderson, C. R. (1973), Sire evaluation and genetic trends, in *Proceedings of the Animal Breeding and Genetics Symposium in Honor of Dr. Jay L. Lush* 10-41, Amer. Soc. Animal Sci. - Amer. Dairy Sci. Assoc. - Poultry Sci. Assoc., Champaign, IL.

Henderson, C. R. (1975), Best linear unbiased estimation and prediction under a selection model, *Biometrics* 31, 423–447.

Henderson, C. R., Kempthorne, O., Searle, S. R., and von Krosigk, C. N. (1959), Estimation of environmental and genetic trends from records subject to culling, *Biometrics* 15, 192–218.

Hensen, L. P. (1982), Large sample properties of generalized method of moments estimators, *Econometrica* 50, 1029–1054.

Heritier, S. and Ronchetti, E. (1994), Robust bounded-influence tests in general parametric models, *J. Amer. Statist. Assoc.* 89, 897–904.

Heyde, C. C. (1994), A quasi-likelihood approach to the REML estimating equations, *Statist. & Probab. Letters* 21, 381–384.

Heyde, C. C. (1997), *Quasi-likelihood and Its Application*, Springer, New York.

Hinde, J. (1982), Compound Poisson regression models, in *GLIM 82: Proceedings of the International Conference on Generalized Linear Models* (R. Gilchrist ed.), Springer, Berlin, 109–121.

Hobert, J. P. and Casella, G. (1996), The effect of improper priors on Gibbs sampling in hierarchical linear mixed models, *J. Amer. Statist. Assoc.* 91, 1461–1473.

Jeske, D. R., and Harville, D. A. (1988). Prediction-interval procedures and (fixed-effects) confidence-interval procedures for mixed linear models. *Commun. Statist. - Theory Meth.* 17, 1053–1087.

Jiang, J. (1996), REML estimation: Asymptotic behavior and related topics, *Ann. Statist.* 24, 255–286.

Jiang, J. (1997a), Wald consistency and the method of sieves in REML estimation, *Ann. Statist.* 25, 1781–1803.

Jiang, J. (1997b), A derivation of BLUP—Best linear unbiased predictor, *Statist. Probab. Letters* 32, 321–324.

Jiang, J. (1998a), Consistent estimators in generalized linear mixed models, *J. Amer. Statist. Assoc.* 93, 720–729.

Jiang, J. (1998b), Asymptotic properties of the empirical BLUP and BLUE in mixed linear models, *Statistica Sinica* 8, 861–885.

Jiang, J. (1999a), Conditional inference about generalized linear mixed models, *Ann. Statist.* 27, 1974-2007.

Jiang, J. (1999b), On unbiasedness of the empirical BLUE and BLUP, *Statist. Probab. Letters* 41, 19–24.

Jiang, J. (1999c), On maximum hierarchical likelihood estimators, *Commun. Statist.—Theory Meth.* 28, 1769–1776.

Jiang, J. (2000a), A matrix inequality and its statistical applications, *Linear Algebra Appl.* 307, 131–144.

Jiang, J. (2000b), A nonlinear Gauss-Seidel algorithm for inference about

GLMM, *Computational Statistics* 15, 229–241.

Jiang, J. (2001), Goodness-of-fit tests for mixed model diagnostics, *Ann. Statist.* 29, 1137–1164.

Jiang, J. (2003a), Empirical best prediction for small area inference based on generalized linear mixed models, *J. Statist. Plann. Inference* 111, 117-127.

Jiang, J. (2003b), Empirical method of moments and its applications, *J. Statist. Plann. Inference* 115, 69–84.

Jiang, J. (2004), Dispersion matrix in balanced mixed ANOVA models, *Linear Algebra Appl.* 382, 211–219.

Jiang, J. (2005a), Partially observed information and inference about non-Gaussian mixed linear models, *Ann. Statist.* 33, 2695–2731.

Jiang, J. (2005b), Comment on Song, Fan and Kalbfleisch: Maximization by parts in likelihood inference, *J. Amer. Statist. Assoc.* 100, 1158–1159.

Jiang, J. (2005c), On classical tests in non-Gaussian mixed linear models, unpublished manuscript.

Jiang, J. and Lahiri (2001), Empirical best prediction for small area inference with binary data, *Ann. Inst. Statist. Math.* 53, 217–243.

Jiang, J. and Lahiri, P. (2004), Robust dispersion tests for longitudinal generalized linear mixed models using Jackknife method, unpublished manuscript.

Jiang, J. and Lahiri, P. (2005a), Estimation of finite population domain means - a model assisted empirical best prediction approach, *J. Amer. Statist. Assoc.* 101, 301–311.

Jiang, J. and Lahiri, P. (2005b), Mixed model prediction and small area estimation (with discussion), *TEST* 15, 1–96.

Jiang, J. and Rao, J. S. (2003), Consistent procedures for mixed linear model selection, *Sankhyā* 65, 23–42.

Jiang, J. and Wang, Y.-G. (2005), Iterative estimating equations for longitudinal data analysis with informative missing observations, unpublished manuscript.

Jiang, J. and Zhang, W. (2001), Robust estimation in generalized linear mixed models, *Biometrika* 88, 753–765.

Jiang, J. and Zhang, W. (2002), Distributional-free prediction intervals in mixed linear models, *Statistica Sinica* 12, 537–553.

Jiang, J., Jia, H., and Chen, H. (2001), Maximum posterior estimation of random effects in generalized linear mixed models, *Statistica Sinica* 11, 97–120.

Jiang, J., Lahiri, P. and Wan, S. (2002), A unified jackknife theory for empirical best prediction with M-estimation, *Ann. Statist.* 30, 1782–1810.

Jiang, J., Lahiri, P. and Wu, C. H. (2001), A generalization of Pearson's χ^2 goodness-of-fit test with estimated cell frequencies, *Sankhyā* A 63, 260–276.

Jiang, J., Luan, Y. and Gan, L. (2004), A test of global maximum with dependent data, unpublished manuscript.

Jiang, J., Rao, J. S., Gu, Z., and Nguyen, T. (2006a), Fence method for mixed model selection, submitted.

Jiang. J., Luan, Y. and Wang, Y.-G. (2006b), Iterative estimating equations: Linear convergence and asymptotic properties, *Ann. Statist.*, revised.

Kackar, R. N. and Harville, D. A. (1981), Unbiasedness of two-stage estimation and prediction procedures for mixed linear models, *Commun. Statist.—Theory Meth.* 10, 1249–1261.

Kackar, R. N. and Harville, D. A. (1984), Approximations for standard errors of estimators of fixed and random effects in mixed linear models, *J. Amer. Statist. Assoc.* 79, 853–862.

Karim, M. R. and Zeger, S. L. (1992), Generalized linear models with random effects: Salamander mating revisited, *Biometrics* 48, 631–644.

Kent, J. T. (1982), Robustness properties of likelihood ratio tests, *Biometrika* 69, 19–27.

Khuri, A. I. (1981), Simultaneous confidence intervals for functions of variance components in random models, *J. Amer. Statist. Assoc.* 76, 878–885.

Khuri, A. I. and Sahai, H. (1985), Variance components analysis: A selective literature survey, *Internat. Statist. Rev.* 53, 279–300.

Khuri, A. I., Mathew, T. and Sinha, B. K. (1998), *Statistical Tests for Mixed Linear Models*, Wiley, New York.

Kim, H. J. and Cai, L. (1993), Robustness of the likelihood ratio test for a change in simple linear regression, *J. Amer. Statist. Assoc.* 88, 864–871.

Kuk, A. Y. C. (1995), Asymptotically unbiased estimation in generalized linear models with random effects, *J. Roy. Statist. Soc. B* 57, 395–407.

Lahiri, P. and Li, H. (2005), A fresh look at baseball data analysis, unpublished report.

Laird, N. M. and Ware, J. M. (1982), Random effects models for longitudinal data, *Biometrics* 38, 963–974.

Lange, K. (1999), *Numerical Analysis for Statisticians*, Springer, New York.

Lange, K. (2002), *Mathematical and Statistical Methods for Genetic Analysis*, 2nd ed., Springer, New York.

Lange, N. and Ryan, L. (1989), Assessing normality in random effects models, *Ann. Statist.* 17, 624–642.

Lee, L. F. (1992), On the efficiency of methods of simulated moments and maximum simulated likelihood estimation of discrete response models, *Econometric Theory* 8, 518–552.

Lee, Y. and Nelder, J. A. (1996), Hierarchical generalized linear models (with discussion), *J. Roy. Statist. Soc. B* 58, 619–678.

Lee, Y. and Nelder, J. A. (2004), Conditional and marginal models: Another view (with discussion), *Statist. Sci.* 19, 219–238.

Lehmann, E. L. (1999), *Elements of Large-Sample Theory*, Springer, New York.

Lehmann, E. L. and Casella, G. (1998), *Theory of Point Estimation*, 2nd ed., Springer, New York.

Liang, K. Y. and Zeger, S. L. (1986), Longitudinal data analysis using generalized linear models, *Biometrika* 73, 13–22.

Lieu, T. A., Newacheck, P. W., and McManus, M. A. (1993), Race, ethnicity and access to ambulatory care among U.S. adolescents, *Amer. J. Public Health* 83, 960–965.

Lin, X. (1997), Variance components testing in generalized linear models with random effects, *Biometrika* 84, 309–326.

Lin, X. and Breslow, N. E. (1996), Bias correction in generalized linear mixed models with multiple components of dispersion, *J. Amer. Statist. Assoc.* 91, 1007–1016.

Littell, R. C., Milliken, G. A., Stroup, W. W., and Wolfinger, R. D. (1996), *SAS System for Mixed Models*, SAS Institute Inc.

Luenberger, D. G. (1984), *Linear and Nonlinear Programming*, Addison-Wesley, Reading, MA.

Malec, D., Sedransk, J., Moriarity, C. L., and LeClere, F. B. (1997), Small area inference for binary variables in the National Health Interview Survey, *J. Amer. Statist. Assoc.* 92, 815–826.

Massey, J. T., Moore, T. F., Parsons, V. L., and Tadros, W. (1989), Design and estimation for the National Health Interview Survey, 1985–94, *National Center for Health Statistics, Vital and Health Statistics* 2, 110.

Mathew, T. and Sinha, B. K. (1988), Optimum tests for fixed effects and variance components in balanced models, *J. Amer. Statist. Assoc.* 83, 133–135.

McCullagh, P. and Nelder, J. A. (1989). *Generalized Linear Models*, 2nd ed., Chapman and Hall, London.

McCulloch, C. E. (1994), Maximum likelihood variance components estimation for binary data, *J. Amer. Statist. Assoc.* 89, 330–335.

McCulloch, C. E. (1997), Maximum likelihood algorithms for generalized linear mixed models, *J. Amer. Statist. Assoc.* 92, 162–170.

McCulloch, C. E. and Searle, S. R., *Generalized, Linear and Mixed Models*, Wiley Series in Probability and Statistics, Wiley, New York.

McFadden, D. (1989), A method of simulated moments for estimation of discrete response models without numerical integration, *Econometrika* 57, 995–1026.

McGilchrist, C. A. (1994), Estimation in generalized mixed models, *J. Roy. Statist. Soc. B* 56, 61–69.

Miller, J. J. (1977), Asymptotic properties of maximum likelihood estimates in the mixed model of analysis of variance, *Ann. Statist.* 5, 746–762.

Mood, A. M., Graybill, F. A., and Boes, D. C. (1974), *Introduction to the Theory of Statistics*, 3rd ed., McGraw-Hill, New York.

Moore, D. S. (1978), Chi-square tests, in *Studies in Statistics* (R. V. Hogg ed.), *Studies in Math.* 19, Math. Assoc. Amer.

Morris, C.N. (1983), Parametric empirical Bayes inference: theory and applications, *J. Amer. Statist. Asso.*, 78, 47–59.

National Research Council (2000), *Small-area estimates of school-age children in poverty*, National Academy Press, Washington, DC.

Newey, W. K. (1985), Generalized method of moments specification testing, *J. Econometrics* 29, 229–256.

Neyman, J. and Scott, E. (1948), Consistent estimates based on partially consistent observations, *Econometrika* 16, 1–32.

Nishii, R. (1984), Asymptotic properties of criteria for selection of variables in multiple regression, *Ann. Statist.* 12, 758–765.

Odell, P. L. and Feiveson, A. H. (1966), A numerical procedure to generate a sample covariance matrix, *J. Amer. Statist. Assoc.* 61, 198–203.

Patel, J. K. (1989). Prediction intervals—A review. *Commun. Statist.—Theory Meth.* 18, 2393–2465.

Patterson, H. D. and Thompson, R. (1971), Recovery of interblock information when block sizes are unequal, *Biometrika* 58, 545–554.

Pinheiro, J. C. and Bates, D. M. (2000), *Mixed-Effects Models in S and S-PLUS*, Springer, New York.

Portnoy, S. (1984), Asymptotic behavior of M-estimators of p regression parameters when p^2/n is large, *Ann. Statist.* 12, 1298–1309.

Prasad, N. G. N. and Rao, J. N. K. (1990), The estimation of mean squared errors of small area estimators, *J. Amer. Statist. Assoc.* 85, 163–171.

Prasad, N. G. N., and Rao, J. N. K. (1999), On robust small area estimation using a simple random effects model, *Surv. Meth.*, 25, 67–72.

Press, W. H., Teukolsky, S. A., Vetterling, W. T. and Flannery, B. P. (1997), *Numerical Recipes in C—The Arts of Scientific Computing*, 2nd ed., Cambridge Univ. Press.

Quenouille, M. (1949), Approximation tests of correlation in time series, *J. R. Statist. Soc. B* 11, 18–84.

Rao, C. R. (1970), Estimation of heteroscedastic variances in linear models, *J. Amer. Statist. Assoc.* 65, 161–172.

Rao, C. R. (1971), Estimation of variance and covariance components—MINQUE theory, *J. Multivariate Anal.* 1, 257–275.

Rao, C. R. (1972), Estimation of variance and covariance components in linear models, *J. Amer. Statist. Assoc.* 67, 112–115.

Rao, C. R. and Kleffe, J. (1988), *Estimation of Variance Components and Applications*, North-Holland, Amsterdam.

Rao, C. R. and Wu, Y. (1989), A strongly consistent procedure for model selection in a regression problem, *Biometrika* 76, 369–374.

Rao, J. N. K. (2003), *Small Area Estimation*, Wiley, New York.

Rao, P. S. R. S. (1997), *Variance Components Estimation: Mixed Models, Methodologies and Applications*, Chapman & Hall/CRC, Boca Raton, FL.

Richardson, A. M. and Welsh, A. H. (1994), Asymptotic properties of restricted maximum likelihood (REML) estimates for hierarchical mixed linear models, *Austral. J. Statist.* 36, 31–43.

Richardson, A. M. and Welsh, A. H. (1996), Covariate screening in mixed linear models, *J. Multivariate Anal.* 58, 27–54.

Robins, J. M., Rotnitzky, A., and Zhao, L. P. (1995), Analysis of semiparametric regression models for repeated outcomes in the presence of missing data, *J. Amer. Statist. Assoc.* 90, 106–121.

Robinson, D. L. (1987), Estimation and use of variance components, *The Statistician* 36, 3–14.

Robinson, G. K. (1991), That BLUP is a good thing: The estimation of random effects (with discussion), *Statist. Sci.* 6, 15–51.

Sahai, H. and Ageel, M. I. (2000), *Analysis of Variance: Fixed, Random and Mixed Models*, Birkhäuser, Cambridge, MA.

Satterthwaite, F. E. (1946), An approximate of distribution of estimates of variance components, *Biometrics Bull.* 2, 110–114.

Scheffé, H. (1959), *The Analysis of Variance*, Wiley, New York.

Schrader, R. M. and Hettmansperger, T. P. (1980), Robust analysis of variance based upon a likelihood ratio criterion, *Biometrika* 67, 93–101.

Schwartz, G. (1978), Estimating the dimension of a model, *Ann. Statist.* 6, 461–464.

Searle, S. R. (1971), *Linear Models*, Wiley, New York.

Searle, S. R., Casella, G., and McCulloch, C. E. (1992), *Variance Components*, Wiley, New York.

Sen, A. and Srivastava, M. (1990), *Regression Analysis*, Springer, New York.

Shao, J. (1993), Linear model selection by cross-validation, *J. Amer. Statist. Assoc.* 88, 486–494.

Shao, J. and Tu, D. (1995), *The Jackknife and Bootstrap*, Springer, New York.

Silvapulle, M. J. (1992), Robust Wald-type tests of one-sided hypotheses in the linear model, *J. Amer. Statist. Assoc.* 87, 156–161.

Smith, H. F. (1936), The problem of comparing the results of two experiments with unequal errors, *J. Council Sci. Indust. Research* 9, 211–212.

Song, P. X.-K. (2000), Multivariate dispersion models generated from Gaussian copula, *Scand. J. Statist.* 27, 305–320.

Song, P. X-K., Fan, Y., and Kalbfleisch, J. D. (2005), Maximization by parts in likelihood inference (with discussion), *J. Amer. Statist. Assoc.* 100, 1145-1158.

Speed, T. P. (1991), Comment on Robinson: Estimation of random effects, *Statist. Sci.* 6, 42–44.

Speed, T. P. (1997), Restricted maximum likelihood (REML), *Encyclopedia of Statistical Sciences* 1, 472–481.

Thall, P. F. and Vail, S. C. (1990), Some covariance models for longitudinal count data with overdispersion, *Biometrics* 46, 657–671.

Thisted, R. A. (1988), *Elements of Statistical Computing—Numerical Computation*, Chapman and Hall, London.

Thompson, W. A., Jr. (1962), The problem of negative estimates of variance components, *Ann. Math. Statist.* 33, 273–289.

Ting, N., Burdick, R. K., Graybill, F. A., Jeyaratnam, S., and Lu, T. F. C. (1990), Confidence intervals on linear combinations of variance components that are unrestricted in sign, *J. Statist. Computation and Simulation* 35, 135–143.

Tukey, J. (1958), Bias and confidence in not quite large samples, *Ann. Math. Statist.* 29, 614.

Verbeke, G. and Molenberghs, G. (2000), *Linear Mixed Models for Longitudinal Data*, Springer, New York.

Verbeke, G., Molenberghs, G., Bickel, P., Diggle, P., Fienberg, S., Krickberg, K., Olkin, I., Wermuth, N., and Zeger, S. (2000), *Linear Mixed Models for Longitudinal Data*, Springer, New York.

Verbyla, A. P. (1990), A conditional derivation of residual maximum likelihood, *Austral. J. Statist.* 32, 227–230.

Wald, A. (1947), *Sequential Analysis*, Wiley, New York.

Wald, A. (1949), Note on the consistency of the maximum likelihood estimate, *Ann. Math. Statist.* 20, 595–601.

Wei, G. C. G. and Tanner, M. A. (1990), A Monte Carlo implementation of the EM algorithm and the poor man's data augmentation algorithms, *J. Amer. Statist. Assoc.* 85, 699–704.

Weiss, L. (1975), The asymptotic distribution of the likelihood ratio in some nonstandard cases, *J. Amer. Statist. Assoc.* 70, 204–208.

Welch, B. L. (1956), On linear combinations of several variances, *J. Amer. Statist. Assoc.* 51, 132–148.

Welham, S. J. and Thompson, R. (1997), Likelihood ratio tests for fixed model terms using residual maximum likelihood, *J. Roy. Statist. Soc. B* 59, 701–714.

Ying, Z. (2003), An asymptotic Pythagorean identity, in *Development of Modern Statistics and Related Topics*, H. Zhang and J. Huang eds., 20–37, World Scientific, New Jersey.

Zeger, S. L. and Liang, K.-Y. (1986), Longitudinal data analysis for discrete and continuous outcomes, *Biometrics* 42, 121–130.

Zeger, S. L. and Karim, R. M. (1991), Generalized linear models with random effects: a Gibbs sampling approach, *J. Amer. Statist. Assoc.* 86, 79–86.

Zheng, X. and Loh, W.-Y. (1995), Consistent variable selection in linear models, *J. Amer. Statist. Assoc.* 90, 151–156.

Zhou, L. (1997). Nonparametric prediction intervals. Ph.D. dissertation, Univ. of Calif. at Berkeley, Berkeley, CA.

Index

Springer Series in Statistics *(continued from p. ii)*

Lahiri: Resampling Methods for Dependent Data.
Le Cam: Asymptotic Methods in Statistical Decision Theory.
Le Cam/Yang: Asymptotics in Statistics: Some Basic Concepts, 2nd edition.
Le/Zidek: Statistical Analysis of Environmental Space-Time Processes.
Liu: Monte Carlo Strategies in Scientific Computing.
Manski: Partial Identification of Probability Distributions.
Mielke/Berry: Permutation Methods: A Distance Function Approach.
Molenberghs/Verbeke: Models for Discrete Longitudinal Data.
Mukerjee/Wu: A Modern Theory of Factorial Designs.
Nelsen: An Introduction to Copulas, 2nd edition.
Pan/Fang: Growth Curve Models and Statistical Diagnostics.
Politis/Romano/Wolf: Subsampling.
Ramsay/Silverman: Applied Functional Data Analysis: Methods and Case Studies.
Ramsay/Silverman: Functional Data Analysis, 2nd edition.
Reinsel: Elements of Multivariate Time Series Analysis, 2nd edition.
Rosenbaum: Observational Studies, 2nd edition.
Rosenblatt: Gaussian and Non-Gaussian Linear Time Series and Random Fields.
Särndal/Swensson/Wretman: Model Assisted Survey Sampling.
Santner/Williams/Notz: The Design and Analysis of Computer Experiments.
Schervish: Theory of Statistics.
Shao/Tu: The Jackknife and Bootstrap.
Simonoff: Smoothing Methods in Statistics.
Sprott: Statistical Inference in Science.
Stein: Interpolation of Spatial Data: Some Theory for Kriging.
Taniguchi/Kakizawa: Asymptotic Theory for Statistical Inference for Time Series.
Tanner: Tools for Statistical Inference: Methods for the Exploration of Posterior Distributions and Likelihood Functions, 3rd edition.
Tillé: Sampling Algorithms.
Tsaitis: Semiparametric Theory and Missing Data.
van der Laan/Robins: Unified Methods for Censored Longitudinal Data and Causality.
van der Vaart/Wellner: Weak Convergence and Empirical Processes: With Applications to Statistics.
Verbeke/Molenberghs: Linear Mixed Models for Longitudinal Data.
Weerahandi: Exact Statistical Methods for Data Analysis.